Office Automation and Information Systems

Office Automation and Information Systems

ARNOLD ROSEN
Nassau Community College

Merrill Publishing Company
A Bell & Howell Information Company
Columbus   Toronto   London   Melbourne

Cover Photo: Merrill Publishing/Larry Hamill

Published by Merrill Publishing Company
A Bell & Howell Information Company
Columbus, Ohio 43216

This book was set in Times Roman

Administrative Editor: Vernon Anthony
Production Coordinator: Carol Sykes
Cover Designer: Cathy Watterson

Copyright © 1987, by Merrill Publishing Company. All rights reserved. No part of this book may be reproduced in any form, electronic or mechanical, including photocopy, recording, or any information storage and retrieval system, without permission in writing from the publisher. "Merrill Publishing Company" and "Merrill" are registered trademarks of Merrill Publishing Company.

Library of Congress Catalog Card Number: 86-62495
International Standard Book Number: 0-675-20557-3
Printed in the United States of America
1 2 3 4 5 6 7 8 9 — 91 90 89 88 87

To my wife, Estherfay, and my son, Paul

It is not really necessary to look too far into the future; we see enough already to be certain it will be magnificent. Only let us hurry and open the roads.

Wilbur Wright

The Merrill Series in Computer Information Systems

Chirlian:	UNIX for the IBM PC: An Introduction	20785-1
DeNoia:	Data Communication: Fundamentals and Applications	20368-6
Harriger and Agrawal:	Applesoft BASIC Programming	20287-6
Harris and Kay:	Introducing Pascal: Workbook and Exercises	20454-2
Horn:	Micro Payroll System: Student Manual	20494-1
Ingalsbe:	Business Applications Software for the IBM PC	20476-3
Ingalsbe:	Business Applications Software for the IBM PC 2.0	20760-6
Ingalsbe:	dBase II® for the IBM PC	20612-X
Ingalsbe:	dBase III® and dBase III-Plus® for the IBM PC	20732-0
Ingalsbe:	Lotus 1-2-3® for the IBM PC	20548-4
Ingalsbe:	Lotus 1-2-3-®, with Version 2.0, for the IBM PC	20759-2
Ingalsbe:	WordStar for the IBM PC	20611-1
Keim:	Business Computers: Planning, Selecting, and Implementing Your First System	20286-8
Mellard:	Introduction to Business Programming Using Pascal: A Structured Problem-Solving Approach	20547-6
Moriber:	Structured BASIC Programming	20106-3
Reynolds:	Introduction to Business Telecommunications	20108-X
Richards and Cheney:	COBOL: A Structured Approach	08041-X
Rosen:	Office Automation and Information Systems	20557-3
Spear:	BASIC: Programming Fundamentals and Applications	20553-0
Spencer:	Computers: An Introduction	20559-X
Spencer:	Computers and Information Processing	20290-6
Spencer:	Computer Science Mathematics	08650-7
Spencer:	Data Processing: An Introduction, Second Edition	09787-8
Spencer:	Data Processing: An Introduction with BASIC, Second Edition	09854-8
Spencer:	The Illustrated Computer Dictionary, Third Edition	20528-X
Spencer:	An Introduction to Computers: Developing Computer Literacy	20030-X
Spencer:	Introduction to Information Processing, Third Edition	08073-8
Spencer:	Learning BASIC for Microcomputers: A Worktext	20437-2
Spencer:	Learning BASIC for Microcomputers: A Worktext for Apple II, IIc, and IIe	20435-6
Spencer:	Learning BASIC for Microcomputers: A Worktext for the IBM PC, AT, and XT	20436-4
Spencer:	Learning BASIC for Microcomputers: A Worktext for the TRS-80	20438-0
Spencer:	Learning Turbo Pascal: A Worktext	20694-4
Spencer:	Principles of Information Processing	20410-0
Sutcliffe:	Introduction to Programming Using Modula-2	20754-1
Thierauf:	Effective Management Information Systems: Accent on Current Practices, Second Edition	20745-2
Thierauf:	Systems Analysis and Design: A Case Study Approach, Second Edition	20229-9
Thierauf, Klekamp, and Ruwe:	Management Science: A Model Formulation Approach with Computer Applications	20006-7
Thierauf and Reynolds:	Effective Information Systems Management	09988-9
Thompson:	BASIC: A Modular Approach, Second Edition	20280-9

PREFACE

Information Age Education

A goal of curriculum development has always been to keep pace with the changes taking place in the real world. The onslaught of microcomputers has led the white-collar masses into the technological mainstream. As the corporate world is moving toward integrating information systems, more information executives and white-collar workers are becoming computer literate. The language barrier between technical and business people is disappearing, and the two cultures are drawing even closer together as the mystique of computers fades.

Consider the following facts:

- By 1990 the number of electronic keyboards in American businesses will equal the number of white-collar workers.
- Installed computing capacity, measured by instruction processing ability, doubles every two to three years. By 1990 there will be roughly 1,000 times as much computing capacity as there was in 1970.
- Circuits and memory are continually getting cheaper, lending to the increasing computing capability of desktop devices. Today's $5,000 device has more computational power than the minicomputers of the 1970s and the mainframes of the 1960s.
- Advances in printer technology, image processing, color and high resolution displays, and especially disk storage have added exponentially to the usability of desktop systems. A small business, or a department in a large business, can automate for less than $5,000 today.
- The number of vendors marshalling workstations into the office is increasing daily. For instance, the number of personal computer vendors has grown from 30 in 1978 to 130 today.

- By 1990 a high level of connectivity will be achieved. Through X.400-compatible products, message interconnection standards will be supported by virtually every computer manufacturer. Computer systems of one worker will be interconnected with virtually every other computer screen on the planet, making the task of sending an electronic letter, a telex, a facsimile, or a voice message as easy as international dialing is today.

Such developments are clear indications that changes in curriculum are long overdue.

Need for This Book

For the past ten years, word processing has been the cornerstone of office automation curriculums. But now a fresh approach is needed, since word processing has expanded to encompass a broader scope of information systems.

Present courses and textbooks offer a limited approach to integrated office information systems: courses geared to secretarial and business education emphasize word processing and office automation, and books and courses designed for data processing and computer science are written for the computer science major, often in technical, hard-to-understand language. *Office Automation and Information Systems* replaces textbooks that focus on a single entity and provides a balance between office automation and management information systems.

Outstanding Features of This Book

Among this book's features are:

- Comprehensive coverage of all aspects of office automation and information systems.
- An easily understandable level of writing.
- A rich assortment of photographs and line drawings.
- End-of-chapter activities and projects that challenge the student to critical thinking and analytic problem solving.
- A complete glossary that includes up-to-date terms in office automation and computer systems.

Although this book is primarily designed for majors in secretarial and office technology programs, it may be used by those in business management and computer science or as an introductory text in other academic areas.

Blending the Best of Two Worlds

As office automation systems, personal computers, and even telephone systems increasingly tie into corporate computers, a new management challenge is emerging. Technical skills, computer science, systems analysis, and programming are no longer the only requisites for successful careers for computer professionals. Now, management and interpersonal communication skills are also necessary.

Office Automation and Information Systems offers a comprehensive approach within a changing business environment. Some of the key topics presented are:

- The planning and implementation of office information systems.
- New technologies and their applications, communications, networks, compatibility, service, and support.
- How users are automating office tools and integrating office technologies.
- How vendors are now offering not only workstations but integrated systems.
- The transition of the office, including the coordination of hardware and software products, word processing, graphics, and spreadsheets.

- New areas of interactive computing, telephony, and integration.
- The linking of micros to corporate mainframes and commercial data banks to access vast stores of data and process information instantaneously.

Recent books may include some of these aspects, but no current book includes a comprehensive presentation of all of these technologies and applications.

Today's educational training is moving away from the highly trained technician toward the well-rounded, computer literate professional. Among the trends that will shape the direction of our future educational programs are:

- The integration of the computer into the entire curriculum.
- The partnerships between business people and educators that are forming in many school districts.
- The growing recognition that the nation's educational plan must go hand in hand with plans for economic and social development.

The future of office education is intricately linked to computer technology. Change is never easy, whether it occurs in a business or an educational environment. Every change, however, represents an opportunity and brings a new and powerful challenge to education. If courses, curriculums, and resources are planned wisely—and if students and teachers face these inevitable changes realistically—the educational experience can be an exciting and profitable venture.

ACKNOWLEDGMENTS

Writing a book of this scope could not have been accomplished without the help of many devoted and talented people. My first contact with the publisher was by way of Dr. Richard Abel, Administrative Editor of the College Division at Merrill. He believed in the project at its inception and was instrumental in its acquisition, development, and continued support. I wish to thank Carol Sykes, who handled the many phases of production and who was always available for advice and professional guidance. There are many other people at Merrill who work behind-the-scenes to help create a successful book. I would like to acknowledge and thank them for their contribution and commitment to this project. Working with the people at Merrill became a pleasant and creative partnership.

I was very fortunate in having a group of reviewers whose critical evaluations were of great value during the preparation of the manuscript. I wish to thank Hal Babson, Columbus Technical Institute; Becki Chaney, University of Arkansas; Leona Gallion, Indiana State University; Mary Ann Hicks, Texas Women's University; Kenniston Lord, Oregon Technical Institute; Vincent Lum, Naval Postgraduate School; Barbara Nichols, Columbus Business University; Kim Staton, Super X Drugs Corporation; and Ardell Terry, Catonsville Community College.

Additional appreciation is extended to the members of the "information society"—the vendors, consultants, executives, and information professionals who have contributed their materials, ideas, and expressions of good wishes.

The foundation of my success in preparing this book was the support and tolerance of those closest to me—my wife, Estherfay, and my son, Paul. It was they who tolerated the disarray of a room full of computers, books, papers, and long periods of isolated work. They offered me the love and encouragement I needed to carry me through the rigors of writing. To you, Estherfay and Paul, I say thank you.

CONTENTS

Part 1   Introduction   1

1.   The Information Age Arrives   3

   Introduction   4

   The Beginning of the Information Age   4

   The Concept of Information Systems   5

   The Information Workplace and Work Force: The Changing Patterns of Occupational Choices, Technology, and Environment   6

2.   The Office in Transition   11

   Moving into Automation   12

   Is Change Necessary?   12

   Start with a Business Plan   12

   Why Plan?   13

   What Are the Elements of Change?   13

   Information   13

   Technology   14

   People: The Most Important Resource   14

   Strategic Planning for Office Automation and Information Systems   14

   Preimplementation   17

   Implementation   17

Part 2   Management Mandates: Information and People   23

3.   Information Processing   25

   Information: What Is It?   26

   Information Stages   26

   Information Processing   26

   Managing Information   31

4.   Organizational Structure and Management   35

   The Structure of an Organization   36

Management Principles Affecting Organization Structure 36

Patterns of Organizational Structure 37

Organizing for Office Information Systems 42

The Evolution of Technology and Its Impact on Today's Organization Structure 42

5. **Personnel: Managing Strategies in a Changing Office**   50

The Role of the Information Manager 51

Motivating Employees 54

Theories of Motivation 54

Human Resource Management in a Changing Society 57

Environmental and External Forces Affecting Personnel Policies 60

Laws and Regulations in the Personnel Field 61

6. **Planning for Integrated Information Systems**   67

Linking the Building Blocks 68

Developing a Strategic Plan for Integrated Systems 68

Part 3 The Information Processing Cycle: Systems and Hardware   75

7. **Voice Processing**   77

Automating the Input Cycle 78

Machine Dictation/Transcription 78

Digital Voice Processing 87

Voice Mail 88

Advanced Voice Technologies 89

8. **The Computer**   94

The Evolution of Processors 95

Computers and Computer Systems 97

Classification of Computers 99

The Personal Computer 101

The Microprocessor 103

Chips and Bits: The Measure of Computer Power 105

Auxiliary Storage 105

Printers 107

Personal Computers Move into the Business World 113

9. **Processors: From Electronic Typewriters to Supercomputers**   117

Electronic Typewriters 118

Word Processors 120

The Executive Workstation 121

User-Friendly Aids 126

Supercomputers 133

Strategies for Acquiring Computer Systems 135

10. **Image Processing**   140

Copiers 141

Copier Features 143

Computer-Interfaced Copiers 146

Phototypesetting 146

Desktop Publishing 148

Optical Character Recognition (OCR) 148

Graphics 150

11. **Word Processing**   157

Features of Word Processing 159

Industry and Professional Applications 165

Word Processing Is Still the Cornerstone of Office Automation 169

12. **Telecommunications**   171

Divestiture: Reorganization of an Institution 172

Telecommunications: A Large "Umbrella" of Technologies 172

Long-Distance Service 175

Electronic Mail 178

Computer-Based Electronic Mail 178

New Technologies—Expanding the Telecommunications Umbrella 181

13. **Computers and Communications: The Micro-Mainframe Link** 193

Expanding Information Resources 194

The Micro-Mainframe Link 194

Part 4 Software and Storage: Components and Technology 203

14. **Software** 205

What Are Software Packages? 206

The Emergence of Software 206

Impact of Personal Computers on Software 206

Types of Microcomputer Software 207

CAD/CAM 211

Integrated Software 216

Software Selection 217

15. **Storage and Retrieval** 221

A Paper-Based Society 222

Coping with the Avalanche of Papers 223

Automated Records Management Systems 225

Micrographics 225

Computers and Micrographics 228

New Technologies in Storage and Retrieval 231

Data and Information Backup 235

Strategies for Implementing a Records Management Program 236

16. **Supplies** 241

Care and Protection of Diskettes 244

Protecting and Storing Other Supplies 246

Part 5 Safety, Security, and Environment in the Office 251

17. **Preventive Maintenance, Service Plans, and Disaster Recovery** 253

Preventive Maintenance 254

Environmental Hazards to Computer Systems 255

Service Plans and Maintenance Contracts 257

Checklist for Computer Maintenance 259

Computer Catastrophes 259

18. **Information Security** 265

Computer Crime and the Law 267

Methods to Ensure the Security of Information 269

Software Piracy: Illegal Copying of Software Programs 272

Terrorism and Sabotage 273

Finding Solutions: A Total Effort 275

19. **Ergonomics** 278

Designing the Workstation for More Comfort and Greater Productivity 279

Video Display Terminals 280

Ergonomic Furnishings 283

Environmental Controls in the Workplace 289

Climate Conditions of the Office 292

Ergonomists—A Career Area Emerges 296

Part 6 Personnel and Training — 299

20. Personnel Selection, Job Descriptions, and Career Options for Information Workers — 301

Sources for Employment 302
Alternate Work Styles 307
Job Titles and Descriptions 308
Career Options for Office Information Systems 309

21. Training, Salary Administration, and Measuring Job Performance — 315

Establishing a Training Program 317
Training Methods 320
Training and Career Opportunities for the Disabled 322
Salary Administration 323
Job Performance: Measurement and Review 325

Part 7 The Future — 329

22. Toward the Twenty-first Century — 331

Glossary 341

Index 351

About the Author 363

PART ONE
Introduction

CHAPTER 1

The Information Age Arrives

After reading this chapter, you will understand
1. The meaning of the information age
2. The changes in the work force and workplace
3. The relationships among information processing, office automation, and other components that comprise a total information system environment

Introduction

Humanity's greatest invention is not the wheel, as many believe; it's the written word, the alphabet. The ability to create, record, pass on knowledge and even improve upon it is quite an achievement.

People have been striving to better use the written word, which has taken us, in just 5000 years, from the sands of the Sahara to the powdery base of the moon's surface and beyond.

The curiosity and inventiveness of the human mind and spirit motivate us to find ways to record information more accurately, store it more efficiently, and, above all, pass it on to others more swiftly.

The *written word:* It's not always written anymore. It has come a long way since the clay tablet. Information today takes many forms. We still collect books, pamphlets, brochures, business cards, and business-related documents. Today, the information industry is more diverse. It has electronic data bases with billions of entries. They are all automated and available in a variety of formats that serve the needs of information users throughout the world.

Innovative formats like microfiche put into the palm of your hand volumes of printed text that are retrievable at the speed of light. Magnetic tapes and disks that store billions of characters are tailored to a myriad of computer systems. They range from room-size behemoths to lap-size micros.

Today, the tools that help enhance office productivity are so numerous that most companies are overwhelmed by their options. A wave of technological change is poised to sweep over the information industry. Faster and cheaper computers will allow users to do things they cannot do today. For example, people will be able to operate a typewriter by voice. The stakes are high: the new technology will affect all information age businesses, ranging from electronic mail networks to computer manufacturing. Consider some of the events that lie ahead:

- Fifty-four million electronic typewriters, word processors, personal computers, and multiuser systems will be installed in the United States alone.
- Ninety percent of the electromechanical telephone switches will be replaced by digital or computer-controlled switches.
- Almost $120 billion will be spent annually on office automation equipment in the United States. Services and software will add another $80 billion.

The Beginning of the Information Age

The information age is less than a generation old. Its future appears to have no limits. When did we suddenly enter this "information age," and what statistical event marked the occasion? It began one day in 1956 when, for the first time in American history, the number of white-collar workers outnumbered that of blue-collar workers. At that moment, the information age was born. The United States began a long and continuing development into an economy based on the manipulation of information rather than on the production of goods.

Information is the raw material for many business activities shaping this new era, just as coal and steel were the basic commodities in the dawn of the industrial age. And, just as coal fueled the transformation to an industrial society, so microelectronics is powering the rise of the information age. Knowledge replaced physical objects as the important resource.

The transition into the information age not only changed the makeup of the work force and workplace but also transferred the control of resources. Those who control the corporate information resources hold the key to the suc-

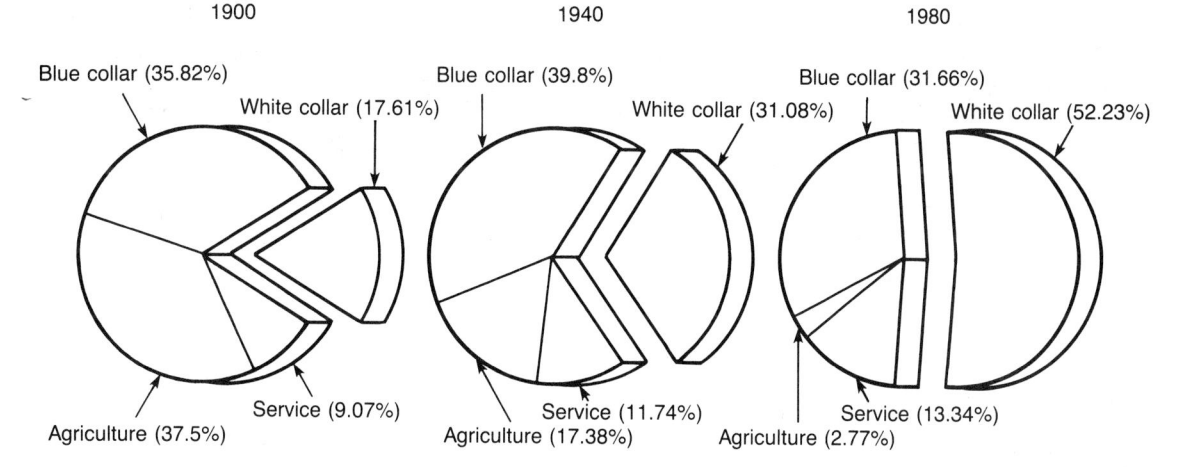

1900
Blue collar (35.82%)
White collar (17.61%)
Service (9.07%)
Agriculture (37.5%)

1940
Blue collar (39.8%)
White collar (31.08%)
Service (11.74%)
Agriculture (17.38%)

1980
Blue collar (31.66%)
White collar (52.23%)
Service (13.34%)
Agriculture (2.77%)

Figure 1-1. The changing structure of the work force.

Source: U.S. Congress, Office of Technology Assessment, *Summary: Automation of America's Offices*, OTA-CIT-288 (Washington, DC: U.S. Government Printing Office, December 1985).

cess of any organization. Over time, control has passed from farmers to merchants to white-collar workers.

Figure 1-1 illustrates the way that work done by people has changed with the introduction of technology, first through the industrial revolution and then at the threshold of the information age. These charts graphically trace the shift of the work force to a period where more than half the work is done by information workers. By the end of this decade, more than two-thirds of the work force will be completely information-dependent.

The Concept of Information Systems

The dawn of a new industry brings more than just the trappings of technology and a changing work force. It carries with it a new set of language, explanations, and definitions. As we move deeper into the information age, the familiar explanations that used to clarify events no longer suffice.

What can we call the new age of information? We need to select a term to identify a complete processing system rather than one that identifies only isolated parts. In the early phases of the information age, the term *information processing* was used synonymously with *data processing*. As word processing grew in importance and secretaries began to use equipment that combined various functions, information processing came to include both data and text processing. Thus, for a brief period of time, word processing and data processing had a common identity. However, information processing has expanded to include not only data and text, but also the integration of graphics, voice, and image processing. Information processing comprises devices, software, skilled people, correct procedures, and the appropriate environment.

The components and technologies have merged in the concept of *office automation* (OA). To some the term still conjures up images of word processing and electronic typewriters. The word *automation* (rightly or wrongly) has a negative connotation. Clearly, office automation has gone far beyond that. It includes integrated office systems with voice and mail capabilities, compatibility among vendors, personal computers, intelligent work stations, high-speed voice and data networks. Communications and other technologies were placed under a single OA umbrella. Unfortunately, key people in the information industry did not take to the term *office automation* and searched for a new comprehensive, contemporary, and professional identity. Vendors, consultants, and key users seem to favor the term *office information systems.*

> Office information systems comprise applications and technologies that serve the information needs of business organizations in the most productive and effective way. The technologies include word processing, data processing, personal computing, telecommunications, and information and storage retrieval.

This change of name was brought about by a number of factors. Office information systems (OIS) is not just a set of applications or a checklist of technologies. It is that, of course, but it is also a change in the way people work. It is a change that identifies the areas of information from a broader perspective than word processing or office clerical tasks. The new name reflects what is actually happening in the marketplace. Word processing and data processing are becoming more a part of information systems than separate and distinct areas.

A new name, however, may not be all that is needed to define the concept of information systems. It is an important starting point. And, as the segments that make up total office information systems are described, the primary definition begins to take on a sharper focus.

Because the nature of the information industry is evolving, definitions must also be flexible enough to reflect the changes within society. Office information systems concepts are not "written in stone." The tools of office technology encourage one to imagine an expansion of office applications. The evolution of communications technology, for example, is an additional element in our changing perception of OIS. Networks and micro-mainframe links (Chapter 13) are part of the OIS framework and are now in the vocabulary of most information workers. Clearly, these trends demonstrate exciting new dimensions for the growth and sustenance of a robust, productive, and cost-effective information systems industry.

The Information Workplace and Work Force: The Changing Patterns of Occupational Choices, Technology, and Environment

What Are Information Workers?

Information workers are professionals who create, process, and distribute information. Jobs in information occupations have seen the most dramatic increase of all occupational categories in recent decades.

Today more than 60 percent of us work with information as programmers, teachers, clerks, secretaries, investment counselors, publishers, editors, accountants, and lawyers. Most Americans spend their time creating, processing, or distributing information.

The Job Market and New Occupational Choices

The implications of the expanded information economy are perhaps most profound in employment. The information age, like the industrial age before it, is creating new jobs while rendering others obsolete. History shows that as machines took over the physical labor, the mental efforts of workers became even more important. These workers moved to jobs that machines could not perform. Today's office environment does not offer such opportunities. Office technology offers more efficient and less expensive alternatives to office workers. Automation changes the nature of work, and corporate planners are slowly learning to examine office technology investments in light of their effect on overall productivity.

New Jobs, New Titles, New Descriptions

The addition of new machines and office technology has changed the things we do and the way we do them. Hordes of office workers, replete with new functions and responsibilities, have suddenly emerged without titles or job descriptions.

Organizations dislike creating a new job category every time new hardware shows up in the office systems market. When technology first appeared in offices, personnel departments scurried to find new ways to describe old positions. Some used equipment to identify a job function. Job titles and descriptions should be designed around the job function, not the tools used to accomplish a job.

In the absence of progress in redefining job descriptions by government and industry, the Association of Information Systems Professionals (AISP) gathered data from more than 2400 companies in the United States and Canada. AISP developed fifteen narrowly defined job categories, descriptions, and titles, ranging from *word processing trainee, WP operator, specialist, supervisor,* and *staff analyst* to *information manager.* The information industry is going through a transition. In many cases, the job descriptions that do exist, such as word processor, will no longer be applicable five years from now.[1]

Whatever title descends upon the information worker, technological innovation has brought forth a new breed of office worker. These people come to their jobs equipped with a basic knowledge of text processing, computing, and communications. Some are recent graduates from community colleges and business schools who are prepared to enter the high-tech electronic work force with good skills and computer literacy.

New Technology

In the past, office technology changed the way people worked. Sometimes, it helped them do the same things they had always been doing, only faster and better. New technologies appeared and old ones were discarded. New systems were introduced and older, "time-proven" ways of doing things vanished. New jobs were created. Old jobs and even old professions were eliminated.

In recent years, office automation consisted of little more than some stand-alone word processors that were operated by specialists who had previously been secretaries. What had been the typing pool emerged as the word processing center, and the office manager became the word processing supervisor.

As the evolution continued, the stand-alone word processors, whether automated typewriters or cathode ray tube– (CRT-) based machines, were replaced by shared-logic systems, executive workstations, and personal computers. The latter was significant because it was the first time professionals were presented with a computer of their own. It freed them from the "tyranny" of data processing. It allowed them to gather and maintain their own data base through micro-mainframe communication links. During the early 1980s, personal computers and executive workstations were considered one and the same. As technology evolved, the two products and their markets began to move in different directions.

Today the office is changing again. New technologies have become entrenched in organizations throughout the country. Equipment has been tested, enhanced, and replaced.

---

[1] Word processing no longer accurately describes a total job function. The dedicated word processor is on the brink of disappearing. Word processing is an application of computing, not a machine per se. Most workers use an integrated workstation capable of word processing, computing, and communicating functions.

Now executive workstations are evolving into voice/data terminals used primarily to gather information crucial to executive decision making. These systems are tied into the corporate communications network. Personal computers (PCs) are also vying for a place in the office. The personal computer industry did not exist ten years ago and now is estimated to be a $15 billion a year business. This office device has the potential to serve business needs, but it has barely met its potential. The trend of shipment of personal computers into American offices will continue to rise for the rest of this century. The PC marketplace offers devices that feature a bewildering maze of options, storage, and computing power. Vendors are stressing the communication aspects of PCs, not just number crunching.

Whether the executive workstation or the PC is the centerpiece of the modern office, both devices will function as communication links to the basic unit of American business—the work group. The work group consists of people working together in small units. They are managers and information professionals who create, communicate, and share information.

The pace of technological innovation in the industry has been phenomenal. One dollar now buys 200 times the processing power it did thirty years ago. To the credit of creative vendors in a flourishing free market environment, the information industry has grown at an extraordinary rate.

The Information Age Landscape

The transformation into the information age has brought about another dramatic change. The landscape across America has been changing from inner-city factories and lofts to office towers and corporate centers. Cities and suburban centers are becoming white-collar towns as a result of a steady migration of office buildings and a white-collar work force. Office towers are springing up in the cities, and corporate centers are emerging in the suburbs.

The demand for information workers is increasing. Many corporations are decentralizing operations and are forming "back offices" that are split away from corporate headquarters. They are migrating to the suburbs, smaller cities, and even the countryside.

For example, Long Island, like so many other areas of our country, is developing office towers and corporate parks. When the Mitchell Field project in Nassau County is complete, 25,000 office workers will be working within a two-mile radius of Nassau Community College in multiunit office parks. After years of steady migration, Stamford, Connecticut, has become the center of one of the nation's largest concentrations of major corporate headquarters. Wave upon wave of workers has moved to gleaming new office buildings, making Stamford less a suburb than a newly rebuilt city with its own urban amenities. The Xerox Corporation, GTE Corporation, Champion International Corporation, Singer Company, and Pitney Bowes Inc. are among the more than a dozen huge corporations based within Stamford's city limits.

Long Island, New York, and Stamford, Connecticut, are microcosms of the transformation that is taking place across the country. Corporate centers are springing up in Ohio, Maryland, Massachusetts, and California and throughout the Sunbelt. Offices will soon be filled with machines and people ready to function in an "electronic office environment."

Telecommunications has been the force to prompt more and more corporate offices to move from central cities to suburban enclaves.

It used to be that all the support services were in the central cities. Now, a multinational corporation like IBM or Pepsico does not have to be located on Wall Street or in midtown Manhattan to get its work done. Thanks to data communications, computers, and global networks, corporations can "talk" to their branch offices electronically, over long distances, instantly and effortlessly.

As more and more cities and states try to

reorient their economies from smokestacks to computers, the new information industries spawn offshoot industries. They include service industries such as law offices, accounting firms, and personnel services. The information age is thriving with a positive regeneration of jobs and businesses.

The Information Age Moves into Higher Gear

The beginning of this chapter presented a brief historical account of the information age. The time span is brief, yet within that short period the rate of change has been very swift. Product cycles no longer relate to depreciation cycles. Vendors merge, buy each other out, spin off, and go in and out of business at the blink of an eye. The market for new technologies is as orderly as a Keystone Kops movie chase. Yet, where we are today is just a foreshadow of what will take place during the next ten years. The beneficiaries of this era will be the people who use the tools and technologies. Computers will multiply as prices drop and performance increases. These devices will become easier to use as the software becomes better and more powerful. They will be linked by growing networks and information systems. We may be the last generation with manual mail delivery systems, newspapers, books, and stand-alone computers. Not only will the industry continue to reach out and be a integral part of millions of people's lives, but it also has the potential to become the most important industry in the world.

Lessons from History

Can we rely on history to chart our technological destiny? Some changes will not occur. Some predictions by prominent analysts and consultants simply will not materialize. The computer revolution has happened so fast that perhaps the only way to cope with the changes is to take them for granted.

One of my educator colleagues was bewildered by the changes occurring. "I am not going to make any curriculum changes just yet," he said. "I am going to wait until the dust settles." But it doesn't seem likely that the dust will ever settle in this technological sandstorm. It just keeps swirling around.

There is no ideal world, and educators as well as today's information systems professionals must realize that there will not be one for some time to come. What one must do to build an office information system successfully is try to implement tactical changes in the office, while not straying from a longer-term strategic plan. Chapter 2 explores these goals in more detail. Being overly cautious and sitting on the fence as everyone around you is installing office automation can be counterproductive.

The information age is in its beginning stages now. The past decade has been an exciting one to live through. History will ultimately record that the computer revolution, like language, writing, and printing, was a profound event in the progress of civilization. But an even more exciting future lies ahead for those who participate and shape its destiny.

Summary

The information age began when the number of information workers (white-collar) surpassed factory and manufacturing workers (blue-collar).

Information workers (also referred to as *knowledge workers*) are professionals who create, process, and distribute information. The transition into the information age changed the makeup of the work force and workplace. Knowledge replaced physical objects as the important resource.

The term *information processing* was used synonymously with *data processing*. As word processing grew in importance, information processing came to include both data and text processing and expanded further to include graphics, voice, and image processing. Thus, information processing includes machines, software, skilled people, correct procedures, and appropriate environment.

Office automation is a component of information processing that pertains to the use of computers to conduct business in the office.

Office automation is part of a larger entity, known as *information systems.* Information systems comprise applications and technologies that serve the information needs of business organizations in the most productive way. The technologies include word processing, data processing, personal computing, telecommunications, and information storage and retrieval.

The personal computer has played a significant role in the evolution of the information age. Personal computers are allowing all workers to come into contact with computers to create, process, store, and distribute information. The personal computers are quietly making operations more efficient, more productive, and more competitive.

The landscape across America has been changing from inner-city factories to office buildings and corporate centers. Cities and suburbs are becoming "white-collar" towns as a result of a steady migration of a white-collar work force.

Telecommunications and global networks have provided the means for corporations to expand their operations. Through computers and communications, organizations can "talk" to their branch offices electronically, over long distances, instantly and effortlessly.

Review Questions

1. What forms will the recording of modern history take?
2. Describe a significant event that marked the beginning of the information age.
3. Why was the term *office information systems* selected to identify the concepts of information systems? What were some of the objections to the term *office automation?*
4. Define an information worker. List some specific job titles.
5. What are the changes taking place that affect job descriptions? Describe such a change in the area of word processing.
6. State some of the specific job requirements and skills needed by information workers entering the job market.
7. How did the personal computer (PC) change the work pattern of the executive?
8. What is a work group?
9. What is a white-collar town?
10. What is the key technology that has been a positive force in encouraging corporate offices to move from urban centers to suburban sites?
11. In the midst of continuous technological changes and new products, when is it ideal to restructure an organization's information system? Describe, in general terms, the best formula for such an approach.

Projects

1. Collect the latest data and statistics pertaining to information systems projections. Summarize these projections into catetories: devices, software, employment or work force projections. List your findings in units or dollar amounts and chart the trend in a table or pie chart.
2. Select a vendor who sells devices that comprise a work group concept. Describe the components that make up the work group and draw a diagram that illustrates the way the devices are connected.
3. Write a report that compares executive workstations with personal computers.
4. Chapter 1 points out that a fundamental change of strategy is taking place among vendors competing in an evolving market. Part of this change is the result of the rapid convergence of computer and communication technologies. What strategic plans or steps should vendors take to survive in such a complex market?

CHAPTER 2

The Office in Transition

After reading this chapter, you will understand
1. Why it becomes necessary for organizations to automate their office operations
2. What a business plan is and what purpose it serves
3. The aspects of strategic planning, the way it should be presented, and the way it should be implemented
4. The importance of training and orientation during planning and implementation

Chapter 1 presented the story of the information age. It traced the external and internal forces that shaped the way our society conducts business. The information age has bestowed upon us the power of the computer. As consumers and providers of information, we are gaining control over our own end products. Soon we will be able to tie our desktop devices into a companywide network. We will share sales projections, retrieve financial data from distant computer banks, and instantly exchange memos. No longer will information workers be dependent on others for the access and control of information. This is the real meaning of office (automation) information systems, as explained in Chapter 1. *Office information systems* has not yet developed into a household term. Corporate America may not fully understand it. However, there is enough perception about the promise of improvement in production and increased efficiency that people are beginning to listen.

Moving into Automation

Office automation is not merely buying more computers. Most big companies already use an adequate number. They have multimillion-dollar mainframes for managing payrolls, minicomputers for specialized tasks, desktop computers, and word processors. Acquiring new hardware does not necessarily provide an automatic solution to the problems of today's office.

Problems persist because office work has not been reorganized. The introduction of new technology into an organization has brought about a rethinking of the work procedures. It has forced us to adopt a new set of rules. People responsible for bringing about this change must have a thorough knowledge of the work procedures that office automation entails. Also, they need extensive technical knowledge of the equipment.

Moving into automation must be guided by a set of intelligent and logical principles. The principles should determine whether or not an organization will be automated, and in what way. Automation is not the solution for everyone or every business problem. If you bring in new equipment and do not change the work procedures, it may have little or no effect on office productivity. Automate the status quo and you will end up with automated problems.

Is Change Necessary?

"If it works, don't fix it" is an old cliché. But if something does not work, a change is needed. Organizations change to improve productivity, reduce costs, and achieve efficiency. Changing management is not a new concern of organizations. However, the importance of implementing change effectively can determine the success or failure of a business.

Changing an organization's way of conducting business concerns not only a single department, but also the organization as a whole. Companywide changes affect many levels at once. The person or persons advocating change better be sure that the change is necessary. If changes are implemented poorly and do not bring about the suggested benefits, the organization may be worse off than before.

Changing information systems can be risky and the results inconsistent if companies rush into such a change without intelligent planning.

Start with a Business Plan

A business plan is simply a formal process of determining long-range objectives and methods to achieve them. Every business venture begins with someone's idea. The purpose of the business plan is to develop the business idea in writing. It should be completed in sufficient detail that it can serve as a blueprint for building and developing the idea.

Why Plan?

Planning helps to analyze and to think through the entire concept. A venture that sounds good in the idea stage may prove to be inadvisable upon closer analysis. The business plan helps management determine where the organization should be in three to ten years. It identifies the resources it needs to get there, and it previews the mix of products or services at the end of the planning stage. The business plan provides the foundation for the profit plan, short-term forecasts, and management reporting.

Planning for office automation is the same process. It involves human and technological resources. They must be marshalled to meet an organization's business goals and objectives. Thus, we need to focus on developing innovative strategies for office automation, planning, management change, and top management education.

What Are the Elements of Change?

The elements in office information systems are people, information, technology, and the office itself. To understand fully the relationship between individuals and technology, and between organizations and technology, we first need to understand the nature of the elements. First let us examine the *office*. The office itself is not simply a room made up of the physical components of information work—the file folders, memo pads, and reams of paper that clog up the desks of the workers. It is an environment that forms the nucleus of all levels of workers. The environment can include secretaries, chairpersons, sales representatives, researchers, a literary agent, a multinational corporation, a simple telephone, or a room-size mainframe computer. The office exists to serve the primary function of information creation and processing. It can exist to support a primary function such as production, marketing, and finance. In most cases the role of office employees is to assist decision makers, technical personnel, and professional people. In essence, the office facilitates activities. The activities include processing information, handling customer's calls, and preparing reports for managers. The product of the office is service. The underpinnings of the office consist of varied types of information created and processed from computers to mainframes.

Information

With the exception of people, *information* is the most valuable resource of an organization. Regardless of their field of specialization, managers often consider information to be their most important managerial tool. It is also one of the most fundamental and precious elements of the total office system. Information exists in many forms: data, text, voice, pictures, graphics. And, it is stored in many locations. Never before have managers had so much information available to support their efforts. There is a virtual tidal wave of volume. Every minute 1 million pieces of paper are added to the 15 trillion already in circulation in the United States.

The importance of information in organizations can hardly be overstated. Most information workers spend most of their time working with and communicating information. The effectiveness with which this information is managed determines how useful it is. The intelligent management of information means that we can *manage* the flood of information, *find* it, and *focus* on the relevant.

Business today needs more information than ever. But it is also true that it needs less information, in the sense of selectivity; that is, the ability to sort through the unwanted and focus on the relevant. The efficiency with which an organization manages its information is affected by the office environment, people, and technology.

Technology

The explosion of *technology,* more than any other factor, is responsible for the rapid increase in the amount of information available to managers. The new technology does not necessarily create vast amounts of new information. Rather, it makes existing information more readily available, accessible, and convenient to use. The technology that is largely responsible for the increased amount of information is often the same technology that is successfully used in its management.

Achieving real support from information technology for key business plans calls for more than simply building and installing computer-based systems. The installation of mainframes, minis, micros, and software will in itself result in no business benefit.

People: The Most Important Resource

The information age has changed the way people work in the office, but not all of them in the same way. For the executive the desktop computer is becoming a reality. It has filtered down the corporate pyramid to middle management, professional staff, and finally the secretarial/clerical level.

The computerization of the office has helped increase productivity, expedite work flow, and improve the quality and timeliness of decision making. Through work group networks, it has brought the different vertical levels closer. The sharing of information has helped workers form a cohesive office team. It has brought about a tremendous rethinking of job responsibilities at every level of the organization. The very nature of work is changing to accommodate the amount of information and the infusion of dazzling technology. We are moving from simple to complex tasks, from standardized jobs to ones with variety, and from fragmented, isolated tasks to jobs that encompass a complete project. Furthermore, we have moved from socially isolated jobs to ones that require interrelation among departments and individuals.

All of this requires a change in work behavior, habits, thinking, and values. Senior managers of corporations who wish to develop a competitive advantage must encourage and support all of the elements for change. They must encourage a new work behavior that best fits into the new technology and new "office" environment.

To change permanent work behavior, more training in the use of a system is required. Patterns in the flow of work may need to be altered, and changes are needed in reporting relationships. New standards and procedures should be established, and performance criteria revamped.

Finally, major strategic changes often involve the most elusive element of all the firm's traditional business philosophy: its culture. This pertains to the way people relate to each other, the work they do, and the organization. These cultural changes can sometimes be attained only gradually. At other times, they may require a complete break from the past.

The recognition and understanding of the key elements must be clarified before a strategic plan can be undertaken.

Strategic Planning for Office Automation and Information Systems

The *business plan* establishes a formal process of determining long-range objectives for an organization. It is useful for organizations that are beginning operations and need an overall goal for the company. *Strategic planning* applies to narrower goals for shorter periods. It is the next level down from business planning.

In its modern meaning, the word *strategy* means leading an organization to obtain certain stated objectives.

Strategy implies the combination of planning and directed action. Strategic planning begins with setting the direction for action. And, it can only be successful when connected with *implementation*; that is, the action that carries the direction into existence.

Implementation of office automation and information systems covers a broad spectrum of the technologies and elements identified in the preceding section. It also raises many human, structural, procedural, and technological issues. The objective of office automation planning is to identify the current and future information requirements of the company and then to determine the way those requirements should be met.

Planning for office automation and office information systems is unique. The information age has a short history. Computer technology has thrust organizations into a relatively new field, without clear precedence of procedures and responsibilities. We are constantly in a state of change, and the time during which new technology emerges is compressed more and more. In the beginning, there was the stand-alone word processor. Next came the personal computer and then the executive workstation. Now, office networks, fourth-generation PBXs, and voice/data phones are the latest developments in a succession of new products. And there is no indication that the speed and volume of technological change will reverse or subside. On the contrary, many feel that the pace of technological change is growing even faster. The computer industry itself has had to change its own views. Data processing and word processing are no longer considered separate endeavors. They are components in a larger field that has moved from information processing to *office information systems*. These cycles will ultimately migrate into the office and will permeate all levels of the organization. No-where is strategic planning more necessary than in the office. Office automation has come of age. Tomorrow's office is here today. The information systems that will serve it already exist. We can buy them now. But we need strategic planning to determine which ones to buy and the way to install them gracefully.

Steps in Strategic Planning

1. Understand the Organization's Goals and Objectives. The first step in strategic planning is to know where the business or organization is going: what its major objectives are in terms of growth, return on investment, and the nature of products and services to achieve those objectives. The organization should have a firm definition of its role. It should have a broad-based understanding of its mission. The mission should be to accomplish the goals through maintenance of existing functions and adoption of new functions.

2. Establish Support and Commitment from Top Management. As in business planning, it is important to receive the endorsement of senior management prior to undertaking strategic planning. Many top-level executives are ready to be sold. They have been exposed to the frustrations of traditional office support systems. They probably have read the advertisements and understand the benefits of office automation. They must be receptive, however, to exploring the opportunities information technology offers. They must realize that results will occur from changes in the way business is conducted, and not from technology alone.

3. Determine the Way the Organization Is Satisfying Current Objectives. Conduct a survey to determine (define) the present status of office support systems. The survey may take the following forms:

a. *Feasibility Study:* A feasibility study is undertaken to determine whether specific operations can be improved and whether the installation of equipment (computers) and the adoption of revised systems and procedures are economically justified for these improvements. The current system is examined through questionnaires, volume studies, or data collection.

b. *Requirements Analysis:* Early feasibility studies were primarily concerned with typing. Through the evolution of office automation, feasibility studies have expanded into *requirements analysis.* Today, a requirements analysis must measure six application areas: word processing, data-base management, accounting and financial reporting, spreadsheet analysis, graphics, and communications. It also analyzes the internal situation (strengths, weaknesses) and the external forces (economic climate, competition) that may affect the organization's health and prosperity.

c. *Needs Assessment:* This is a survey that may focus on office functions such as productivity, applications, volume, storage, and compatibility of equipment.

The data collected in these surveys are important in the development of a precise statement of the organization's office automation requirements. Then the most appropriate and cost-effective systems can be selected.

4. Develop Specific Office Information Goals. The overall goals of your strategic plan to improve the office support system must be defined. Each organization must develop its own set of goals. Examples of some organizations' objectives and goals in office information systems may be

a. To support and improve the management process.
b. To increase management's flexibility and responsiveness to change.
c. To reduce barriers to effective decision making.
d. To integrate people with systems.

The objective of strategic planning is to identify the current and future information requirements of the company and to determine the way these requirements should be met.

5. Develop Strategies to Meet These Needs. This very well may be the most important step in strategic planning. This step is the actual generation of the plan itself. It is important to understand the total information needs of your organization and to expand your vision to look at all the technologies that are presently available and those that are anticipated in the future. The strategy may involve changing traditional office systems in depth: the delivery systems and the way support people interact with one another—and with the executives. It may require projects that attempt to change the way executives interact among themselves. Four major steps to achieve strategy development are

a. Evaluate alternative solutions.
b. Define discrete projects with deliverable time and benefits.
c. Outline implementation strategies.
d. Establish benchmarks against which to measure progress in the implementation stage.

6. Explore Alternate Choices. Offering a single choice might prove ineffective. In order to develop an effective strategy you must have alternative solutions for your problems. Therefore, strategic planners must know the state of the art of information systems. Planners must present more than hardware and mere dollars. The strategists must also know management and organizational procedures to solve problems.

7. Recommend Solutions. The solution involves more than tangible hardware and human resource allocation. It also involves a blending of the key areas such as people, technology, support, and training. One cannot rely on vendors to help find a solution. Guard against letting the dazzle of hardware lead you into the trap of "solutions" when there are no problems. Just because some feature exists does not mean your organization should use it. Obviously vendors will recommend their own brand of equipment as an ultimate answer for all problems. Independent thinking is essential. The correct piece of hardware or the software package that meets the specific needs of an organization should be part of the recommended solution. The first consideration should be the organization's information requirements, not the availability of the technology.

8. *Present the Plan.* This step involves communicating your plan to key managers and decision makers. It should take the form of a written and oral presentation. The written report should contain a step-by-step action plan. It should be written in clear, easy-to-understand language and should present your plan in a logical sequence. Terms should be defined and a clear set of recommendations on ways to implement the actions should be included.

After management has sufficient time to digest the report, an oral presentation should be scheduled. Do not expect everyone on the management team to read your report. Therefore, your oral presentation may have to be a brief summary of your plan. A question-and-answer session should follow. Management may either approve your plan or request that you reexamine some of the recommendations and return with a modified proposal.

Preimplementation

Once the strategic plan for office information systems has been approved, a few additional steps must be taken before the actual implementation. This preimplementation phase will help to ensure that the plan is carried out successfully.

Select a Team

Team members should be well versed in office functions and knowledgeable in most of the technologies. The implementation team should be composed of progressive people who are highly regarded and qualified to carry out the project. Ideal team members may include personnel managers, department managers (MIS and Office Automation), key operators, and secretarial support personnel. In addition to technical knowledge, the team should have sensitivity to human factors, technical processes, and procedures. Coordinating the process through a group leader with real authority is a necessary approach. The leader should have total responsibility for the project and skills that include people management, financial management, and knowledge of methods and procedures. Schedules should be determined. Realistic goals should be set, and periodic meetings should be scheduled to access progress and bring the members up to date on the overall progress.

Employee Participation

As the plan progresses and moves from behind the planners' closed doors into the workplace, it makes good sense to solicit employee cooperation on all levels. Strategic plans will ultimately affect the information workers, who should know what changes will occur and in what ways these changes will benefit their working lives. They should participate in the ongoing implementation process. Their involvement should be established from the beginning. This includes feedback on systems analysis, vendor selection, training, and procedures. Employee participation will ultimately bring goodwill and high morale to the implementation process.

Once the team has been selected and the employees are willing to participate, the implementation cycle moves into the actual implementation stage.

Implementation

Implementation refers to "doing" or "enacting." This is the "action" part of the plan that makes things happen. Remember, the overall goals toward achieving a successful office information system were spelled out in a strategic plan. The general goals for reorganizing the office information system usually include some or all of the following:

1. Integrating technologies across departmental and functional lines.
2. Delivering information in the proper form to the decision makers.
3. Linking devices through networks and telecommunications technologies to permit sharing and communicating of information.
4. Combining existing stand-alone technologies into cost-effective and capable hybrid systems.

The key in this overall plan is integration. The specific goals and the means to carry them out will be discussed in Chapter 6, "Planning for Integrated Information Systems." At this point, you are ready to put your plan into action.

Although there is no universal blueprint for implementing office information systems, the following general guidelines can help ensure a smooth and successful transition:

1. *Develop a Time Schedule.* This schedule should include the following:

- Equipment selection and purchasing
- Organization and staffing
- Site preparation
- Forms and procedures
- Orientation and training
- Installation and phase-in
- Review and evaluation

Consideration should also be given to any physical changes, training and retraining of personnel, alterations of job descriptions, redefinition of responsibility and authority, and changes in procedures or policies. Each detail and consequence should be considered before actually making changes.

2. *System Selection.* System selection involves addressing the vendor issue. Which vendor is best suited to the long-term office information needs? Who is the best vendor for achieving a smooth transition?

During the planning stage, a needs assessment was made. This data should now be brought forward to help in the selection process. Your judgment in selection of a vendor should be based on the following factors:

- Current state of the art
- Needs versus availability
- Future requirements
- Vendor support

3. *Designing Organizational Structure.* Determining organization and management strategy is a critical step. At this stage a decision about the way to plan for a new organizational structure should be reached. Office automation can change the way many departments are organized. The implementation team's input into the company's organizational plan is important. The variety and alternate choices of organization strategy and structure will be discussed in Chapter 4, "Organizational Structure and Management." Do you intend to have all of your support staff report to one person, with responsibility for performance appraisals, merit increases, promotional opportunities, and so on? The organizational reporting structure for office information systems responsibilities is not clear-cut. Much depends on the personality of the company. There should be a conscientious effort to integrate all of the information resource activities (OA, MIS, Telecommunications, Records Management, and other support services) under one management umbrella.

4. *Site Selection.* Site selection and space planning should be established. Careful consideration should be given to the following:

- Layout and design

- Furniture
- Ergonomic factors such as lighting, heating, ventilation, health, safety, and comfort of display terminals
- Accessories and supplies

Special workstations and equipment may require dedicated and conditioned wires and cables. Special air conditioning may also have to be considered to keep computers "up and running."

5. *Employee Orientation and Training.* Training and orientation take several forms. They may consist of general orientation in a class setting or specific skills training for management and staff employees. The training environment should be designed with comfort in mind. The training team may consist of the in-house implementation team, vendor trainers, or a combination of both. Vendor training sessions are usually provided at the site. Unfortunately, the vendor's office systems support staff is often limited and cannot provide the highly personal and effective "hand holding" needed in the early stages of learning. Users often learn the basic start-up steps and, after that, must prepare to fend for themselves. Experienced in-house users should be available for help as a means of follow-up learning. Training on computers is an elusive experience. The learning curve varies from person to person. The primary employee concerns are likely to include fear of change, technology anxiety, lack of self-confidence, and system adjustments. The implementation team should not expect immediate results but rather should observe a progressively improved rate of production over three to four months.

6. *Phasing In.* The best approach to implementing massive changes is through a gradual phase-in. This can be achieved in small steps over a period of time through careful planning and budgeting. The best order of phase-in is to implement a system at the highest level of sophistication. That will bring critical systems "on-line" and make them operational. This scheme will show immediate results and will have immediate payback. Peripherals and components can be installed in later phases.

7. *Feedback and Flexibility.* During implementation, it is important to solicit feedback from all levels of information workers to make sure you are meeting their needs. Listen to their comments. After all, they are the ones using the system daily. If their suggestions are reasonble and altering the overall plan will not be impractical, do so.

Problems that necessitate changes in plans are expected to occur. As the implementation progresses, you may need to revise the schedule. Although this revision will create a need for continued analysis, you will finally reach a point where most changes have been effected.

8. *Review.* Reviews will help assess how well the equipment and systems measure up in terms of response time, capacity of data storage, and user needs. During reviews, it is important to refer to cost data. Reviews and evaluations also help answer such questions as the following: How is the new plan working? How quickly can users learn the system? Is the equipment more efficient and effective than the old equipment? What changes or improvements are necessary? At this point, there should be a solid relationship between the vendor and the planning and implementation team. They should work together to iron out the rough edges. A spirit of cooperation and accommodation between these parties will contribute to the project's success; also, there will be improved efficiency and effectiveness within the organization.

Summary

The process of strategic planning and implementation is a never-ending task. In a highly technological environment no policy dealing with information systems should be inflexible. Nor can any organization be committed to a single vendor for a lifetime. New products appear within a short time span. Those that promise more efficient and productive service should be considered for implementation.

Successful implementation also involves keeping abreast of new information processing equipment and software as they become available. After the implementation has been completed and the system is "up and running," education and training must continue. Ongoing training sessions should be scheduled to enhance operators' skills. They should learn advanced functions on existing equipment. Periodic information-sharing sessions should be scheduled, to evaluate and consider new equipment, software, and peripherals. Information workers are becoming more computer-literate and astute. They are eager to experiment with newer technologies and systems. Planners, as well as users, must be continually alert and committed to keeping abreast of the market and changing technology. New technology will continue to thrive throughout this decade. The impact of these changes will continue to improve productivity in the office. Adequate planning and proper implementation strategies, coupled with sensitivity to the employees, can make the transition into the automated office a successful venture.

Review Questions

1. What factors determine whether or not an organization should automate office operations? Is automation necessary for all companies? Explain.
2. What is a business plan? What purpose does it serve?
3. List the elements that compose office information systems. Which is the most valuable element (resource)?
4. Briefly describe strategic planning for office information systems. How does it compare with (or differ from) a business plan?
5. What makes strategic planning for office automation so unique? Why is it especially necessary to plan for change in this area?
6. Briefly describe a feasibility study.
7. Cite some typical office information goals within a strategic plan.
8. What role can vendors play in the selection of hardware? What precautions must be observed?
9. How should the plan be presented?
10. List some preimplementation steps.
11. Describe implementation.
12. What is a needs assessment? List some activities involved in this phase of implementation.
13. What forms do training and orientation take? What role, if any, does the vendor play in training?
14. Explain how changes can be implemented in a "phase-in approach."
15. Once implementation is achieved, how can managers and information workers help maintain peak productivity and efficiency in work flow and office operations?

Projects

1. You have recently been appointed to the strategic planning team of your organization. The head of the committee has asked you to research the key elements of office information systems.

   You may draw upon this chapter as a resource and supplement your further research to describe ways that technologies are affecting changes in these key elements. Your findings should reflect the relationships between people and technology and possible effects on organizations about to embark on strategic planning and implementation.

2. As a leader of the strategic planning team you are required to select a team. You have decided to select key members of the Management In-

formation Systems Department and the Administrative Services Department of your organization. These two departments hold diverse views and philosophies regarding the allocation of equipment, flow of information, and manner in which support service should be handled throughout the company.

To compound this problem further, the acquisition of personal computers by executives has decentralized the control of information flow. It has also diminished the function of the MIS Department. There is a growing level of resentment between the staffs of the two departments. You need the expertise of these two departments to help plan and implement a new office information system. How can you bring these divergent groups together to work in harmony for the good of the company?

3. Establishing support and commitment from top management is an important step in any strategic plan. In order to establish this support, many of the executives in your company must understand the benefits of the systems and technologies you intend to introduce into the corporate environment. As part of an education process design a brief table explaining the benefits (strengths) of the following technologies: computers, micrographics, telecommunications, electronic imaging, word processing, and reprographics. Create a two-column table. Use the title "Information Technologies of Office Automation" as a main heading. The first column heading should be "Technology." The second column heading should be "Advantages."

PART TWO

Management Mandates: Information and People

CHAPTER 3

Information Processing

After reading this chapter, you will understand
1. Information as a corporate resource
2. The flow of information within the information processing cycle
3. The subsegments of information processing
   Input
   Processing
   Storage and retrieval
   Output
   Distribution
4. Ways to manage information

The introductory chapters provided the reader with a frame of reference for office information systems. We explored the necessity of developing a strategy that defines the objective and direction of a corporation's information management system. Arranging the key resources of people, information, and hardware within a suitable environment remains the primary mission of most business organizations. Each segment, however, must be examined in isolation. Terms must be defined. Then we can develop a clear understanding of the key resources and their interrelation to the concept of office information systems.

This chapter focuses on *information:*

- Information as a resource
- Information cycle and flow
- Forms of information
- Effective gathering, processing, managing, control, and dissemination of information.

Information: What Is It?

As described in Chapter 2, information takes many forms: data, text, voice, and image. It can be conveyed in a variety of styles and methods.

The basis for today's economy is rapidly moving toward information. The labor force is becoming predominantly a processor of information either by hand or tool.

Information is the lifeblood of an organization. In order for any business to operate, information must be exchanged. People exchange thoughts and ideas through *written communication*. The arrangement of written information is called a *format*. The variety of arrangements of written material are called *documents*. They include letters, memos, and reports.

Information can also be exchanged by people speaking with each other either face to face or over the telephone. To be effective, information must be communicated. In an organization, *communication* is the process by which managers establish a degree of "common understanding" with their employees, clients, customers, and the general public. Thus, communication is the means by which information and human thoughts are exchanged among people.

Information Stages

No matter how individuals exchange information, people respond to information in the following stages (Figure 3-1):

1. We *receive* information by listening to it or reading it.
2. We *organize* information by analyzing it, categorizing it, or manipulating it.
3. We *remember* information by storing it in memory or writing it down on paper.
4. We *distribute* information by telling it or sending it to someone else.

Information Processing

As documents move through the stages just described, they go through an *information processing* flow. Thus, *information processing* can be defined as the movement of a document through various stages to produce a desired result. The processing of information not only applies to the manipulation of text and data but also to integration with other office information technologies, such as image and voice processing. Other factors are involved in information processing: people, procedures, and the environment. Information processing involves a series of functions that are especially unique to an office environment.

Input

Input is the beginning step of information flow. It involves the creation and capture of

1. Receive Information
2. Organize Information
3. Remember Information
4. Distribute Information
Here it is!   Thanks!

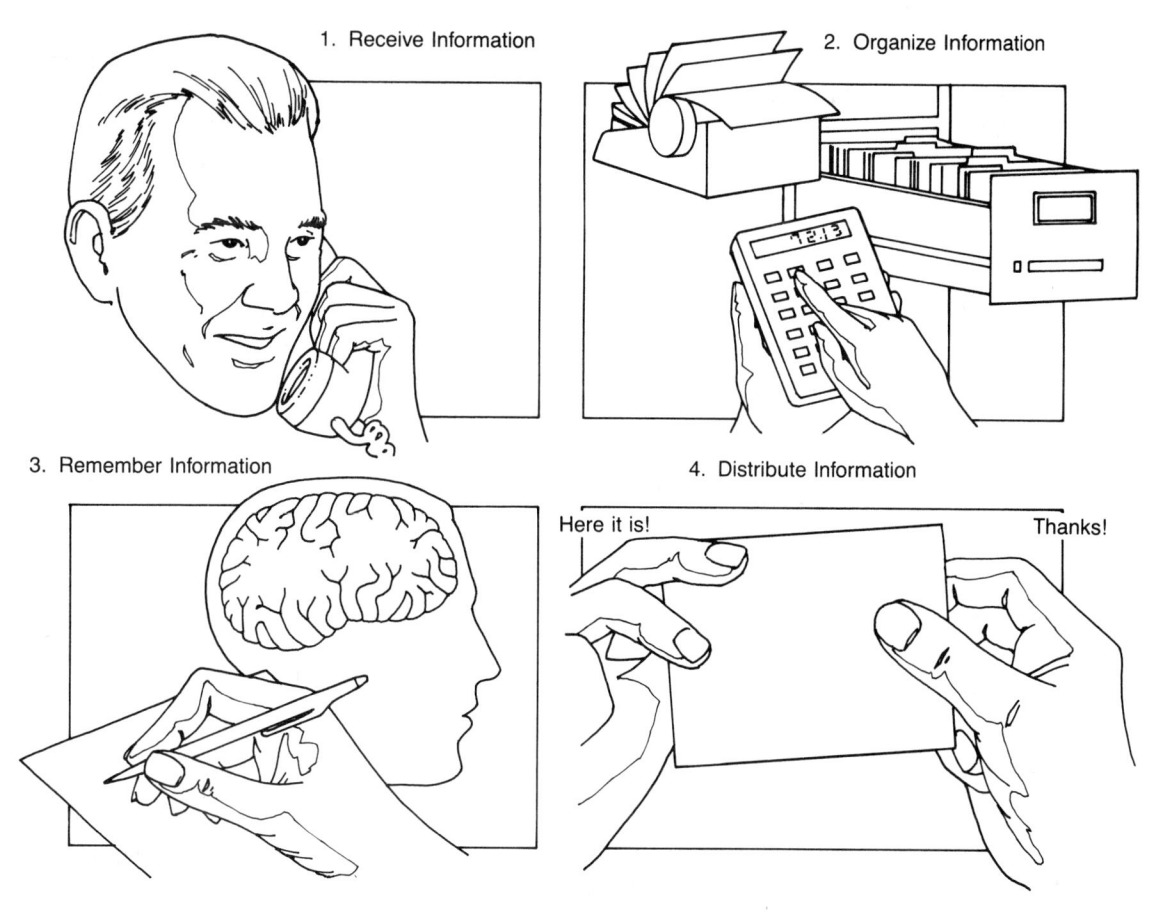

Figure 3-1. The stages of response to information.

information into an acceptable form. Ideas can be captured by writing, keyboarding, or dictating. These forms of information are in raw form; that is, they are not yet refined and processed.

In an automated office, input is usually captured by keyboarding or dictating. The benefits of automating the input function include increased input efficiency when capturing data and minimized retyping or rekeying of previously entered data.

Processing

Processing is the manipulation of previously captured (raw) data to produce information. Processing is sometimes referred to as *throughput.* It may be considered the workhorse of the cycle of information flow. Processing is the stage in which data such as text, numbers, images, and voices are transformed into forms and formats suitable for reading, digesting, and analyzing. The processing cycle uses the technologies of voice, data, word, image, and distributed processing.

Many of the machine functions can be generated or processed from the same piece of equipment, such as multifunction terminals or executive workstations. Integrated voice/data workstations allow originated information to be transmitted from one point to another via telephone wires, private wire networks, and satellites. The benefits of an automated processing cycle include the following:

1. Productive and efficient use of time
2. Immediate access to current information to aid decision making
3. Improved accuracy of computations

Several technologies are used in the processing phase:

Word Processing. Word processing is the creation and development of an idea from its inception to its final form through advanced electronic means. The key benefits of word processing are as follows:

1. *Keyboarding:* The first step in word processing is keyboarding the document. The information to be keyboarded may be in longhand notes or other forms. Keyboarding is like using a typewriter. As the information is keyboarded, it immediately appears on a display screen above the keyboard. However, when using a typewriter, one must set aside each completed page of type and insert another sheet of paper to continue keyboarding. Word processing eliminates this manual task. It electronically stores each page of information inside the system. Then it automatically clears the screen for a new page of text.
2. *Editing:* Once the information is keyboarded, it can be edited, changed, and formatted. Instead of cutting and pasting or making changes with correction fluid, one can simply insert, delete, and rearrange. These can be done electronically without retyping the entire document.

Word processing follows through with the final phases of storage, output (printing), and distribution. They will be discussed as separate functions of information processing.

Data Processing. Data processing is the manipulation and processing of data quickly and efficiently. The information obtained is timely, meaningful, and accurate. Computers are used to read incoming data (input), process it, and produce outgoing information (output).

In the context of data processing, *data* are defined as raw facts that need to be processed so that information is produced. *Information* (as used within a data processing environment) is defined as processed, structured, and meaningful data. Various levels of computers are used to apply these functions. Chapter 9 discusses the technological advances that have occurred in data processing and computer science.

Image Processing. Image processing is the conversion of data into visual and graphic representations: signatures, charts, and electronic photographs. This technology includes everything from one-color to multicolor graphics to high-speed facsimile transmission. *Facsimile* is a process in which a scanning device "reads" the original communication line by line and converts it electronically to digital impulses. The impulses are sent through telephone wires and "reconstructed" as a reproduction of the original document. Image processing is a young technology. It is rapidly emerging as an integral part of office information systems.

Audio Processing. Audio processing in the "processing" function differs somewhat from the "input" function. In the input process, thoughts and ideas were captured by electronically recording one's voice. Voice input technology is usually accomplished by *machine dictation,* a fast, efficient method of capturing initial ideas and data in the input stage. The input stage is the first step in the information processing cycle. *Audio processing* enhances the spoken input by advanced technology such as voice synthesis, voice store and forward, digital compression, voice recognition, and teleconferencing. Through advanced computers, newer technologies enable computers to recognize and respond to speech commands. The answer may be displayed on a screen, or the computer takes the required action. The potential markets for these developments have exciting possibilities in the areas of security, processing, and transcription of information.

Storage and Retrieval

Storage and retrieval are the placement, retention, and recovery of information. *Storage* refers to the systematic preservation of information. *Retrieval* is the recalling of the stored information for reuse. Information can be maintained in traditional paper files and handled manually or stored in electronic files for quick access.

Computers and word processors offer electronic storage and retrieval of information in two ways. The memory capacity within the system can store enormous amounts of information. Also information can be stored externally, or *off-line*. Off-line storage is outside the system, for example, on floppy disks. These two storage methods are described more fully in Chapter 15, "Storage and Retrieval."

The benefits of automating the storage and retrieval function include the following:

1. Minimized storage costs
2. Improved management of documents
   a. Access and handling procedures
   b. Retention and file purification (disposal or long-term storage of outdated files)
   c. Backup and recovery
3. Minimized risk of document misplacement
4. Improved response times for information retrieval

Output

Output is the process of transferring captured information from electronic storage into another form, which people can read and understand. Output may take the form of printing, photocopying, photocomposition, intelligent copiers, and other advanced forms of technology. Output may be alphanumeric characters or graphic images. A brief description of each of these forms follows.

Printing. Special printers are used as a peripheral to word processors and computers. Unlike a typewriter, which prints the document as you type, printers only print the document on command. When you instruct the system to print a specific document that is stored in the system, the printer produces a paper copy of the text. The printed copy, called *hard copy,* is a form of output.

Copiers. Photocopying is a photographic process of making multiple copies of an original. With the introduction of high-speed, plain-paper copiers, the process has greatly enhanced the output function of information processing.

Photocomposition. Photocomposition produces typeset copy directly from electronic data. Through phototypesetting, information is put into type through the process of exposing characters on photosensitive material such as photographic paper or film. This book was phototypeset.

Intelligent Copiers. Intelligent copiers or printers are devices that can be linked to a computer, word processor, or executive workstation. This device produces high-quality copies of the text that appears on the display screen system. Intelligent copiers can control the output of a network by coordinating input from computers and word processors.

Distribution

Distribution is the transfer of information from one place to another. This transfer of information occurs both between the employees within an organization and outside it. Internal distribution can be more flexible when an organization has the services to support it. Benefits of automating the distribution function include the following:

1. Reduced physical handling of documents since some internal information can be sent through in-house networks
2. Reliable and rapid delivery of documents and messages.

The following are forms of distribution:

Data communication. Data communication is the electronic transmission and reception of information among computers or similar remote terminals. Data communication is the heart of "telecommunication" and information distribution.

Electronic mail. Electronic mail is a form of data communication in that it provides for the transmission of information via electronic mailboxes. Electronic mail can be sent by facsimile, communicating word processors, computer-based message systems, and teletype. These methods are discussed in Chapter 12, "Telecommunications."

Teleconferencing. Teleconferencing is another form of electronic distribution of information. Basically, it is a meeting among geographically separated conferees. They are connected simultaneously via telecommunication systems and utilize two-way voice or video communication. The chief benefit of teleconferencing is that it eliminates the need for executive travel.

Telephone. The telephone is still the most frequently used mode of information distribution. It transmits voice simply and efficiently. Though not a new concept, telephony is important to our overall integrated systems approach. Advanced technologies have linked the telephone instrument to the computer.

Together they provide the transmission of data by using combined analog and digital transmission techniques. An outgrowth of advanced telephone systems is the introduction of *voice mail:* the recording, storage, and transmission of voice messages in electronic form using a computer-based system.

Networking. Networking is the backbone supporting all the technologies. The office of the future will certainly require products that form a total corporate distribution system. Many experts feel that networking is the key element within the overall office information system.

A network is a system that links various information processing devices to receive, exchange, store, and reproduce information. Networks have a variety of levels. The first and most important is *local area networks* (LANs). LANs consist of devices communicating to other devices within one building or site. Studies indicate that the majority of information generated from a given location is destined for points within the same location. Memos often take as long to reach a destination across the country. LANs alleviate this inefficiency.

Figure 3-2. Computer makers are starting to work together to make sure that they use the same protocol, or rules that allow computers to communicate with each other.

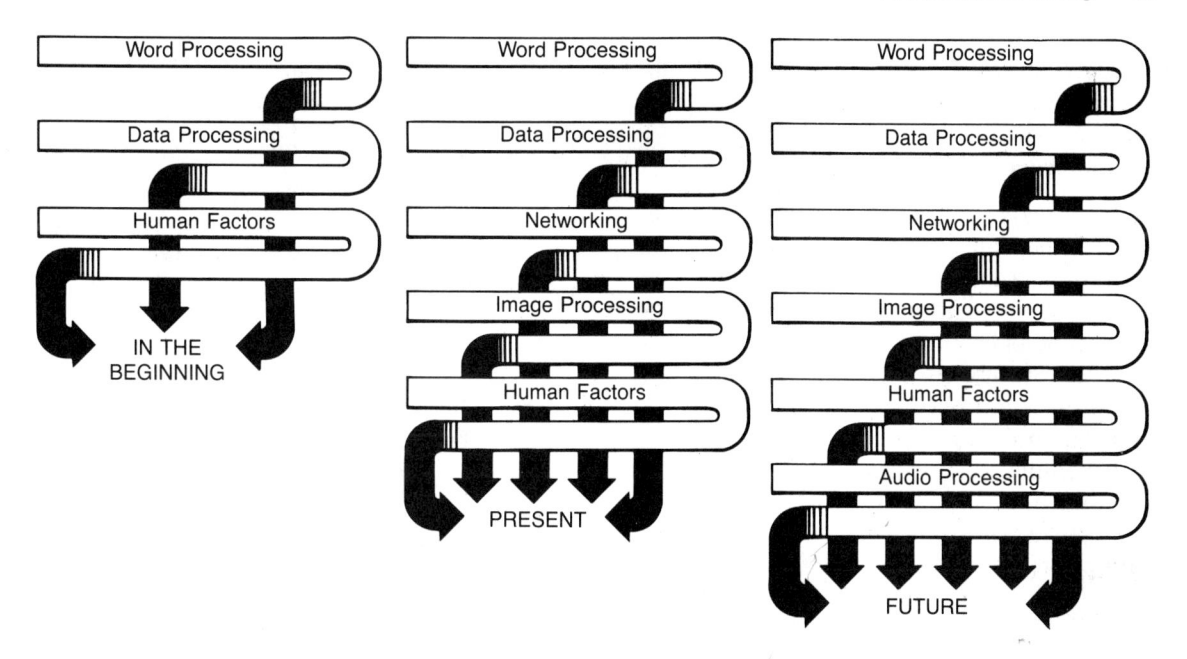

Figure 3-3. The evolution of office automation.
Source: Wang Laboratories.

Remote networking is another networking level. It is composed of devices and products that share resources and communicate between each other from site to site. Electronic mail and computer-based document distribution systems are examples of these applications. *Gateways* (Chapter 13) are a form of networks that allow different products and devices to communicate with each other. Enhancements such as *protocols* (Chapter 13) help these devices speak the same electronic language.

In the next few years, vendors will prosper by providing true solutions to the problem of *connectivity,* particularly those trying to deliver information in the marketplace. All of the crucial issues involving the processing and distribution of data come down to the same things: connectivity and networking.

The establishment of industrywide communications standards may prove to be a utopian concept, but many vendors are striving for this goal. Every computer user wants reassurance that his or her systems will be able to talk to each other (see Figure 3-2).

Processing functions present constant challenge to management on all levels. The human resource factor, although not classified as a separate technology, is crucial to the success of information processing. Document flow and information processing need human intervention. High technology must interreact with the human element. People must activate, touch, and command the devices. There must be a balance between people and technology. Thus, all of the information processing levels are meaningless unless we understand the ways that departments and office workers can extract benefits from each (Figure 3-3).

Managing Information

We have seen that information processing helps improve the flow and availability of information. We are capable of providing increased volumes of information for use in the management decision-making process.

A Tidal Wave of Information

Information is the end product and the key resource, when all engines of the information processing structure are working together. As improvements in computer and communication technologies occur, the volume of information keeps growing. Is communicating all this information necessarily a good thing? Is information the problem or the solution? Are we becoming obsessed with the technology churning out volumes of data? Should we focus our attention on the quality of information? In other words, are we overlooking the ends and giving too much attention to the means?

A Sudden Need to Manage Information

Information is transformed into *power* for the decision makers only when presented in a form that facilitates human understanding and action. The information should be related to a current, particular problem or project. The challenge is to manage the information processing forces so they harness information and assist the knowledge workers relate the data to their current jobs. Relevancy and time also contribute to the usefulness of information. As the amount of information presented increases, the delivery time of the data may be a key factor in management decisions.

Information Management Functions

Information is taking on new dimensions within the structure of today's dynamic organization. Its management revolves around technology and human productivity. Information management will demand management's special attention. It will become an organized business management function, much like personnel, finance, and marketing.

The major forces that caused this sudden need to manage information are the following:

1. The tremendous growth in information processing technologies
2. The need for relevant information for problem solving and decision making
3. The need to reduce *information float;* that is, information in transit
4. The need for access to information to keep abreast with competitors in the business environment
5. The demands for information across the corporate spectrum—marketing, finance, production, personnel, research and development

The major goal of information management is to *get the right information to the right place at the right time.* It requires careful planning, organizing, implementing, and managing the entire information processing function.

Information Resource Management

To manage information effectively it must be clearly understood that information is a valuable resource. Once this concept has been established within an organization, management can set about managing this resource just as it does other resources. The management plan is to design an information delivery system that provides instant accessibility to the data required. Then, they can make the necessary decisions. Redesigning a system necessitates sound and innovative management planning.

Traditionally, organizations did not feel that managing information was as important as technology resources. Unfortunately, information planning was overlooked.

The tide has turned, however. Because of the growing importance of information, traditional management principles are not suitable to meet management's needs. New concepts of planning and managing information are needed. The *human resources* of the organization must be part of the overall plan.

Social, psychological, and managerial disciplines necessary to develop and support information resource planning must be part of a new management approach. *Information resource planning* (IRP) is the process of identifying the fundamental structure of information available to an organization. This process involves the way the information should be accessed, processed, and delivered to support the organizational requirements.

Once an information resource plan has been established, *information resource management* (IRM) can begin to take shape. IRM is an overall approach for managing the information resources within an organization. IRP and IRM will become part of the organizational structure of most large companies. Other traditional management approaches will be inadequate to manage effective information resource environments.

The question of who within the organization will be given the mission to carry this out is partially addressed by the discussion of the chief information officer (CIO) (Chapter 4).

Or a new title may emerge. The person in charge may have the title of *information resource manager* (IRM) and may report directly to the CIO. Traditional MIS or data processing management philosophies dealing with specification, design, and maintenance of the information must be improved. Management functions are not keeping pace with the requirements of modern business. The focus has shifted from technology to people. The new information resources manager must have skill in interpersonal relationships, technical knowledge, and communication skills. The person in charge of IRM must have the necessary organizational position and influence to implement policies and plans effectively. The person should therefore be a member of top-level management.

Companies are beginning to realize that there is an information technology gap. Relying on a hierarchical decision-making structure, with decisions emanating from one powerful chief executive officer, may not be the best way to manage companies in the future.

A change of philosophy and direction is taking place in organizations. Top management people are becoming increasingly aware that their companies need to recognize the value of the information they produce. Many companies see the need to create positions such as information resource manager or chief information officer.

Summary

More organizations will change their management structures if they are educated about information. Then they will be willing to accept the concept that information is a resource. Terms like *decision support, corporate data base,* and *artificial intelligence* are becoming part of the language of information management teams.

There is a need for one person to manage all aspects of information. Right now, no one knows for sure where that person will come from in the organization. Data processing departments and MIS personnel will be given first chance. Although MIS has been successful in managing information in some organizations, the focus in other companies remains on technology. If they cannot cope with the newly emerging environment, the traditional MIS/DP group will give way to the emergence of a more humanistic, organizationally oriented group.

Managing information in the information age is important to the success of all organizations. As the industrial society continues to make the transition into the information society, enlightened organizations are going to do the following:

1. Elevate the function of information to its rightful place on the organization chart

2. Implement an information resource management plan
3. Create a new information title that will provide leadership for their company's future
4. Begin or continue to treat information as one of the major functions of business

Review Questions

1. What is the relationship between information and communication?
2. Compare format with documents.
3. Briefly list and describe the ways that people respond to the stages of information.
4. Define information processing.
5. What forms does input take?
6. Briefly describe the processing function. What hardware devices can be used in it?
7. What are the two key functions of word processing? What other applications can word processing systems perform?
8. Describe the meaning of *information* as used in the data processing environment.
9. What is image processing?
10. What is audio processing? What other advanced technologies relate to audio processing?
11. Compare storage and retrieval.
12. List the various forms of output.
13. Do you see a difference between data communication and electronic mail? Explain.
14. Describe a local area network. What benefits do LANs have that make them attractive to companies?
15. The textbook poses the question, Is information the problem or solution? How would you respond to this question?
16. What makes information valuable and usable?
17. What is information float?
18. Why have some organizations felt that managing information is not as important as managing other company resources?
19. Define and briefly describe information resource planning.
20. Describe the functions of an information resource manager. How does this person compare with a chief information officer?

Projects

1. Ed Whitson, the newly appointed information resource manager, has just completed an intensive information management seminar. It stressed that corporations and top management must be made aware of the need to recognize the value of the information they produce and the need to manage it properly. Ed Whitson, a former MIS professional, a soft-spoken, conservative technician, likes to work in his own unique environment. He often remains in his own MIS work group and is reluctant to engage in office politics. If you were Ed Whitson, what plan would you take to encourage your company to rethink information strategies? How can you, as an information resource manager, do the following:
   a. Elevate the position of information resource managers to the level of other resource managers within the corporation?
   b. Become a catalyst for bringing information systems into the mainstream of corporate life?
2. Draw a graphic or flow chart that illustrates the movement of a written document through the various stages of information processing. Label each stage and use connecting lines and arrows to indicate direction of flow.
3. The position of information resource manager (IRM) is a new approach to managing information within an organization. Research current magazines and publications and gather information on this subject. Present your findings in a report. Include the following:
   a. Personal and professional requirements for position
   b. Functions that must be managed
   c. An organization chart that includes the IRM as a director, vice president, or other top-level management position

CHAPTER 4

Organizational Structure and Management

After reading this chapter, you will understand
1. The structure of an organization
2. Management principles affecting organizational structure
3. Patterns of organizational structure
4. The evolution of technology and its impact upon today's organization structure
5. The way the introduction of a new technology (telecommunications) affects organizational design

The quest for improving information processing begins with the realization that there must be a better way to manage information. As the preceding chapters make clear, information is one of the key resources in today's competitive market. The major purpose of information is to improve the efficiency and effectiveness of the human decision-making process. Strategies, goals, and implementation are the means to develop intelligent information management. Designing new corporate structures that manage information, technology, and human resources is at the heart of bringing about successful change.

The odyssey toward improvement begins with studies to find out how office work is currently being accomplished and whether improvements can be made. Once the facts are known, the statistics are presented, workflow is diagrammed, costs are calculated, and evaluation is complete, the time is at hand to design an organizational structure that integrates these key components into a corporate environment.

This chapter will examine the challenge of designing a new corporate structure that integrates new technologies into an existing corporate plan. It will also discuss management policies and principles that relate to this new corporate environment.

Before we begin to examine new directions in corporate structure, it is necessary to define, explain, and compare traditional forms of organizations. These forms can apply to any enterprise and to nearly every activity found within an enterprise.

The Structure of an Organization

An organization is a group of individuals playing different but interrelated roles to achieve a common goal. Furthermore, the basic function of the organization is to blend together people, procedures, and equipment within a comfortable environment to achieve that goal. Organization structure is the established pattern of order or relationships among people, jobs, and work groups within the firm. The purpose of an organization structure is to provide an environment in which people can perform. The organization's internal structure determines the patterns of authority among its employees. The simpler the structure, the more easily the organization will be understood and the clearer the interrelationships among the workers will become.

Management Principles Affecting Organization Structure

Sound planning and intelligent implementation have a substantial impact on the development of the organization. Once established, the structure comes to life through fundamental principles of management. The following principles provide basic guidelines that are useful in designing and structuring various activities. These principles also play an important role in guiding the organization toward successful achievement of its goals.

1. Objectives. Objectives, or goals, are the end points toward which we aim. Objectives must be clearly defined, understood, and accepted by the individuals working within an organization.

2. Policy. A policy is a broad guidepost to action. It is developed to define and reflect each of the organization's objectives. The purpose of policy is to assure that those who make decisions within a company do so consistently and in the best interest of the firm.

3. Authority, Responsibility, and Accountability. A clear and specific channel of authority and accountability must be established. *Authority* refers to a relationship among the participants in the organization and is not an attribute of one individual. The person has the duty to obey the command. Authority is the right to invoke compliance by subordinates on the basis of formal position and control over rewards and sanctions. It is impersonal and goes with the position rather than the individual. Furthermore, authority and *responsibility* should be directly linked. An individual who is given the responsibility for a task must also be given sufficient authority to assure that the task is completed.

Accountability is associated with the flow of authority and responsibility and is the obligation of the subordinate to carry out his or her responsibility and to exercise authority in terms of the established policies.

4. *Delegation.* In an organization of any size it is impossible for the top manager to keep all the authority and make all the decisions. Authority and responsibility, then, are delegated throughout the organization to lower-level managers. *Delegation* is a function of supervisors who provide for the assignment of meaningful tasks to subordinates.

5. *Participative Management.* Participative management is related to delegation. It is a management process that involves the worker. Participative management ensures that everyone affected by a decision is fully informed about the reason for it. This process allows the workers to join in the discussion about their expected roles. It gives them the opportunity to discuss the best way to reach objectives that have been established.

The benefits of participative management accrue to both managers and workers. It involves the workers in the process of making decisions so as to tap their resources and expertise and arrive at the best possible decisions. Furthermore it provides group members a sense of input for the decisions made. They have a sense of responsibility to meet the defined objectives successfully.

Participative management must be executed with care and common sense. It should not advocate majority rule or democratic voting. Workers should not assume that their decisions will be automatically accepted or even open to negotiation. It is simply an opportunity for workers to voice their opinions and become involved. Involvement does not mean decisions must be unanimous. Neither does it mean compromise. When workers understand these ground rules, participative management can be an effective motivational tool.

6. *Span of Control.* The span of control, or span of supervision, relates to the number of subordinates whom a superior can supervise effectively. It is closely related to the hierarchy structure and departmentalization.

Hierarchy refers to the number of levels of management within the business. It is arranged by degree of authority.

Departmentalization refers to the process of establishing departments. Although there is no rigid formula for determining an appropriate span of control, several factors should be considered, such as (a) the skills of the manager or supervisor and the subordinates, (b) the complexity of the work, (c) the repetitiveness of the task, and (d) the amount of time that the manager or supervisor spends on nonsupervisory work. Figure 4-1 shows two spans of control.

A narrow span of control usually produces more levels of management. Figure 4-1 illustrates a narrow span of control with four workers reporting to one supervisor. This narrow span of control ensures sufficient communication within the group so that the supervisor can guide, evaluate, and "stay on top of" all developments. A wide span of control usually implies few managerial levels. It is practical where the work of a work group is less complicated. Figure 4-1 illustrates a wide span of control with eleven workers reporting to one supervisor.

Patterns of Organizational Structure

The best form of organization is the simplest. In designing an organizational structure, it is also important to provide a situation in which people can perform. Most organizations, regardless of size, will use several structures and

Wide Span of Control

Narrow Span of Control

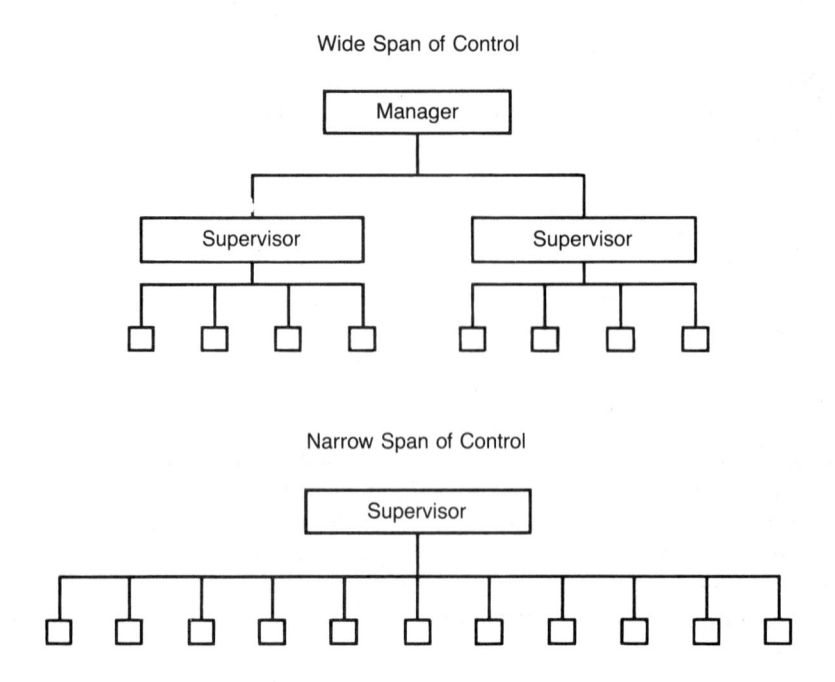

Figure 4-1. Spans of control: wide and narrow

design approaches. There is no single best way to design a company's organization.

Traditional structures are line structure, line and staff structure, functional structure, product structure, service department, committee structure, and matrix structure. Wise business planners will select a pattern of organization that fits their needs rather than just copy that of another firm, no matter how successful the other firm may be.

Organization Charts

An effective way to grasp the meaning of an organization is by using an organization chart. An *organization chart* is simply a graphic representation of an organization. Through boxes and connecting lines it depicts the line of authority, the span of control, and the responsibility for each function or department. By including names and titles, it immediately presents the hierarchical relationships among people. It is useful for identifying lines of authority.

The organization chart is used to identify promotional opportunities for job applicants and new employees. It also identifies areas suitable for training and orientation. In presenting the various organization structures, it illustrates different features.

Line Organization

A line organization is the oldest and simplest of the structures (Figure 4-2). There are direct and clearly understood lines of authority and communication flowing from the top of the firm downward, with employees reporting to only one supervisor. The direct authority characteristic of the line organization structure makes each employee directly responsible for the performance of designated duties. In othe words, the performance of duties is directly traceable to a worker and the one immediate superior in command.

Line and Staff Organization

In the line and staff organization, a line of command runs down the structure, as in the

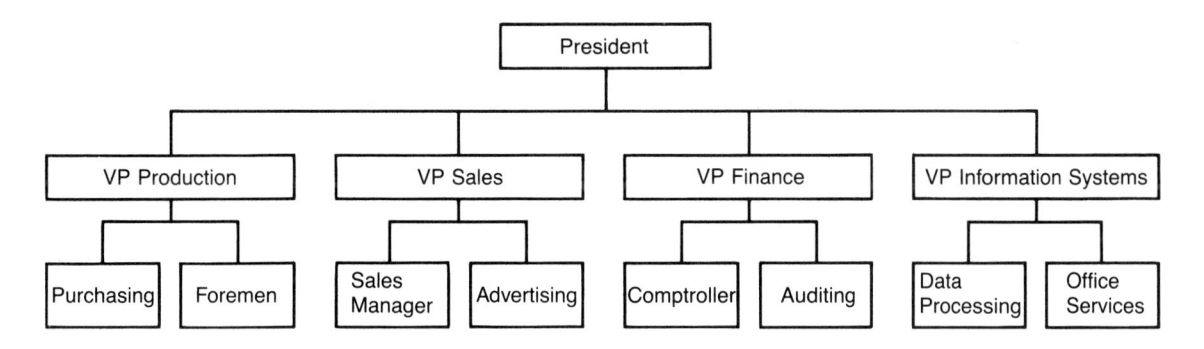

```
                        President
        ┌──────────┬──────────┬──────────┐
    VP Production  VP Sales  VP Finance  VP Information Systems
```

Figure 4-2. Line organization chart

case of the line organization, but other staff personnel are available to advise and consult on decisions (Figure 4-3). Some departments may be referred to as *line departments* and others as *staff departments*. Thus staff author- ity, based on expert knowledge, is the right to advise, assist, and support line managers. Staff managers usually have the necessary line authority to supervise their own departments. Their basic functions are to provide expert ad-

——————— Line Authority
- - - - - - - Staff Authority

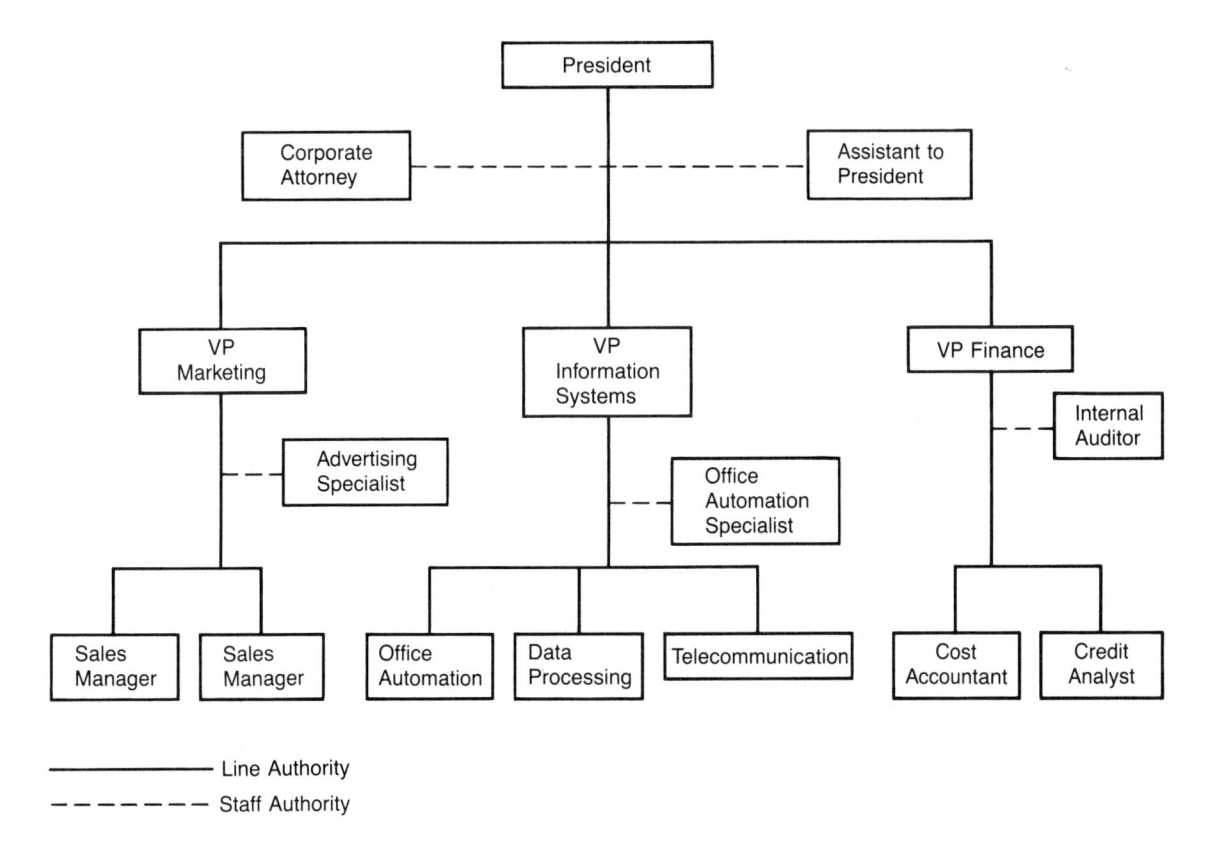

Figure 4-3. Line and staff organization chart

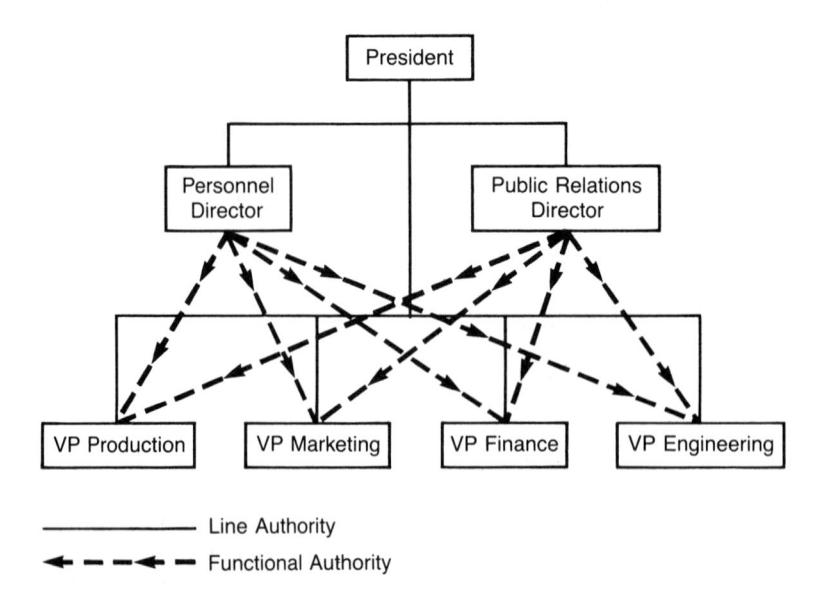

Figure 4-4. Functional organization chart

vice and to render services for the line managers.

Functional Organization

A functional organization organizes a company according to the major operations that the company performs (Figure 4-4). In a temporary help company, they may be office support, factory workers, health services, and transportation. In a manufacturing company, they are likely to be production, marketing, finance, and engineering. Each functional area is managed by a specialist who is likely to have the title of vice president. These managers have both line authority and functional authority.

Departments that are primarily staff in nature are often given limited authority over people who do not report to them. This is also

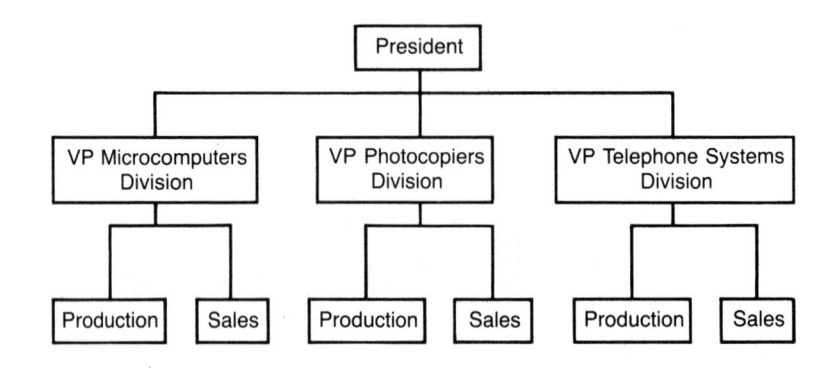

Figure 4-5. Product organization chart

true of many service departments. For example, in Figure 4-4, the functional authority of the personnel director or public relations director is really a part of the president's authority.

Product Organization

A product organizational structure is fashioned according to company product lines (Figure 4-5). Each major product is given a category or division status and a top-level executive heads the unit. General Motors Corporation has a product organization structure. Divisions, such as Oldsmobile, Buick, Cadillac, and Chevrolet, are separate entities. Product organization structure may be found in divisions within a company. The marketing divison, for example, may have people in charge of various product lines, and so may the engineering division.

Functional and product organizations are examples of *departmentation* grouping. The basis on which grouping is done varies. In addition to functional and product grouping, departmentation can occur by *customer type* or *clientele* or *geographical location* (see Figure 4-6).

Committee Organization

The committee structure serves the needs of an organization by allocating authority and responsibility among a group of individuals rather than a single manager. The shared ideas and views of the committee members are frequently accepted more readily than the recommendations of one individual.

By Customer or Type of Clientele

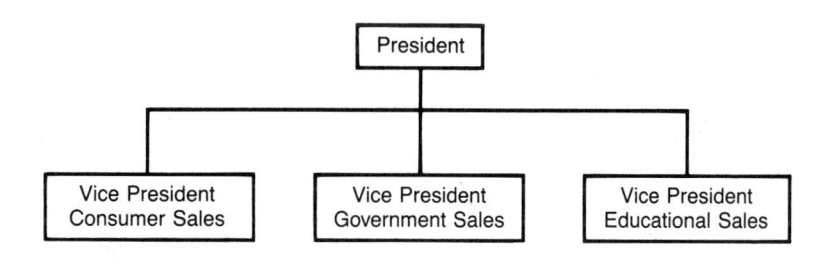

```
                President
        ┌──────────┼──────────┐
  Vice President  Vice President  Vice President
  Consumer Sales  Government Sales  Educational Sales
```

By Geographic Location

```
                President
        ┌──────────┼──────────┐
  Vice President  Vice President  Vice President
  Eastern Area    Western Area    Southern Area
```

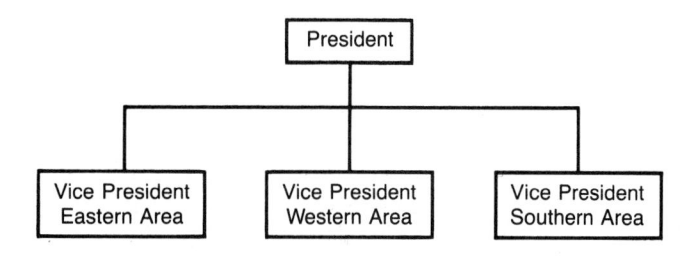

Figure 4-6. Department organization chart

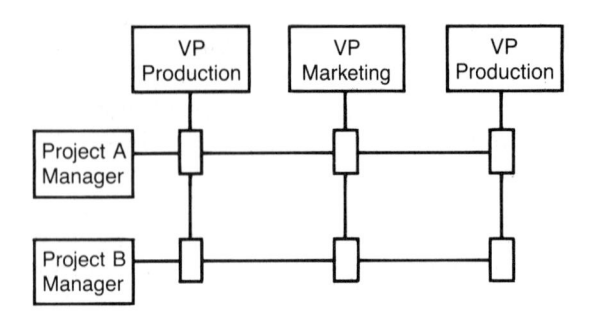

Figure 4-7. Matrix organization structure

Matrix Organization

Matrix structures are comparatively new and have been developed by firms in high-technology industries, such as computers, aerospace, and telecommunication. The essence of a matrix organization, as we normally find it, is the combining of functional and product patterns in a single structure (Figure 4-7).

The matrix structure helps an organization meet the special needs presented by a variety of projects. Thus, matrix management represents a kind of compromise between functional and product organizational arrangements. This structure integrates both vertical and horizontal relationships into a temporary new unit, called a *project*.

This form of structure is well suited to companies that are primarily interested in the final product or project. For example, vice presidents of production are in charge of production schedules for each project under them (vertical authority). But, at the same time, a special supervisor with technical expertise may be in charge of that entire project (horizontal authority).

Organizing for Office Information Systems

Technology is changing the "look" of traditional organizational structures. We see evolution and turmoil as the information systems industry moves into one phase and out of another. There are new relationships, lines of authority, and hierarchies. All of the structures may apply in the information organization, or none of them. There is no one correct way to organize office systems and integrate them into an existing company. Implementation occurs at different times and with varying degrees of intensity. Lines of authority and changes in tasks must be introduced gradually. Teamwork is essential. Yet, no one style or structure works for all companies.

The Evolution of Technology and Its Impact on Today's Organization Structure

Forces Affecting Organization Strategy

There are trauma, confusion, and excitement in the quest to create a structure that accommodates integrated office systems. The cycle of change has moved through the industry with intensity. Alfred Lord Tennyson said, "Let the great world spin forever down the ringing grooves of change" in 1842. He should observe what is going on now.

Strategies implemented during the last century to deal with the industrial world of Tennyson's day are certainly outmoded today. We are thrust into the electronic world of computers and communications. The pace of technological change has accelerated to a point where it seems out of control. What are some of these forces? How do they apply to redesigning organizational structures?

Decentralization

In the early stages of office systems design, centralization (Figure 4-8) of office and information systems became prevalent. Management felt that centralizing information flow and access was cost-effective. In a centralized

CENTRALIZATION

```
                    Corporation
                         |
                        MIS
   ┌─────┬─────┬─────┬──┴──┬─────┬─────┬─────┐
 User  User  User  User  User  User  User  User
 Dept. Dept. Dept. Dept. Dept. Dept. Dept. Dept.
```

DECENTRALIZATION

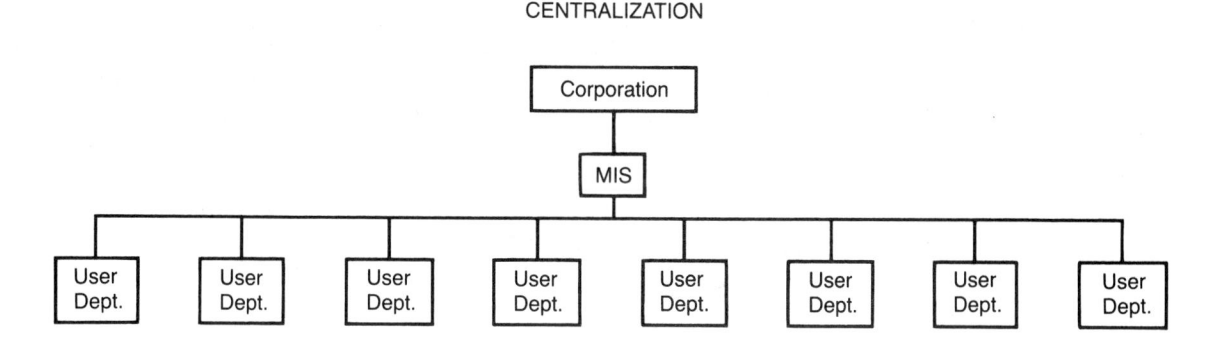

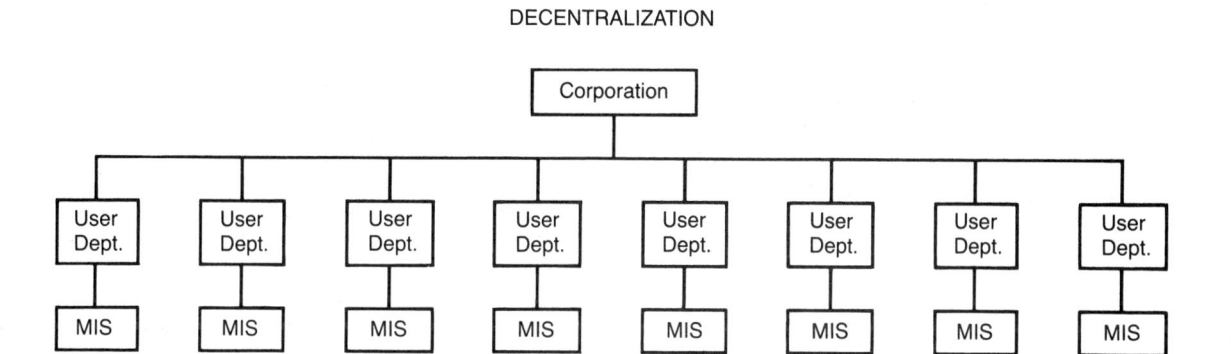

```
                    Corporation
   ┌─────┬─────┬─────┬──┴──┬─────┬─────┬─────┐
 User  User  User  User  User  User  User  User
 Dept. Dept. Dept. Dept. Dept. Dept. Dept. Dept.
  |     |     |     |     |     |     |     |
 MIS   MIS   MIS   MIS   MIS   MIS   MIS   MIS
```

Figure 4-8. A corporation with a centralized MIS department has many user departments that use MIS. When MIS is decentralized, each user department gains its own MIS work group, which handles only the department's needs and whose expenses come solely from the department's budget.

environment, large computers were located at a single site with a large and complex support staff organization.

To retrieve information from the centralized system, one needed to access the central facility. The centralized approach was the first approach used because no other method was economical at the time. *Decentralization,* on the other hand, allows several departments or individuals located throughout the organization to control their own activities. There are small computers, word processors, and photocopiers in departments or work groups with communication links into the main computer.

As the price/performance ratio of microcomputers improves, corporate managers and planners find themselves facing an ever-expanding array of tools. Decision support systems, end-user computing, and personal computers have assumed permanent positions in the modern, large corporation. One effect of these tools has been to change the way people do their work. A more subtle effect has been to change organization structure to accommodate decentralized users.

Thus, the current trend is toward decentralized or distributed systems.

Personal Computer: Key to Decentralization

The merging of distributed data processing (DDP) and office automation (OA) played a key role in decentralization. Both were seen as taking computing power from the data center and bringing it closer to the actual users. Each was a partial solution: Office automation focused mainly on administrative personnel; and DDP remained largely a traditional data processing function, but on a smaller scale.

What managers of information did not foresee was the impact of the personal computer, which placed unprecedented computing power and flexibility into the hands of individual users. Those users leaped at the chance to automate their own little projects. As personal computer prices dropped, MIS managers lost control over information access and flow.

MIS Departments Regain Technological Expertise

As organizations stepped up their bulk purchases of PCs, MIS directors decided micros would help them more than hurt them. Survey data tend to support this view of growing MIS dominance.

This is the conclusion of a study that contends that office technology decisions are quickly becoming the realm of the MIS/DP manager and not the office manager. The study "Technotrends '85: Inside the American Office" was published by the Omni Group, Ltd., who conducted 315 interviews with members of planning committees, administrative services, department representatives, and MIS managers in Fortune 1000 companies. Seventy percent of the companies surveyed said MIS is the most influential department for personal computer purchasing. MIS is responsible for deciding which software packages will be bought for personal computers in 74 percent of the firms surveyed.

Another survey, by *Computerworld*, a weekly trade magazine, found that 82 percent of the responsibility for acquiring and controlling micros was with the MIS managers. MIS support for personal computers has planted the seed for some innovative approaches such as information centers, micro centers, and company stores.

The Information Center

The information center is not a place, but rather a concept. It is designed to offer support for end-users. It is the area within a business organization where end-users can go to learn how to use computer resources and get the data they need. It makes users work to solve their own problems.

Originally, information centers were rooms in which end-users could gain immediate access to management-information-systems (MIS) data. Staffed by professionals trained to formulate searches of the mainframe data bases, these early centers provided a means of satisfying growing demands for data without increasing the number of hands-on terminals.

The information center of today still fills this role and has expanded its support services to include all aspects of information processing. With the tremendous growth in the use of personal computers and office automation devices, the information center's staff must be prepared to answer questions concerning a range of technologies beyond MIS and PCs: questions about electronic mail, network file transfers, advanced data bases, and word processing.

The objective of setting up an information center should be that it will be a corporate command post for all aspects of information. These centers should be managed by their own personnel. They should be separate from traditional data processing functions. However, they might work closely with them and fall under them on the organization chart.

The staff of the information center must be patient, diplomatic, organized, and good at communication. They must also have a working knowledge of the business and a broad understanding of the tools to be used.

The Chief Information Officer

Information centers are having a major impact in reshaping corporate structures. The concept is also causing some rethinking among corporate executives as to who shall manage this new corporate unit and at what level of authority. With the importance of the information asset in a growing number of companies, there is a sudden need to establish an enterprise-level resource manager called the *chief information officer* (CIO).

The CIO is a generic title for an upper-level executive (usually a senior vice president), who participates on an equal basis with other corporate officers. Together, they formulate company policies.

The chief information officer really understands the interconnection between the information flow and the business. On an operational level, the CIO has implied (dotted-line) authority and functional control over telecommunication, word processing, and data processing (Figure 4-9).

CIOs need to relate to both management and its functions. They need to have enough technical knowledge to translate management's ideas into the systems they want. The chief information officer's job has acquired prestige and urgency in decentralized companies. The CIO usually reports directly to the chief executive officer or similar corporate general manager. This change is important. The elevation itself is central to the concept. The new reporting relationship underscores that the CIO now has a companywide responsibility and orientation.

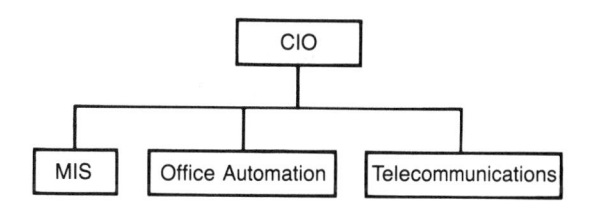

Figure 4-9. An information department headed by a chief information officer

Perhaps Thomas Peters and Robert Waterman were thinking about the emergence of the CIO in their book *In Search of Excellence,* when they describe an effective leader. "An effective leader must be the master of two ends of the spectrum: Ideas at the highest level of abstraction and actions at the most mundane level of detail."

The Company Stores

There is a need to support users in their quest for using micros and other information tools. The concept of a company store has emerged in several organizations. In-house company stores, or *microcenters,* started because managers recognized the need to introduce micros to employees in a way that was less intimidating. Company stores are central locations where employees with an interest in microcomputers can go to select hardware and software and to receive advice and training. Company store staffing and equipment can range from one microcomputer with a part-time staffer to a specially designed and constructed microcomputer store within the company. Unlike a retail computer store, for a company store education is the key to the whole idea. Physically, company stores may be a combination of a display area and a training facility. Research and technical review of new hardware and software options are carried out by company store staff members on a continuing basis. In addition, a library of evaluated software is maintained for users to try when they are searching for a new application.

Information centers and company stores can offer organizations excellent opportunities to support users in building an awareness and acceptance of microcomputers. The MIS department's control of mainframe computers will soon give way to these newer organizational concepts. Computing has steadily moved out of the computer room, first with minicomputers and now with microcomputers. Software that once ran only on mainframes now runs on minicomputers and is migrating to microcomputers. We are entering a new age of computing, better served by information centers and computer stores as new entities within a corporate structure.

Telecommunication and Its Effect on Organizational Design

Perhaps no other technology will affect the reorganization of a corporate structure more than telecommunication. The pace of change in this area proceeded very slowly for the first ninety years. All of a sudden, technological advances burst forward in a torrent of events. The most profound are competition and technological breakthroughs. First there was the Carterfone legal decision in 1968, which led the courts to question the monopoly held by AT&T. That was followed, in the early 1970s, by opening the floodgates to competing long-distance services. Technological breakthroughs—in signal processing, transmission, switching, software, and space launch—are contributing to the world's telecommunications growth. And finally 1/1/84—"Divestiture Day."

The Trauma of Divestiture

The trauma of divestiture and deregulation has sent shock waves throughout corporations. All of a sudden there was an urgent need to restructure companies' business communication resources.

Before divestiture, AT&T ran the networks and discouraged the establishment of non-telephone company networks. Telecommunication management was relegated to an unimportant role in many companies. Departments were small, with very little budget and little political clout. Those in charge simply responded to end-user demand for circuits and telephones.

When divestiture came along, corporations suddenly found themselves in charge of resources that AT&T used to control. Yet many organizations did not have the technical know-how to deal with this new responsibility and freedom.

Telecommunication Departments: A Bright Star on the Organization Chart

Now organizations must look on the bright side of this postdivestiture communications-department world. They must rebuild internal management structures and revamp procedures to fit the changing environment. As networks become integrated, long-distance service, hardware, and organizational issues must be decided. Who should be in charge: perhaps the CIO, the MIS manager, or a separate telecommunication manager? In many cases it seems to depend on the personality, experience, political strength, and business savvy of the individuals involved.

Summary

As we move deeper into the information age, there is a new imperative for managers to redesign their traditional organizational structures. For years, the pattern of organizations—the reporting structures, scope of authority, and span of control—worked well in American companies. But the proliferation of computers and the broad dissemination of information are changing all that.

The old style of management is giving way to a networking style referred to as *participa-*

tive management, in which every employee participates in decision making. Desktop computers, communication, new information processing departments, and decentralization are some of the technologies and techniques that will affect organizational structure and management.

This is a time of transition in organizational structure. New trends in interactive computing, office automation, and telecommunication have brought the functional components closer together. Each area has its own set of specific technologies. A wide range of solutions is available. New technologies are beginning to alter the shape of traditional organization structures. No one formula works for all companies. The lack of role models has led organizations to be resourceful. They must implement structures that are innovative.

In larger organizations, for example, a line manager may be responsible for each area. In smaller organizations, some of these functions may be combined under fewer departments. The organizational reporting structure for information systems may not be clear-cut for every company. Much depends upon the corporate "personality" of each. We can, however, make some assessments of the trends:

1. Companies must rebuild internal management structures and revamp procedures to fit the changing environment.
2. As technology advances, separate telecommunication departments must be established. They must be equal to other information resource departments in terms of prestige, budget, and personal training.
3. A new breed of manager must emerge to lead these information-intensive departments. The job should be at the director or VP level. It has the overall responsibility for the organization's information resources. The job holder should be responsible for interpreting the group's mandate, for presenting its proposals to top management, and for meeting the objectives established for the unit's operation.
4. The position of chief information officer should be established to manage and co-ordinate the total information responsibilities of the organization.
5. Support for computer users should be established in the form of information centers and company stores.
6. Office technology decisions are becoming the realm of the MIS/DP manager, who is becoming the most influential resource for personal computer acquisition and guidance.
7. The office of the future will be decentralized. It will consist of small work groups. Traditional hierarchical structures will give way to informal networks.
8. Finally, the human factor must be considered in planning, implementing, and redesigning corporate structures. User participation implies help in providing data to every active participant.

Office information systems are no different from other management functions. The traditional, formal organization chart found in college textbooks (or even in the figures of this chapter) probably will not be what you will find in the real business world. The analogy might be compared to the formal structured lesson plans of teachers who schedule every task into neat forty-one-minute packages. Things do not always work according to a preconceived plan. There must be flexibility and resourcefulness. Organizations must have structure, lines of authority, and reasonable management policies. And they must not be inflexible. As systems become friendlier and are installed with software that is flexible and easy to use, organizations will find the various aspects of the technology shifting. The lines that connect boxes on an organization chart may shift and change as users' needs evolve.

The challenge for management is to try matching corporate needs and human resources. With respect to office productivity, once management has achieved this formula, the organization will be in tune with today's environment.

Review Questions

1. What is the structure of an organization? Why is it necessary to establish a structure within an organization?
2. Authority and responsibility should be directly linked. What does this mean?
3. Briefly describe delegation. How does proper delegation benefit the company? The worker? The supervisor?
4. What ground rules must management make clear to workers when engaging in participative management?
5. Describe span of control. Which type would be useful to a manager who must supervise complex and highly creative work?
6. How does the line of authority flow in a line organization?
7. What is the role of staff personnel in an organization?
8. Briefly describe the product organization style. Can you cite an example of a major corporation (besides General Motors Corporation) that uses this form of structure?
9. Define an organization chart. How would an organization chart help new or prospective employees?
10. Briefly compare centralized office operations and decentralized office operations. What key factor has led to the increase of decentralization in most companies?
11. The textbook indicates that organizations look for leadership, guidance, and advice from MIS personnel in selecting computers. Why do you think this is so?
12. What is an information center? Why do you suppose it was created?
13. Describe the duties of a chief information officer. What qualifications are necessary to become a CIO?
14. What are the similarities between an information center and a company store?
15. Describe the role of the telecommunication manager within the corporate environment prior to the divestiture of AT&T.
16. Which key departments fall under the realm of office information systems?

Projects

1. Ethel Zalinsky, a manager of office information systems, is planning to upgrade her existing terminals to personal computers. She has asked her employees to participate in deciding which model to select. For the past month, she has suggested that members of her staff visit trade shows, observe demonstrations, and present their selections based on the applications of the work group.

    After reviewing the choices from the members of her staff and consulting with the MIS department, Zalinsky has made a selection. Victor Nebre, one of the members of the staff, was furious that his choice was not selected and expressed his disappointment in a meeting after the selection. He indicated that this exercise was a farce and that workers' input was useless. He suggested that the manager knew exactly what brand to select before the workers were brought into the picture. His accusations were made in front of the group. How would you handle this situation? What would you say to the group pertaining to participative management? What would you say to Victor Nebre?

2. For many years, the CIO position has been more idea than reality at Comptel Computer Corporation. It was a dotted box sketched onto a corporation's management hierarchy with no name to fill it. The company organization consists of a president and the following reporting relationships: a chief financial officer, to whom the following horizontally equal departments report: director of MIS, director of office auto-

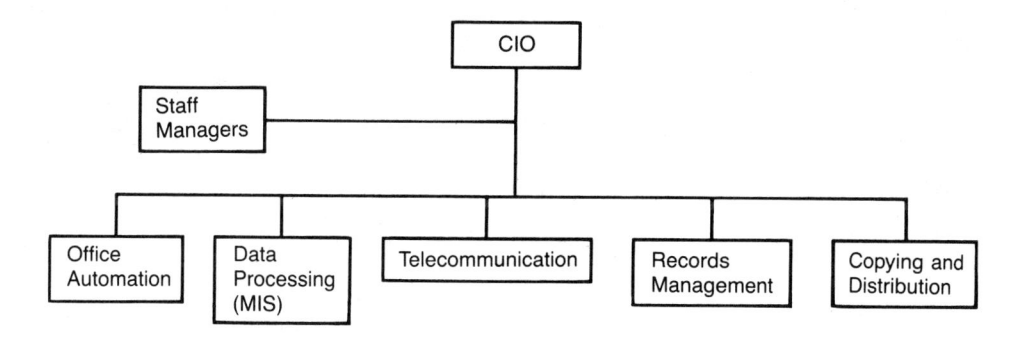

```
                          ┌──────┐
                          │ CIO  │
                          └──┬───┘
        ┌─────────┐          │
        │ Staff   ├──────────┤
        │ Managers│          │
        └─────────┘          │
    ┌──────┬──────┬──────────┼──────────┬──────────┐
┌───┴───┐┌─┴────┐┌───────────┴──┐┌──────┴───┐┌─────┴──────┐
│Office ││Data  ││Telecommunica-││Records   ││Copying and │
│Automa-││Proc- ││tion          ││Management││Distribution│
│tion   ││essing││              ││          ││            │
│       ││(MIS) ││              ││          ││            │
└───────┘└──────┘└──────────────┘└──────────┘└────────────┘
```

Figure 4-10. Project 3 corporate structure

mation, and director of telecommunication. Draw an organization chart of a decentralized line structure that incorporates the new position of chief information officer into the hierarchy.

3. Your organization has redesigned the corporation structure reflected in the organization chart in Figure 4-10. As chief information officer you must assign the following information-processing functions to one of the following operational level departments. Complete the organization chart and indicate the placement of each of the following functions:

Word processing
Batch processing
Telephone systems
Local Area Networks
Information Center
Archive Maintenance
Mailing services
Microfilm
Information retrieval service
Secretarial support

CHAPTER 5

Personnel: Managing Strategies in a Changing Office

After reading this chapter, you will understand
1. The role of the information manager
2. The way an information manager can motivate employees
3. The ways that personal problems such as stress, alcoholism, and drug abuse affect job performance of workers
4. The effects of environmental and external forces on personnel policies
5. The influence of laws and regulations on information workers and the work environment

We learned in Chapter 4 that there has been an evolution in the structure and design of organizations. Decentralization, integrated information systems, desktop computers, and new job titles are some of the forces shaping these changes. Information technologies, such as word processing, data processing, and communication, have reshaped organizations and brought together workers from different backgrounds. New titles for workers, supervisors, and managers that are more in tune with the modern corporate structure have emerged. The title of chief information officer, for example, has emerged to distinguish a high-level resource manager who brings together a diverse mix of people and technology in a new work group. What are the attributes that characterize a successful supervisor or manager? What are the responsibilities and challenges of the "new information manager"?

We have explored some of the job titles, responsibilities, and requirements for information workers. Now, let us turn our attention to the person who must lead these workers.

The Role of the Information Manager

Any management position entails generally accepted standards that apply to all such positions. Managers basically have the responsibility to direct, control, plan, and implement in order to obtain the best results. The process of managing extends from top management through middle management to the supervisory management level. Our discussion in this chapter will be devoted to examining the information manager at the middle- and lower-level positions.

These positions typically deal with the daily handling of people and operations. At this level the information manager plans and organizes work experiences in such a way that both the organization's needs and the employees' needs are met. Beyond these broad areas of responsibilities, the role and duties of the information manager are changing to accommodate new procedures, advanced technologies, and a new computer-literate work force. The change revolves around working with a team approach. The old notion that secretaries seek change and managers seek to maintain current roles and strengthen the established distances between managers and workers has been replaced by a new philosophy.

Today's information manager seeks to narrow this organization gap. The manager knows that there is a tremendous benefit to be gained through the effective integration of the entire work team. Managers are finding that their staff can contribute significantly. The manager fosters a oneness of thought and purpose. In this environment, automation is viewed by both workers and managers as a positive force to improve work quality, hasten the pace of routine tasks, and provide more time for creative work.

Qualifications

Where does an information manager come from? Who will manage the integrated work group? What are the qualifications for the new information manager?

Because many organizations have established integrated information systems, managers must have technical knowledge in office automation, data processing, and telecommunication. The people occupying these positions today may have been recruited from the data processing ranks or, in some cases, from a telecommunication background.

Ideally, managers should have a bachelor's degree or several years of appropriate work experience, including supervision, and thorough knowledge of the area to be managed.

Titles

The advent of technology has dramatically changed the structure and reporting relationships within organizations. New departments, new titles, and new leaders are emerging. Legions of office personnel with new functions, labels, and responsibilities have sprung up—without benefit of formal job titles or descriptions. Managers and supervisors of information fill relatively new positions within organizations. Information managers' titles vary: They may be called *information processing managers, micro managers, office service managers, office information system coordinators,* or *information services managers.*

Departments using microcomputers as the predominant processing tool have developed their own set of supervisory titles. The managerial candidate becomes the in-house microcomputer authority. His or her title may be *director of microcomputer operations* or *microcomputer support specialist.*

Duties

The job titles and the descriptions of duties that follow serve as general guidelines only. The dynamic nature of office information systems makes it difficult to establish a uniform set of responsibilities that apply to all managers. Duties of an information manager may include some or all of the following:

- Responsibility for overall operation of department or work group
- Design and implementation of future information systems
- Coordination and assignment of work
- Design and implementation of a management reporting system
- Responsibility for budgets, overall production reports, and coordination of services
- Management of the operation of document preparation, photocopying, printing, mailing, graphics services, MIS, and telecommunication monitoring projects
- Potential management of other major information services, such as records and retention and microfilm services
- Interviewing, selection, and training of new personnel
- Evaluation of performance and production of employees
- Understanding of corporate goals and responsibilities

Characteristics of the Successful Manager

The information manager is first and foremost an administrator. This type of person must think and act with an open mind. The information manager must provide the leadership required to bring together all the talents and skills necessary to form a successful team. Choosing a leader is, perhaps, one of the most important management decisions.

Leadership, planning, and imagination are basic requirements for any managerial role. Integrated work groups and restructured information centers are relatively new to organizations. The new manager may have relatively little background, experience, and credentials to bring to this new position. Some are thrust into their new positions almost overnight. This is where a wise and perceptive selection policy can pay off. Insight into a person's ability to adapt to new environments, respond to situational factors, and exhibit sensitivity to the employees' human needs is a hallmark of success. Just like the president who grows in stature when undertaking the many duties connected with the presidency, so the new information manager grows in stature when undertaking the many duties connected with an integrated office. The following are some of the characteristics and traits that a manager should possess.

Leadership. Leadership is one of the most significant characteristics of an effective manager. It includes the capacity to direct the work of others and to organize and integrate their activities to attain the goal of the work group. The leader must know how to handle people at the operating level and must be able to get along with both men and women. Work must be done efficiently and rapidly. Getting sufficient work done without too much grumbling and resentment is a measure of leadership skill.

Communication. Messages to employees should be clear and direct. Do not take anything for granted. Hidden agendas and messages are barriers that create difficulties for the uninitiated. The effective manager must realize that there is a great deal more to communications than simply words. Nonverbal, verbal, and written communication and feedback are an integral part of effective communication. Effective communication is a two-way process requiring at least two people—one who sends information and one (or more) who listens, receives, and acts upon information. The *listening* aspect of communication may be the part we do least well. The effectiveness with which you communicate ideas and instructions will play a large part in determining your success as a manager.

Fairness. Fairness is creating a climate of equity toward employees. It is a psychological climate perceived by an open employer-employee relationship. Managers must be able to deal with the everyday problems that, at times, threaten to disrupt the efficient functioning of their departments. They must be ready and willing to see both sides of problems and to solve them fairly and reasonably.

In a fair and open climate, people are prepared to express their feelings and anxieties without fear of recrimination. Openness tends to foster a better understanding of the people involved and may well lead to a better solution for the problem.

Trust. Workers who trust managers will have confidence and respect. Good supervisors will keep all their promises and support their cooperative workers, even when the supervisors are dealing with other departments and with top management. Trust is the basis upon which relationships are established and maintained.

Discipline. Discipline is a common responsibility of the manager who must deal with employees who do not perform as expected or behave in acceptable ways.

Discipline involves more teaching and positive persuasion than reprimanding or punishing. To be effective, discipline should be constructive so that employees can learn from their mistakes. Consistency is an important aspect of discipline. The manager's reactions to like situations should be similar. The manager's reaction to employee A who has failed to comply with an order should be the same as that for employee B's failure to comply with the same order.

If the manager is not feeling comfortable with the actions of a particular individual, the most satisfactory way of handling this discomfort is to discuss and resolve the area of conflict as soon as possible. Generally, reprimands are necessary and they should be made in private.

Delegation. Delegation is a positive and useful process that entrusts meaningful tasks to workers. It transfers to a subordinate, on a temporary basis, something that the supervisor or manager would ordinarily be obliged to perform. Delegation is not simply a straightforward mechanical operation, in which the manager "hands off" some selected function to a subordinate. Along with the delegation should go sufficient authority to get the job done. The instructions should be communicated precisely. The authority should be given to a worker who has demonstrated ability.

Managers who cannot resist doing the work themselves are unskilled in delegation and often make poor leaders. Many managers find it easy to do the work but encounter problems when they have to delegate it to someone else. Competent information managers soon develop dependable people to whom definite tasks can be assigned. These people are necessary in any work group and can help solve some of the routine problems. Intelligent delegation can mean a challenge for workers as they begin to realize that their manager is willing to give them an opportunity to learn. It also helps extend the supervisor's capabilities, encourages team work, and results in higher productivity.

Flexibility. A manager in the dynamic field of office automation requires a great deal of supervisory flexibility. The office is a place where new practices and systems are emerging at a rapid pace. Information managers must be alert to new developments and flexible enough to incorporate those operations that can be of help. Flexibility also means having the good sense to abandon equipment or procedures that are inefficient.

Education. Vendors of information systems are constantly introducing new and upgraded equipment. Using such equipment effectively requires men and women who possess greater training and skills. New methods of operations and software applications may require more than just an in-house training session by a manufacturer's representative. It may demand upgrading one's education by attending seminars or enrolling in college courses. Computer literacy, management skills, and government regulations are areas of knowledge that managers must keep up to date.

Professional Growth. In addition to attendance at seminars and enrollment in college courses, membership and participation in professional associations can help a manager's career. The Data Processing Management Association, the Association of Records Managers and Administrators, the Association of Information Systems Professionals, and the International Communications Association are a few examples. Membership in a professional group is a form of recognition and a great opportunity to influence your profession. Increased status and self-esteem may be the results.

Motivating Employees

Of all the characteristics for a successful manager, the skill of *motivation* may be the most vital. To motivate employees, the manager is responsible for creating conditions that enable workers to use their abilities and to realize their potentialities. Motivation creates the desire in employees to perform effectively. Threatening, cajoling, or teasing may get people to do what you want, but it is certainly not an effective way to motivate them.

The first stage of motivation must be to create an environment in which employees can function and interact more effectively. It must be an environment in which workers are enthusiastic and have a desire to work. Motivation has been the key to successful management and productivity for organizations throughout the world. Lee Iacocca rejuvenated the dying Chrysler Corporation through employee motivation. The Japanese practice a continuous, day-by-day process of motivation in their factories and offices. It has become a national management code of conduct. Whatever the reason, organizations are beginning to take a variety of approaches toward enhancing motivation.

Theories of Motivation

Assessing a person's motivational level involves some subjectivity, however. Although certain elements are required, there are often no right or wrong answers. Theories of motivation have been formulated throughout the history of the labor movement. The following are some of the motivational approaches in dealing with workers.

McGregor's Theory X and Y

Douglas McGregor expressed concern over the lack of opportunities for higher-level-need satisfaction in many jobs in business and industry. He reasoned that management traditionally has tended to direct and control closely the actions of its employees because of false assumptions about human behavior. McGregor has stressed the point that the "assumptions we have about people affect our behavior toward them." According to McGregor, there are two different types of assumptions we can make about employees that will affect how we behave toward them on the job. These two different sets of assumptions he called *Theory X* and *Theory Y*.

Theory X. *Theory X* is a point of view that sees people as having tendencies toward cruelty, hate, and destruction, with the need for close control and regimentation. An extreme example of this approach would be "If I held a .37-magnum to your head, the likelihood was that you'd go out and do what I asked."[1]

Theory X makes certain assumptions:

1. The average person has an inherent dislike for work and will avoid it if he or she can.
2. Because of this dislike for work, most people must be coerced, controlled, directed, or threatened with punishment to induce them to put forth adequate effort toward the achievement of organizational objectives.
3. The average person prefers to be directed, wishes to avoid responsibility, has relatively little ambition, and wants security above all.

---
[1] Colby, Wendelin, "Motivation in Motion," *Infosystems* (August 1985) page 81.

Theory Y. Theory Y, on the other hand, suggests that effort invested in work is as natural as effort spent in play or recreation. It is possible, and in fact necessary, to provide sufficient scope to a job (latitude, responsibility, freedom, and so on) that employees can integrate their personal goals with the goals of an organization. Theory Y originates in the belief that employees typically use only a small fraction of their intellectual and creative ability to work. If employers develop the right organizational climate, employees will unlock their hidden potential and become both happier and more productive.

Theory Y's assumptions are as follows:

1. People do not inherently dislike work.
2. Because work can be as natural as play, people do not have to be checked on to see whether they are working.
3. People will assume responsibility and willingly direct their efforts toward the objectives of the organization, provided that in doing so they are able to satisfy their own higher-level needs.

A manager holding Theory Y's assumptions would, therefore, concentrate on satisfying the employees' needs, which are closer to Abraham Maslow's views on worker behavior and motivation.

Maslow's Theory on Hierarchy of Needs

According to Dr. A. H. Maslow, a person's wants are always increasing and changing. Once an individual's basic (primary) needs have been satisfied, other (secondary) needs take their place. To satisfy these needs, people expend energy. However, once a need has been somewhat satisfied, it no longer acts as a motivating force. Individuals then begin to invest their energies in higher-level needs (Figure 5-1).

Maslow's theory of motivation stresses that people are motivated to satisfy many needs, some of which are more pressing than others.

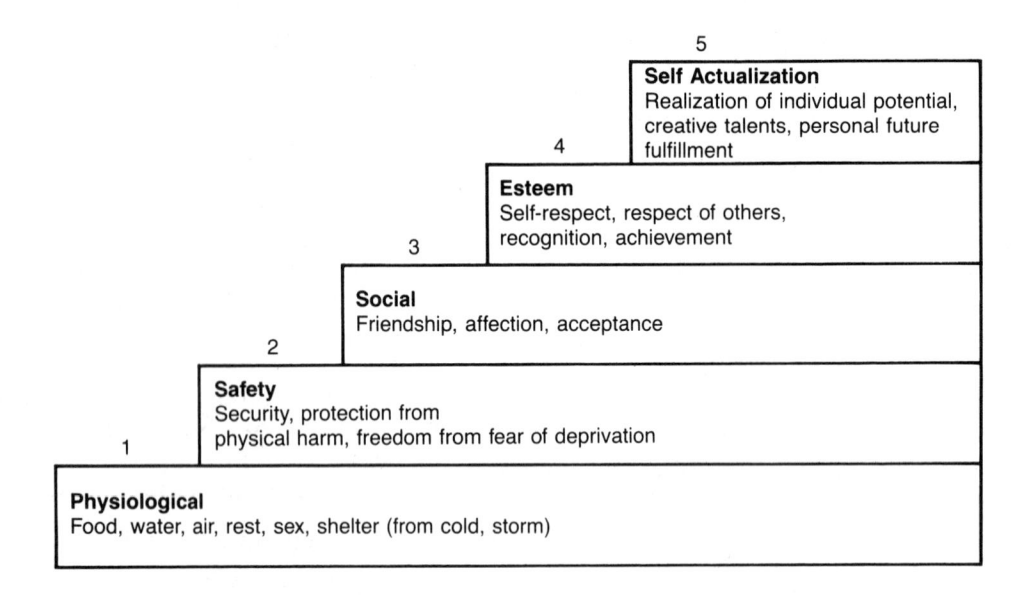

```
                                          5
                                    ┌──────────────────────────────────────┐
                                    │ Self Actualization                   │
                                    │ Realization of individual potential, │
                                    │ creative talents, personal future    │
                              4     │ fulfillment                          │
                          ┌─────────┴──────────────────────────────────────┤
                          │ Esteem                                         │
                          │ Self-respect, respect of others,               │
                     3    │ recognition, achievement                       │
                      ┌───┴────────────────────────────────────────────────┤
                      │ Social                                             │
                  2   │ Friendship, affection, acceptance                  │
                   ┌──┴─────────────────────────────────────────────────── ┤
                   │ Safety                                                │
                   │ Security, protection from                             │
              1    │ physical harm, freedom from fear of deprivation       │
                ┌──┴────────────────────────────────────────────────────── ┤
                │ Physiological                                            │
                │ Food, water, air, rest, sex, shelter (from cold, storm)  │
                └──────────────────────────────────────────────────────────┘
```

Figure 5-1. Maslow's hierarchy of needs

If a number of needs are unsatisfied at any given time, the individual will move to satisfy the most pressing one(s) first. Maslow identified five levels in his hierarchy of needs (see Figure 5-1).

Physiological needs are the most pressing. Once our physiological needs are largely satisfied, the next level of needs in the hierarchy begins to emerge. These are *safety needs*, among them the avoidance of physical harm, illness, economic disaster, and so forth. In a similar manner, satisfaction of the safety needs gives rise to the emergence of *social needs,* then *esteem needs,* until the satisfaction of all these leads the individual to be primarily concerned with the highest-level needs, those of *self-actualization.*

Maslow believes that all levels of needs probably exist to some degree for the individual most of the time. Rarely, if ever, is any one need completely satisfied, at least for very long. Our hunger, as a simple example, may be fairly well satisfied after eating lunch, only to emerge again before dinner time.

New Breed of Manager

The new breed of enlightened information manager will bring to the job a greater understanding of motivational skills and technical solutions. The new manager can learn from past motivational theories and workers' behavior patterns. The new manager will create an environment in which employees can function effectively. Effective managers can motivate workers by talking to them, being open with them, and accepting the challenge of finding the right jobs for the right employees.

A good manager has to know how to schedule people, how to motivate them, and how to get things done. Technical competency must be blended with human resource skills. New technology offers tremendous challenges for the future.

An even greater challenge is the management of human resources necessary to plan, develop, and incorporate this technology. Successful managers must be able to harness the full potential of the staff. To do so, they must develop a climate of openness and trust. If their team is prepared to question the status quo, they must be creative and learn to take risks.

Human Resource Management in a Changing Society

In addition to understanding the principles of effective management within the office, there exists a rapidly changing corporate environment that must be met and dealt with. Forces at work in the regulatory, economic, and technological climate have contributed to this revolution in both structure and scope. In addition, personnel problems and practices commonly found in organizations will be discussed in this section.

An understanding and awareness of personal problems should be the concern of the information manager. If not handled and resolved, these personal problems may interfere with the employee's job performance.

For example, environmental changes, personnel, and human relations problems are among the various stresses that affect interpersonal relationships and the performance of groups of people.

These forces are exerting a profound impact on personnel trends within organizations and at times these forces overlap. The following sections describe some of the human resource issues found in the workplace.

Stress

There is a quiet crisis taking place in the office workplace. The growing level of office automation can be a blessing to some and a curse to others. Clearly, the growing computerization of the work force has brought its own form of stress. The machines can perform miraculous feats of storing and calculating. They can spew forth productivity reports with the stroke of a key. They can make some tasks so simple and repetitive that the boredom factor overrides any gains in productivity that may have resulted.

The opposite end of boredom is anxiety, trauma, and stress. The sudden conversion to new technology places everyone on a new learning curve. Each person in the work group—manager, supervisor, and staff—learns at different levels. This may increase the potential for complicated and unpredictable results.

Some employees have a difficult time making a genuine adaption to new technology. They experience emotional upsets at having to learn so many new tasks so quickly. And computers can cause deep insecurity for workers who simply do not catch on to their use as quickly as some of their peers do. The information age has brought with it a new kind of computer psychology. It pertains to the question of whether human beings have the biological and social capabilities to adapt to the advanced technological environment that they are creating. Some experts have labeled this condition technostress.

Technostress may be defined as a condition resulting from the inability of an individual or organization to adapt to new technology. Stress in and of itself may not be harmful. It can challenge and energize. It can bring individuals to a surge of vitality and prompt them to pursue tasks with renewed enthusiasm. However, stress can be harmful and counterproductive when it places overwhelming pressure on individuals. The results can be loss of productivity, poor attendance, costly errors, and levels of depression. Computer technology has placed undue stress on information workers who cannot cope with major changes in a healthy manner.

Information managers must take an active role in working toward a solution to this problem. They must find a graceful path for the proper integration of the computer into the workplace and into the lives of the workers.

Strategies must be formulated to introduce new technologies gradually, in a series of phases. A humanistic approach is the key to technological phase-in. Each individual reacts differently to change. Some accept it more readily than others. The manager must understand individual workers: meet with them, orient them to the new system, and stress the positive virtues of automation and the new technology. Effective training and mastery are other phases that encourage employees and build confidence. These phases will be discussed in Chapter 21.

Outside help can be brought in to complement the manager's efforts to alleviate technological stress. Tech psychologists are specialists in computer psychology and may help to identify computer-caused problems and to offer solutions.

Stress affects executives as well as workers. Although many executives claim to prefer work to leisure, vacations are essential to combat stress inherent in a fast-paced business life. Many so-called workaholic executives confess to not taking full vacations for years.

Specialists in stress recommend to such executives, "Rather than kill yourself or burn out, learn to relax. Take regular nonworking vacation breaks to recharge your batteries."

Technological stress, like other forms of stress, should be identified early. The stress brought about by a sudden introduction of automation and change can exist throughout most offices. Developing humanistic programs can usually solve or minimize stress problems. By alleviating the trauma and anxiety of stress, information workers can prosper in a technological world.

Another kind of stress is creeping into the workplace. It is not caused by a sudden influx of new machines or technology. It is taking place in corporate boardrooms, where high-level company mergers, takeovers, and buyouts are occurring with increased frequency. These corporate maneuvers are having a disturbing effect on workers and their commitment to the company.

Working within an environment of uncertainty and job insecurity can cause levels of stress for the employees. Workers need a workplace that has a reasonable level of security and predictability. What is needed is a positive employee-boss relationship.

Corporate mergers and takeovers are and will continue to be a fact of business life. There are no easy solutions. Management can lessen the disruptive effects on workers by better employee-boss relationships. Workers should know what lies ahead. They should be informed about ways that their working conditions will be affected by corporate mergers, takeovers, and other changes. Both outgoing and incoming corporate management should share in an obligation to safeguard the nation's human resources.

> Without work all life goes rotten. But when work is soulless, life stifles and dies.
> —Albert Camus

Alcoholism and Drug Abuse

Alcoholism and drug abuse, like stress, can drastically affect job performance. Alcoholism is a progressive disease characterized by the uncontrolled consumption of alcohol. It has reached epidemic proportions in our society and has permeated into the office work force. Workers suffering from alcoholism are victims of a progressive illness that must be recognized, accepted, and treated. Otherwise, slow deterioration and eventual disaster will occur in all aspects of their work careers and personal lives.

Because managers are often close to the workers in their departments, they may be asked by corporate officials to help identify people who are abusing alcohol. Since there are degrees of alcohol abuse, supervisors and managers should exercise extreme tact when approaching employees. The manager who perceives that an employee is drinking excessively may attempt to discuss the problem and offer advice on how to cut down. Certainly, a manager should not accuse a person of alcoholism. Most organizations have special programs and policies for dealing with alcoholics. An *alcoholic* can be defined as a person who is powerless to stop drinking and whose normal living pattern is seriously altered by drinking. When this person has been identified, the supervisor may take the following steps:

1. Review policies with your immediate supervisor.
2. Determine whether your company has a counseling service.
3. If your organization does not have a counseling service, identify and contact an agency or consulting group that provides such assistance.

New attitudes toward alcoholism have helped these troubled employees rebound from their affliction. Alcoholism can be treated. Progressive companies have established in-house treatment centers or work with professionals who attempt to rehabilitate the alcoholic worker.

Drug abuse is a condition in which the abuser exhibits strong psychological dependence on drugs. As a result of this drug addiction, employees cannot function or perform at their peak efficiency. A drug user is often preoccupied with getting the next "fix." Cocaine is the newest and most devastating drug. It is estimated that 22 million Americans have tried cocaine and 4 to 5 million are regular users.

Crack is a concentrated form of cocaine that causes addiction among users much more quickly than regular cocaine. The quick high it produces, the depression afterward, and the yearning for more of the drug also are more extreme. The use of crack and the spread of cocaine addiction is rising among middle-class and professional people. Despite the ravaging effects of the drug, and its high addictiveness, crack users can be rehabilitated. The problem is to get white-collar addicts to acknowledge that they are hooked. Often, there is also a high level of denial on the executive level of addiction among professional and white-collar workers.

As for alcoholism, progressive attitudes must prevail in treating these debilitating diseases. Today, organizations believe there is value in helping employees take responsibility for solving their drug and alcohol problems. Managers must be concerned with the physical and mental well-being of their workers. They must have a positive attitude and adopt a supportive, corrective philosophy.

Other Personal Problems Found in the Office

Other forms of mental and emotional illnesses less severe than drug abuse and alcoholism are conditions that plague a productive work group. A variety of mental health problems such as anxiety, low self-esteem, and jealousy can occur. They may stem from stress, job insecurity, unpleasant work conditions, and hazardous work. Harmony in the office can be an elusive goal. The complexities of human behavior and interpersonal relations are beyond a detailed discussion in this section. Managers must, however, be on the lookout for a worker's conflict, unrest, and other issues that affect job satisfaction. Among these are the following:

>   Excessive tardiness and absenteeism
>   Marginal employees
>   Nepotism
>   Smoking in the office
>   Sexual harassment
>   Office romances
>   Petty theft
>   Foul language or racial slurs
>   Release of confidential information

Some of these personal problems have been regulated through federal laws. They will be discussed in detail in a later section of this chapter.

Acquired immune deficiency syndrome (AIDS), which is usually fatal, is caused by a virus that attacks (depresses) the body's immune system. Its victims are most often homosexuals or drug users. Although the Center for Disease Control has said that it cannot be spread by casual contact, corporate managers face a critical problem when a worker with AIDS is identified. At present there is no known treatment and doctors cannot speculate when they might find a cure. This is the dilemma that AIDS represents. The disease, which has now spread to all continents, was first identified in the United States in 1981. Public hysteria connected with the disease frequently has made persons with AIDS societal outcasts. Because of this groundless public prejudice, a U.S. law has been enacted to protect those with AIDS on the job. Under the law on employment, a handicapped person is deemed to be qualified if an individual analysis determines that he or she can, with "reasonable accommodation," perform the essential functions of a job.

In most cases, the disease will not present a safety or health risk that would justify exclusion, segregation or otherwise discriminatory treatment of persons with AIDS or antibodies to the virus.

Environmental and External Forces Affecting Personnel Policies

Women in the Workforce

The changing makeup of the work force has created new policies relating to hiring, training, salaries, and promotions. The male-female composition of the labor force changed dramatically during the 1980s (Figure 5-2). More women are in the work force than ever before in our country's history. Projections show that we will soon reach the point where women will represent over 46 percent of the workers. Economic and social conditions have prompted women to enter the work force as a second wage earner in families with children. Divorces have left many women as breadwinners, often the sole or primary support of a family. The number of divorced, widowed, or separated working women with children under eighteen increased by 75 percent in the 1980s.

Figure 5-2. As of 1986, women outnumbered men in the professions.

Source: Photo courtesy of Computer Consoles, Inc.

Not only are more women entering the work force, they are aspiring to more management positions. The path toward wide gains for women in management has been frustrating and slow. Traditional attitudes and values are difficult to change. Degrees have helped women enter corporations at the same levels as men, but diplomas have not ensured comparable careers. Numerous studies have revealed that women managers and administrators earn on the average only half of what their male counterparts do. Some analysts indicate that many of the old barriers for women managers persist. Mary Ann Devanna, former manager of Columbia's Career Development Center, says, "People get ahead easier when they are similar to the people evaluating them. Women will progress more slowly because they're different, by definition, from the existing power structure."

Many companies have begun to overcome entrenched attitudes about women, partly because of legislation and partly because presidents and chief executive officers have daughters and granddaughters in business schools.

Unions and the Information Worker

During the early period of the labor movement, little effort was expended to unionize office workers. Now the unions are increasing their efforts to attract the millions of white-collar workers. The nation's unions are adopting new methods of organizing. They are emphasizing new issues that are geared to the health, safety, security, and professional concerns of the office worker such as autonomy and input to decision making.

Unions are also wooing more young workers. They are trying to recruit part of the 52 million American workers under the age of thirty-five, who represent half of the nation's labor force.

White-collar (clerical, professional, technical, and managerial) workers offer a new field for unionization. These individuals seek the same goals as their blue-collar counterparts in the industrial sector—more money, greater security, and satisfactory working conditions. Periods of continuing inflation, technological changes, and routinization of work have stimulated interest in unions among office workers.

Office workers may not respond, however, to traditional organizing methods. Individuals in this category usually have more formal education and present new challenges to unions. In addition to the basic needs of money and better working conditions, they care about flexible schedules, computer technology, and career development—issues that do not concern traditional blue-collar unions.

Unionization of white-collar workers continues to meet with resistance from management as well as the workers themselves. Many young office workers oppose unionization for the following reasons:

1. They are concerned that unions are stuck in their old ways.
2. They feel that unions create inefficiencies that could eventually ruin a company.
3. They perceive that unions create unrest. They feel that at times unions are involved in illegal activities.
4. Many young workers take jobs in service industries such as insurance, banking, and computer programming. They feel that unions have little success in these areas.

Unions, however, are trying to counter these objections. They are doing more to educate workers in the advantages of unions and hiring more organizers than at any time in the past three decades. Aggressive unions have sent organizers to enlist workers into corporate offices.

A new breed of information worker has presented new challenges to unions and has raised new policy questions for the information manager dealing with a unionized work force. Today's workers are looking for growth, upward mobility, and training in new skills. Union leaders are beginning to recognize the needs and desires of the information worker. They are developing strategies to organize white-collar industries as well as high-tech companies.

Laws and Regulations in the Personnel Field

The job of the information manager must go beyond technology, leadership, and human relations skills. Today's managers must know the *law*. New job titles, skills, and a new work force have created much legislation dealing with personnel issues. The following important laws and regulations affect information workers and the workplace.

Civil Rights Act

The Civil Rights Act was passed in 1964. This act deals with equal employment opportunities (EEO) for persons who may have been treated unfairly in the past. One section, Title VII, affects the hiring practices not only of business firms but also of unions and employment agencies, and it defines certain actions as unlawful employment practices. The act deals with fairness in hiring, evaluating, promoting, and dismissing employees. Some firms have developed specific goals of hiring a certain percentage of persons from minority groups and women. These goals are often part of an affirmative action plan.

Affirmative Action

Affirmative action involves an employer's setting goals and taking positive steps to guarantee equal employment opportunities for everyone. The Office of Federal Contract Compliance Programs (OFCCP), a branch of the Labor Department, enforces equal opportunity in companies that do business with the federal government. The requirement specifies that government contractors adopt goals and timetables for hiring minority people and women. Although no universal affirmative action plan is mandatory throughout the private sector, corporations have initiated affirmative action programs on their own. Debate continues about whether or not affirmative action achieves realistic benefits for minority workers. Some feel that employers can achieve true fairness in hiring standards only by ignoring race and sex altogether—not by favoring women, blacks, and Hispanics over white men. Edward Meese, attorney general in the Reagan administration, compared supporters of racial employment quotas to Americans who once argued "that slavery was good not only for slaves but for society." In a speech attacking job quotas, delivered at Dickinson College in Carlisle, Pennsylvania, on September 17, 1985, Mr. Meese said,

For all intents and purposes, a new version of the separate-but-equal doctrine is being pushed upon us. Those who advocate quotas in hiring and promotion as a means of remedying past injustice will tell you that whatever discriminatory features such policies employ, that discrimination is benign; that it is benevolent.

But you should not forget that an earlier generation of Americans heard from some that slavery was good not only for the slaves but for society. It was natural, they argued, it was a kind of benevolence. Counting by race is a form of racism. And racism is never benign, never benevolent. The idea that you can use discrimination in the form of racially preferential quotas, goals and set-asides to remedy the lingering social effects of past discrimination makes no sense in principle; in practice, it is nothing short of a legal, moral and constitutional tragedy.

Others feel that affirmative action is necessary to provide opportunities for minority people and women. Civil rights activists have long argued that numerical goals are necessary to overcome past injustices against minorities.

Supporters of affirmative action say that it is designed to offset negative action. The Reverend Jesse Jackson defends affirmative action programs by saying, "No one can deny that slavery and segregation and sexual discrimination disfigured justice in this country. And judges—white, male judges, if you will—made a conclusion after considerable study and debate that corrective surgery was needed." Other legal and personal issues are the following.

Age Discrimination

The Age Discrimination Act of 1967 was designed to promote the employment of older persons on the basis of ability rather than age. The act prohibits arbitrary age discrimination in employment and helps employers and workers find ways of meeting problems arising from the increasing population of older workers.

Mandatory retirement has been raised from sixty-five years of age to seventy. Some companies have promoted earlier retirement programs by offering a one-time incentive for people with a specified number of years of service to retire as early as age fifty or fifty-five.

Salaries

Equal pay for equal work legislation provides salary equity among workers. This provision helps to ensure that salaries are similar when persons of different race, sex, national origin, or religion perform essentially the same work in the firm.

Sexual Harassment

Sexual harassment may be defined as unwelcome sexual advances, requests for sexual favors, and other verbal or physical conduct of a sexual nature. The Equal Employment Opportunity Commission (EEOC) provides guidelines that state that harassment on the basis of sex is a violation of Title VII of the Civil Rights Act and that the employer has an affirmative duty to prevent and eliminate sexual harassment. Many organizations have developed programs to train managers on how to identify and respond to sexual harassment. Managers must be alert to warn workers against making sexist comments or engaging in any activity that constitutes sexual harassment. Sexual harassment in the workplace should be condemned at all levels.

Smoking

The legal issue of smoking in the workplace has resulted in several lawsuits filed by nonsmokers against employers who failed to control the work environment to the detriment of the nonsmoker. The problem of employees' smoking in the office and its effect upon productivity must be the concern of supervisors and managers. Although smoking in the workplace started out as a social and health issue, some local communities and cities have enacted laws in this area.

Many nonsmokers, who once saw cigarette smoke merely as an annoyance, now view it as a threat to their health. Environmental tobacco smoke (ETS) has raised a controversial issue pertaining to the respiratory effects of involuntary exposure to smoke.

Many companies are convinced that employees who smoke increase operating costs—for example, heating and air conditioning, workers compensation, and disability insurance. As a result, some companies encourage employees to give up smoking by giving them a chance to win prizes or cash bonuses.

The Washington Business Group on Health compiled a list from employers who use a variety of ways to entice employees to stop smoking. This list (Figure 5-3) offers a wide range of creative ways to discourage smoking in the workplace.

In the absence of laws, organizations have set up a variety of policies regarding smoking on the job. Some companies have followed the example of federal regulatory agencies that require no-smoking sections within government buildings. There are strong feelings for and against smoking in the workplace. Although no universal laws regarding smoking exist, it is likely that within a decade many companies will be required to provide nonsmoking areas in their offices or to ban smoking completely.

Nepotism

The practice of *nepotism* pertains to favoritism in the employment of relatives. It is estimated that 96 percent of American corporations are family-owned or family-dominated. Thirty-five percent of the Fortune 500 companies are family-dominated, or at least have had successive generations of the same family

Figure 5-3. Worksite Stop Smoking Incentives

Sometimes people need just a little added push or reward to motivate them to stop smoking. Many businesses have recognized this need and are responding with an inventive variety of incentives designed to take some of the sting out of the process. Employers also have devised ways to reward long-term employees and new hires who already are nonsmokers.

To stimulate thinking about what added reward or gimmick might entice employees to consider stopping, the following list illustrates some incentives being used by businesses today. The key, of course, is to find the right combination of policy, cessation programs, information, environmental changes (elimination of cigarette machines and sale of cigarettes, improve air filtering), and/or incentives that will turn on your employees to turn off smoking.

- Offer nonsmokers a differential rate (discount) on health and life insurance (Provident Indemnity Life Insurance, Norristown, PA).
- Offer free or reduced-rate cessation programs (Boeing, Seattle, WA; Provident Indemnity).
- Pay for all or a portion of the cost of cessation programs taken in the community (Radar Electric, WA and OR; MSI Insurance, Arden Hills, MI).
- Provide cessation programs on company time or on shared time with employees (Provident Indemnity).
- Offer cessation programs to family members, as well (Bonne Bell, Lakewood, OH).
- Offer cash payments to quitters after six or twelve smoke-free months (Mahoning Culvert, Youngstown, OH; Bonne Bell).
- Incorporate disincentives for quitters who revert to their smoking habit (Bonne Bell).
- Hold drawings for prizes for quitters (Sentry Insurance, Stevens Point, WI; MSI Insurance).
- Provide equal incentives to long-term employees and new hires who are smoke free (City Federal Savings & Loan, Birmingham, AL).
- Reward nonsmokers who adopt-a-smoker and encourage him or her to quit (MIS Insurance, Sentry Insurance).
- Participate in the national quit smoking day (The Rhulen Agency, Monticelo, NY; Arkansas Department of Health, Bass Shoe Company, Farmington, ME; Texas Business Magazine).
- Select your own twenty-four hour period and encourage smokers to quit for the day (Hartford Insurance Group, Hartford, CN).
- Hold a stop smoking fair for local vendors from all types of community stop smoking programs (Hartford Insurance).
- Conduct stop smoking competitions among volunteer teams of employees or with neighboring companies with prizes (Sentry Insurance).

Source: Washington Business Group on Health. Reprinted by permission.

somewhere in top management. Companies have traditionally defended the practice of nepotism. Firms that practice nepotism believe that employing a relative, compared with a nonrelative, provides a loyal and dependable employee. The benefit of having a clan sitting atop the organization and others scattered throughout is an almost familial feeling that binds folks together.

Organizations vary regarding their policies with regard to hiring married couples. Nepotism should not necessarily be classified as good or bad. The practice still remains widespread. No federal or state laws ban nepotism. The only legal aspect pertains to equal treatment for married workers. Rules must be applied equally to both sexes. Therefore, it is illegal for a company to have a policy against hiring married women unless the same rule is applied to employing married men.

Summary

Today's information manager is working in a dynamic, continuously changing environment. Rapid chages in office operations, technology, laws, and regulations are placing increased demands on supervisors and managers. A new breed of employees has joined the office work force. These information workers bring a new set of values, attitudes, and beliefs into the workplace. Their beliefs differ drastically from those of their predecessors a generation ago. They want more say in establishing working conditions and a voice in decision making.

External forces have entered the picture to help workers address their needs. Society's treatment of women, minority people, disabled, and older workers has been changed by legislation, labor unions, and special interest groups. Some progressive organizations are taking the initiative entirely independently of legislation and outside pressures. They are providing their employees with opportunities to participate in decision making. They are sensitive to employees' personal problems and have developed programs to help those who have become victims of the stress and complexities of a fast-paced technological world.

Many corporations have conceded that they have an obligation to provide career opportunities for individuals. Chapter 21 discusses the efforts companies are making to provide training, career development, and career planning for workers. Opportunities for the disabled worker and salary issues will also be discussed.

To sum it up, new directions in personnel policies have taken root within the organizations. The voice of the employee is being heard. Workers' attitudes are beginning to influence personnel policies. Companies cannot ignore this new voice. They must continue to keep the channels of communication open and be responsive to the changing aspirations of both men and women.

Review Questions

1. Describe discipline as it relates to interpersonal relations. Are reprimands necessary? Explain.
2. Why do some managers find delegation difficult?
3. How can a manager keep abreast of new technological developments?
4. Briefly, compare Theory X to Theory Y as motivational approaches to management.
5. Can an understanding of past motivational theories help a manager in the role of dealing with workers? Explain.
6. What is technostress? Why do some information workers suffer from it although others do not?
7. Why should managers be concerned with workers who suffer from stress?
8. Who do corporate mergers cause stress to workers?
9. What steps can a manager take when a person in a work group has been identified as an alcoholic?
10. What factors have hampered the increase in the number of women in management positions?
11. List some of the reasons why white-collar workers resist union attempts to organize them.
12. In what ways have unions changed their recruiting tactics to organize white-collar workers?
13. Some government and corporate officials feel that employers can achieve true fairness in hiring standards only by ignoring race and sex altogether—not by favoring women, blacks and Hispanics over white men. Can you explain why they feel this way?
14. What is mandatory retirement? How are companies promoting early retirement of workers?
15. What is nepotism? Why are there no federal regulations in this area?

Projects

1. It has been announced that Sheila York has been appointed the new information manager for the Information Services Group of Ozark

Electronics Corporation. Fred Daily, a senior microcomputer operator for the department, is disappointed at this new appointment. As a man, he is upset about having a woman as his superior. He has worked in the department for three years and expressed his nervousness and anxiety to his present (male) manager. As his present manager, what can you do to resolve this problem?

2. Over the years, IBM has resisted the attempts by unions to represent their employees. Research and write a report about IBM and their relations with labor unions. Describe the history of union attempts to organize IBM workers. Why have these attempts at IBM and other high-tech companies met with failure?

3. Ed Foster has been described by his information manager, Fred Delgado, as a "marginal employee." Ed does not perform badly enough to warrant dismissal, but he does not measure up to the department's expectations or his own potential. Ed's coworkers have been annoyed with his performance, and this has caused a level of dissention within the department. Complaints have been voiced both from upper management and other employees. They expect Fred Delgado to resolve this problem. If you were Fred, how would you handle the situation?

4. What is the corporate commitment to affirmative action within your community? You are to explore this question by conducting a small survey. Design a brief questionnaire that deals with such questions as the following:

- Does your organization have an affirmative action program?
- Has it been initiated to satisfy corporate objectives? Government regulations? Good business relations?
- What are the numerical objectives in employing women and minorities?
- Do you plan to continue to use numerical objectives to track the progress of women and minority people in your corporation, regardless of government requirements?

CHAPTER 6

Planning for Integrated Information Systems

After reading this chapter, you will understand
1. The meaning of integrated information systems
2. The elements that comprise the model integrated office system
3. The way to develop a strategic plan for integrated systems

In corporate boardrooms, executive offices, seminars, and conferences, decision makers are continually exploring ways to enhance their current office information systems.

Some of the reasons are obvious. Organizations are vying for business as never before. They are seeking new ways to enhance customer satisfaction. They want to create new offerings to fulfill unmet needs and expand from traditional markets to new products and service areas.

The tools and devices have been improved and refined. Never before has such an arsenal of hardware and software been at the disposal of corporate managers. It helps them fight paperwork, enhance production, and speed internal communications. Fortunately, organizations already have many parts of the "technological superstructure" in place. But in order for these tools to work effectively and intelligently, the scattered pieces must be blended. They must harmonize through integrated software, universal workstations, micro-mainframe links, shared networks, and "open door" information centers.

Basically, the term *integration* means uniting with something else. In the context of office systems, integration goes beyond the mere combining of function. It means that all the products offered by one vendor, such as word processors, personal computers, and voice communication, work together. The ability of products to work together is only part of the definition of office integration. The other important aspect is compatibility. *Compatibility* goes one step beyond integration; in this sense, it refers to the ability of different types and makes of equipment to work together.

Currently, machines made by different companies have difficulty communicating. A document typed on a Wang word processor, for instance, cannot easily be sent to and edited on an IBM personal computer. An engineering diagram stored in a Harris/Lanier minicomputer cannot be viewed and modified by someone with a Digital Equipment Corporation (DEC) workstation.

The tools offered by vendors must have the capability to coexist with those of other vendors.

The achievement of integration requires a commitment from vendors. They must define industry standards and work together to survive in a multivendor environment.

From the user's standpoint, purchasing decisions will be made on a vendor's ability to offer integrated technologies. This user-oriented philosophy will give executives more buying power. They will be looking at technology according to its efficiency in providing solutions to business problems. Information workers are beginning to realize that new software and hardware products are great when they can be made to work efficiently and with practical effect.

Linking the Building Blocks

Achieving integration also implies linking together the building blocks of office functions. These basic functions (some of which will be discussed in detail in later chapters) are input, processing, storage, retrieval, output, and distribution. This chapter will discuss some of the other elements that the model integrated office system comprises. These elements are the following:

1. Interactive computing
2. Integrated workstations
3. Shared network systems
4. Micro-mainframe links
5. Data integration

Developing a Strategic Plan for Integrated Systems

When developing a strategy for integrated information systems, the factors of the equation are people, technologies, and applications. The office applications that must be included in any integration plan are word processing, data processing, communications, telephony, and personal computing. The last two are exciting new growth areas that are vital to any integrated system.

One of the first steps in planning a change is to look at the current environment and the information systems already in place.

Management Issues in Integrating Office Systems

The internal politics of organizations have obscured the objective of developing a workable office information system that everyone can use. Top management often selects the technically oriented people to map out the strategies and make the final decision on new office information systems. Data processing (DP) departments are often viewed dimly by nontechnical employees, who claim that programmers do not speak English, do not understand rudimentary business applications, and cannot meet software deadlines. But DP professionals will be the ones with the technical expertise who will tie the systems together.

This failure to understand one another seems ironic in an era when communication technology can circle the earth and penetrate outer space. Yet, the pervasive growth of the same technology has certainly contributed to the communication gap between those who know the vocabulary of "technobabble" and those who do not.

Bridging the Communication Gap

Different companies have adopted different approaches to bridging this gap successfully. In some cases, top-management people have become familiar with the technology. They have learned to accept it, feel at home with it, and understand ways it can be creatively applied. But sometimes the managers have to rely heavily on a senior executive who can translate the technical jargon into understandable language.

Today, more information executives are practicing "computer-age" management. They are becoming computer-literate, or at least they know the language. Terminals are appearing on executive desks. As computer skills are learned, the language barrier between technical and business people is minimized. In addition, the two segments are being brought much closer together as the mystique of computers fades. On the other side of the coin, the systems people must also strive to learn the business vocabulary.

One of the first steps in building an integrated information system is to allow people to communicate. Organizations must encourage an understanding of information technology on all levels. In the final analysis it is not just knowing the language that counts: it is most important to know what to do with it.

Defining the Goals

Before integration takes place, the people in an organization must perceive the office as an "information system" (Figure 6-1):

The goals of a totally integrated information system network are the following:

1. To obtain hardware and software products that work together efficiently
2. To upgrade equipment so that it is multifunctional and has the ability to communicate
3. To provide a network that links small workstations to large computers with access to vast stores of data and instantaneous computing power
4. To extend the network to link customers, suppliers, and other individuals who can communicate data on a domestic as well as global scale

Figure 6-1. Integrated systems provide for the transfer of information among multiple hardware.

Source: Courtesy of Sperry Corporation.

5. To train and encourage workers to use the hardware and software.

The drive toward a totally integrated information system is a long and tedious journey.

Finding the "Right" People

More often than not the task of weaving overall business strategies with an automated environment falls on the shoulders of the management information systems (MIS) professional. But so many people within an organization are affected by information technology that it is not possible for one individual or a group of individuals operating in a vacuum to make the decisions. Thus, the first priority is to establish a committee that will not only plan the integrated systems but will also continuously review them in the future.

Even if the office automation choices are made by a committee and all voices on the committee are equal, the technical voices may carry the most weight. It is therefore crucial to maintain a spirit of cooperation between the organization's end-users and the MIS professionals.

It takes top management to begin the process and companywide cooperation to foster it. However, more resources than exist in one organization may be needed to implement it.

Technology Issues

The changing pattern of technology may have helped to foster a spirit of cooperation between end-users and MIS professionals. These technology issues relate more to when integrated office systems will arrive than the way they will appear to the user. A new philosophy and direction are taking shape in offices today. These events are helping to accelerate this change within the information work force:

1. *End-user computing* is unilaterally changing the MIS function. The computer industry is changing its marketing and sales directions from large, mainframe computers to the distribution and support of desktop microcomputers. As a result, the role of MIS changed from computer tender to end-user coach. A new breed of MIS employee emerged, one who welcomes and guides end-users rather than one who tends the computers.

One method being used by MIS departments to help end-users who want to develop their own applications is through the *information center.* This concept provides a setting where MIS staffers offer support to end-users. End-users are taught the procedures to develop applications with high-level programming language and other design tools.

Another method is through the *microcenter,* or in-house "computer store." Here MIS departments actually buy and support hardware and software for end-users. Information centers and microcenters were discussed in detail in Chapter 4.

2. The mystery is being taken out of computers. More end-users are getting their hands on microcomputer keyboards and like what they can do. New professional-quality software, characterized by high functionality and easy mastery, are attractive inducements. As employees get to know what computers can and cannot do, they also get a better understanding of what data processing can and can not do.

3. Computer software companies modeled their programs after those used for bigger or more dedicated machines. Word processing programs, for instance, are now approaching the capabilities of expensive typesetting equipment. No special programming expertise of the user is necessary. All packages are friendlier, and help is usually at the user's fingertips. What is displayed on the computer's screen is what is printed. There is no manipulation or reformatting.

4. A *decision support system* (DSS) allows managers and other information workers direct and speedy access to the numbers and data they need to analyze a current problem or to make projections. A decision support system consists of a combination of hardware and software. It permits managers or users to access and manipulate information from various sources in a variety of ways. They can retrieve and combine data from data bases within or outside an organization. Also, they can develop and analyze forecasting models, run statistical tests, and display data in various reporting and graphic formats. Decision support tools are readily available in software packages for microcomputers. The most common decision support software are spreadsheet packages. Calendaring, forecasting, and data-base management programs are examples of other practical decision support tools. Decision support software packages will be reviewed in Chapter 14.

5. New personal computers (PCs) with generic options (described in Chapter 8) are emerging. Learning time is minimal. Executives are learning about new products on-line, bypassing bulky training manuals. *On-line* refers to equipment and peripherals that are directly connected to a large computer. Manipulating information on-screen is now highly visual. *Icons* (graphic symbols), pull-down commands, and pop-up menus make operating easy. Modern features such as *mice, touch screens,* and multi-windowed displays handle simultaneous tasks. (These features are part of the modern PC and are defined and discussed in Chapter 8).

Expanding Information Resources

Personal computers have lived up to most of their claims. They are proficient tools to manipulate information. But creating and processing information at a single workstation do not utilize the full potential of interactive computing. Managers at all levels in the corporation are making business decisions in an increasingly competitive environment. They must secure the information required for their decision making. Personal computers provide these managers with an invaluable information analysis tool, but they are limited to the information stored within their memories. Managers need to tap other sources of information.

This need is the essence of information transfer. It is the ability to load corporate data into a personal computer, use the data for decision-making processes, and then analyze the alternatives. Indeed, information stored on a corporate mainframe computer is essential to a manager's analysis, but it is often difficult or impossible to access. In addition to the information stored in the corporate mainframe computer, other data critical to corporate decision making are being generated on desktop computers, which are scattered throughout the organization. A consistent, secure method is needed to channel this information when and where it is needed.

In most large organizations, corporate data reside on one or several large computer systems, in a variety of formats. To obtain these data, information workers must ask mainframe programmers to extract the information from one or more sources. The sources include various corporate data bases, indexed files, and sequential files. This process is expensive, time-comsuming, and hopelessly backlogged. Think of how a network of personal computers in the corporation, all having access to the corporate mainframe, could speed up interoffice communications.

Guidelines for an Effective Integrated Information System

Today's information professionals must realize that there is still no ideal single system that works for every organization. What information users must do to build an integrated information system successfully is try to implement tactical changes in the office, while not straying from a long-term strategic plan. We have just discussed some of the issues involved in building an integrated information system. The following summarizes the principles that are important to the achievement of integrated information systems in the long term while realizing benefits in the short term:

1. Select vendors who offer integrated systems, not just workstations.

2. Emphasize applications and tools. Select hardware and software products, word processing, graphics, and spreadsheets that work together (are compatible) and are appropriate to your applications and needs.

3. Seek to encourage a spirit of cooperation between personnel from different departments. Frequently, the personal trauma associated with moving through the stages of office integration and change is painful. However, it is important for organizations to climb the office technology curve because the benefits of integrated office systems are too great to ignore.

4. Bridge the communication gap between managers, technicians, and workers. Establish seminars and workshops, information centers, and computer stores to teach computer skills and the language of information technology.

5. Do not collect incompatible personal computers and word processors. The most difficult aspect of moving into integrated systems will be the conversion of existing files and procedures and the training of users.

6. Establish flexible planning procedures. Do not overplan. Do not expect 100 percent integration. Do not plan to use one system from one vendor throughout. Approach systems integration with levels of compatibility. Flexible planning procedures and early warning systems for plans

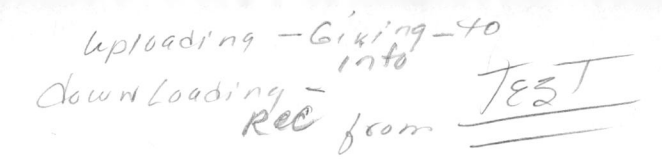

that go awry have the best chance of surviving the zigzagging technological change. In addition, flexible plans for office systems accommodate the introduction of improved products as they become available. Then, users are able to experiment with the equipment and develop their own applications.

7. Establish procedures for data and information security in a distributed environment. Such procedures should limit the use of personal computing devices to noncritical applications. Isolate the facilities to prevent unauthorized use and restrict their operation to authorized personnel only. Provide an intelligent balance between security and access without imposing seemingly "unreasonable" constraints.

8. Establish an integrated micro-mainframe communication link that allows personal computers to download, manipulate, and upload information with the central data base.

9. Establish an integration plan that provides for data integration, transactional integration, and operational integration.

Each of these planning principles must be considered in relation to the effective organization and management of office information systems. Further, it is necessary to respond to all of the factors rather than select only a few. Although the convergence of information systems began over a decade ago, within the past several years the pace of technological change has accelerated to staggering proportions. To understand the speed at which technological change is occurring, we can observe the following comparisons: It is happening about ten times faster than the accommodation of society to the automobile, and one hundred times faster than the accommodation of society to mechanical power.

Moreover, the changes taking place are enormous. Any information systems manager who has been out of touch for a year or two is in for a great shock. It is crucial, therefore, that managers and decision makers keep abreast of the changes in the marketplace and in the technology.

The next five to ten years promise to be even more intriguing than in the past ten. In this environment, strategic planning will become an ongoing office information systems (OIS) practice. Maturation of the OIS function, beyond the stage of integration, will demand a comprehensive functional approach and a commitment from the highest levels of each organization.

Summary

The term *integrate* means "uniting with something else." The integration of office automation and information systems goes beyond the mere combining of functions so all the products offered by one vendor—word processors, voice communications, electronic mail, facsimile equipment, and so on—work together. Integration combined with compatibility will achieve the degree of technological unification sought by users and vendors of office information systems.

Communication is the one essential goal of office automation. Managers face a continuing challenge to integrate cost-effective office automation systems with communication systems.

Emphasis will also focus on computer-literate personnel. Information workers should not have to understand the technology of an office automation system in order to use it. They should know how to apply the technology that is now available. Managers must stay abreast of both the hardware and the methods of managing it.

Review Questions

1. Name the office applications that must be included in any total integration plan.

2. What is the main impediment to developing a workable integrated office information system?

3. Briefly explain "computer-age" management.
4. List the goals of a totally integrated information system.
5. Which types of people can management look to for guidance in developing office automation?
6. What is end-user computing?
7. What improvements have been made in computer software?
8. Briefly describe decision support tools. Give specific examples.
9. What are the generic options in personal computers that make them user-friendly (easy to learn and easy to operate)?

Projects

1. The text refers to communication and personal computing as two of the exciting growth areas vital in any integrated system. Expand and justify this statement.
2. The lack of standards has frustrated computer users who want to connect all the computers in their offices and factories into networks to exchange information. Many users and information executives are frustrated by a lack of standards and compatibility. What efforts are being made by computer makers to develop standards that will allow users to mix and match machines and components from different vendors?
3. The General Motors Corporation, one of the nation's largest computer users, developed its own standards for connecting computers, machine tools, and other electronic gear in a factory. They developed a system know as *manufacturing automation protocol,* or MAP. Briefly describe this system. Can you see ways that a similar system can apply to the office as it now does for factory devices?

PART THREE

The Information Processing Cycle: Systems and Hardware

CHAPTER 7

Voice Processing

After reading this chapter, you will understand

1. The role of voice processing in the overall information processing cycle
2. The meaning of *input*
3. Characteristics and advantages of machine transcription
4. The variety of voice processing devices and media
5. Advanced voice processing technologies, such as voice recognition, voice verification, and speech synthesis

Voice processing is part of the first stage in information processing. It is a method to use the human voice to originate and communicate ideas and thoughts. There are various choices for creating input, including handwriting, keyboarding, person-to-person dictation, and machine dictation. *Machine dictation* is a process whereby a person dictates into a recording device; in *machine transcription* recorded dictation is converted to written information. The automated approach to machine dictation/transcription has evolved into *voice processing* which encompasses the tools that enable the human voice to create, process, and input. Voice processing goes beyond the traditional concept of machine dictation and transcription. Advanced technologies allow people to speak to computers and computers to respond to the spoken word. This chapter will concentrate on the *voice* and *voice processing* aspects of information processing. We will also discuss the following:

Historical aspects
Dictation/transcription cycle
Voice processing techniques
Equipment
Advanced technology

Automating the Input Cycle

The process of written communication two centuries ago consisted mostly of handwriting. Today, advanced technology and sophisticated equipment assist the flow of communications between an originator of words and the person who transcribes such words onto paper.

We have discussed the so-called tidal wave of information sweeping down upon office workers. This paperwork explosion has brought about an increase in labor costs. As a result, management cannot afford to conduct business an usual. In order to survive in a competitive market, organizations are insisting on high-quality, cost-effective document production. As a result, a greater emphasis is being placed on the use of machine dictation as an input *medium.*

There are several reasons for this greater reliance on machine dictation:

1. The cost of producing a business letter continues to rise. A large portion of this increase is directly attributable to the higher salaries paid office employees.

2. The introduction and acceptance of word processors and personal computers have resulted in a much greater utilization of dictation/transcribing machines.

3. New digital dictation systems available for today's users are technically superior to older models in terms of appearance, construction, voice quality, and accessibility.

Machine Dictation/Transcription

Document Flow

The machine dictation/transcription cycle is the most efficient approach to creating original input. It should be remembered that *input is information or ideas in raw form. Input does not become information until it is processed.*

Figure 7-1 illustrates the flow of information processing using a machine in the dictation/transcription method. The originator, or author, dictates into a machine and the process begins. The dictated material is sent to the transcriber for the output phase. The material is transmitted electronically through voice mail, or through a central dictation system. Skilled, proficient transcribers who follow systematic procedures can greatly enhance the productivity at this stage. The completed document is sent back to the originator for final proofreading and distribution.

```
Input
  ├── Handwriting
  │   Keyboarding
  │   Person-to-person dictation
  │   Machine dictation
Processing
Storage and Retrieval
Output
Distribution
```

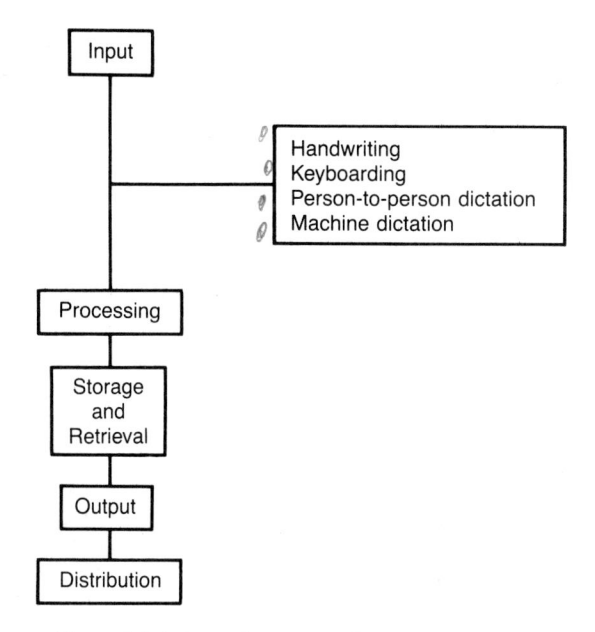

Figure 7-1. Information processing stages.

Benefits of Machine Dictation

The use of machine dictation/transcription as the first stage of information processing offers a number of advantages over longhand and shorthand input.

1. Speed. Speaking is a quick and natural process. Dictating to a machine is faster than writing, keyboarding, or dictating to a stenographer.

2. Efficient Use of Personnel. Machine dictation is more efficient than shorthand because it eliminates the need for a second person in the recording process. This time-saving feature also produces a great cost saving to the company.

3. Convenience. Machine dictation enables the executive to dictate an "immediate response," regardless of the availability of a secretary. The dictator can dictate to a machine at any place and any time.

4. Simplicity. Modern dictation and transcription equipment is easy to learn and easy to use. For the dictator, buttons and settings are as simple to use as a television. For the transcriber, it is just as easy. Transcription is less frustrating than dealing with the erratic speed of live dictation. Speed is controlled by activating the foot pedal to start, stop, review, and forward-space without the operator's hands leaving the keyboard.

The Beginning of Recorded Sound

When Thomas Alva Edison invented the phonograph in 1877 (see Figure 7-2), he did not think of it as a device simply to reproduce music. The word *phonograph* means "sound-writer"; it comes from two Greek words—

Figure 7-2. Thomas Alva Edison demonstrated his tin-foil phonograph before President Rutherford B. Hayes and the National Academy of Sciences in April 1878. This photograph was taken by Matthew Brady, the Civil War photographer.

Source: Courtesy of the U.S. Department of the Interior, National Park Service, Edison National Historic Site.

Figure 7-3. The first practical recording device.

Source: Photograph courtesy of Dictaphone Corporation, Rye, New York.

phono, which means "sound," and *graphon,* which means "to write." To Edison, it was a device to give people time and money. It permitted individuals to write with sound rather than pencil and paper.

The first practical recording device was invented by a three-man team: Chichester Bell, a chemical engineer; Charles Sumner Tainter, a scientist; and Alexander Graham Bell. Their goal was to try to find a practical way of recording and reproducing sound for use with the newly invented telephone.

On October 17, 1881, the two Bells and Tainter deposited in Washington's Smithsonian Institute the fruit of their research (Figure 7-3). It was a device with a rotating cardboard drum coated with a fifty-fifty mixture of beeswax and plain paraffin. The mouthpiece had a diaphragm not unlike that in Bell's telephone. Attached to the diaphragm was a concave steel stylus, which vibrated up and down in the same degree as the diaphragm of someone speaking into the mouthpiece. It cut a "hill and dale" record of the human voice into the wax coating of the rotating drum.

Charles Sumner Tainter gave the new invention its initial test. He picked up the funnellike mouthpiece and carefully recited a quotation from Act 1 of Hamlet: "There are more things in heaven and earth, Horatio, than are dreamt of in your philosophy." Thus, recorded sound was born.

Dictation/Transcription Equipment

The earliest machines were recorded on cylinders of tin foil. But foil quickly yielded to reusable wax cylinders. With this recording medium, a needle cut grooves into the paraffin surface (see Figure 7-4). After transcription, the cylinder was shaved and reused. Wax cylinders were very delicate. They could be ruined by listening to them several times. The wax cylinder dictation method has been around for a long time. Wax as a medium gave way to plastic. Again, metal-etched grooves preserved the sound.

Plastic was not much of an improvement since the plastic record or belt could be used only once. It was not possible to erase or rerecord on the same record or belt. Although plastic was not reusable, it was durable. Also, for the price of one wax cylinder, a user could buy several plastic disks.

The emergence of magnetic recording media drastically changed recording methods. It now became possible to make corrections in the dictation. One could review dictated material easily, record for longer periods of time, and reuse the same magnetic tape or belt an almost unlimited number of times.

Today's information worker will find current dictation/transcription equipment improved over the bulky, complicated machines of the past. Present machines are designed with the user in mind. Controls are clearly identified for ease of operation. They are built to last, yet are compact and attractive. The voice quality is rich and clear. These superior features make it easier for the secretary to transcribe dictation quickly and accurately. The variety of equipment available furnishes a machine to fit almost every need.

Dictation/transcription equipment falls into three basic categories (see Figure 7-5):

Figure 7-4. Transcribing from an 1888 Edison phonograph with a dictation attachment.
Source: Courtesy of Dictaphone Corporation, Rye, New York.

portable models, desktop units, and central systems.

1. Portable Units. Portable units are designed for the traveling executive or the person who finds it necessary to dictate away from the office. These units are compact and vary in weight from about ½ pound to 2 pounds. Battery-powered, they are small enough to fit into a shirt pocket or purse.

2. Desktop Units. Desktop units are designed for in-office work. These units are meant to be stationary, located on the executive's desk for convenience and accessibility. Some units serve a single purpose, either dictation or transcription; others incorporate both functions. The dictator plugs in a microphone before dictating; the transcriber unplugs the microphone and attaches a foot pedal and a set of earphones.

Some microphones are shaped like telephones to make the user feel more comfortable with them. All controls are located on the microphone.

3. Central Systems. Central dictation systems consist of central recorders that are

Figure 7-5. The three basic categories of dictation/transcription equipment.

Source: Photographs courtesy of Dictaphone Corporation, Rye, New York.

permanently installed and wired to either handsets or telephones. These systems provide a distributed approach to input.

In a centralized environment, dictation is likely to be given a priority by a supervisor and relayed to individual transcribers through a switching system. This type of system is most often used by firms with large numbers

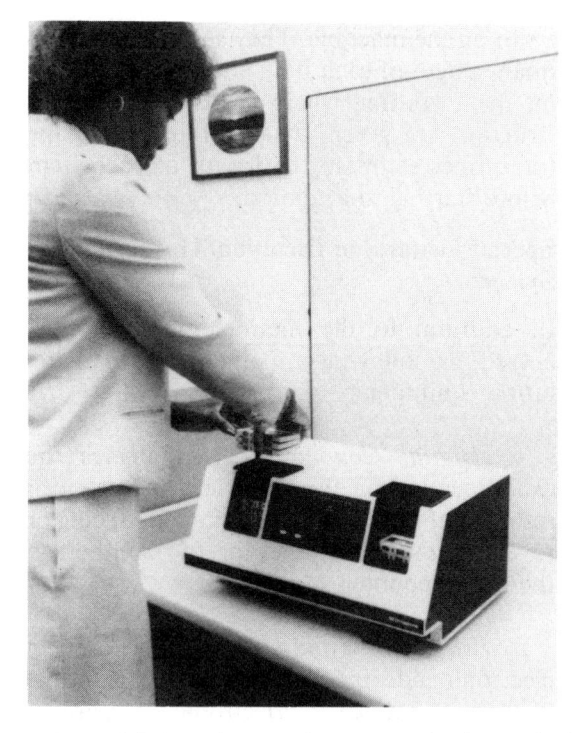

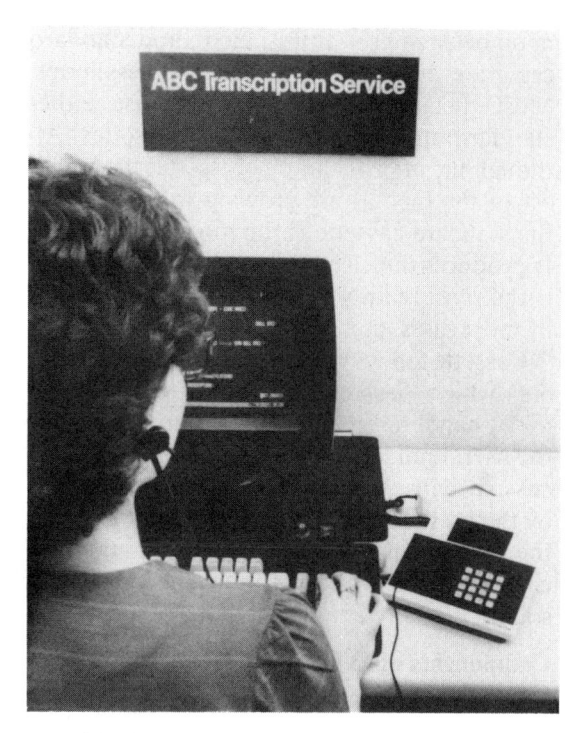

Figure 7-6. A word processing manager loads standard cassettes into the remote transcription system, left, which may be transcribed off premises via a remote transcription terminal, right, which operates over ordinary telephone lines.

Source: Photograph courtesy of Dictaphone Corporation, Rye, New York.

of people who occasionally need transcription services, yet do not require private secretaries (Figure 7-6).

In a central system, the user dictates into a microphone or telephone that is hooked into the central recorder. This is called *remote dictation,* since the recording device is located at some distance rather than on the dictator's desk.

Each system has a central processing unit that uses either *endless loop* magnetic tape or *discrete media* cassette tape. Endless loop dictation systems contain tanks that house the magnetic tape. Because, in this system, the media are not physically handled, it has the advantage of providing many hours of continuous recording. *Multiple cassette* systems use a series of cassettes that are loaded into the system in sequence. When one cassette fills with dictation, the unit automatically begins recording on the next cassette.

With discrete media systems, the supervisor lifts the medium from the recorder and carries it to a transcriber at a compatible transcription station. With endless loop systems, the media are not physically handled. Dictation on a tank must be directed only to one transcription station.

Dictation Media

The emergence of magnetic media was a significant step toward automating the dictation process. It allowed dictators to correct, review, and reuse magnetic tape. Recording media fall into the following categories: magnetic belts; micro-, mini-, pico-, and standard cassettes; disks; tape reels; and endless loops. Most dictation machines today use either standard, micro-, mini-, or picocassettes. Although the micro- and picocassette (the smallest of the media) are growing in acceptance, the standard cassette is the most widely used. It produces the clearest sound of the four and is the most readily available at a price that is, in most cases, competitive and acceptable.

Cassette tapes are housed in plastic containers, which protect the magnetic head and make handling the media safer and easier for the user. Endless loop systems, previously discussed, have eliminated altogether handling by the user. The advantages to the dictator are the convenience of the equipment, the efficiency of the workflow, and the quality of sound.

Components of Dictation/Transcription Equipment

Combination dictation/transcription systems provide dictators and transcribers with versatility in one compact unit. Typical desktop units consist of two components: the main recorder and the microphone. The *main recorder* accepts the recorded media and has various controls for volume, tone, clarity, speed, and index scanning. The *microphone* is attached to the recorder by cable. To convert the unit into a transcribing device, the user simply plugs the foot pedal into the microphone receptacle. The headset is plugged into the headset receptacle on the side of the unit.

The *foot pedal* controls forward and reverse action. It can usually be adjusted to permit the transcriptionist to have the last few words of dictation repeated after resuming playback.

An *ear plug* or *headset* is worn by the transcriber during transcription. Most models have a headset in the form of a chin band with sponge eartips. Since sound is heard in both ears, most outside noises are somewhat eliminated.

The *index slip* serves as a record of dictation. Index slips are strips of paper inserted in a slot on the machine. They show the approximate length of each item and indicate points on the recording where corrections and instruction are given. The markings help the transcriber estimate the length of each item before starting to keyboard.

Special Features of Dictation/Transcription Systems

In addition to the main components discussed, the following are other features of dictation equipment:

1. Digital Counter. This indicates the exact location on the tape. The operator can return to the same position after advancing or reversing the tape by noting the number on the digital counter.

2. Electronic Tone Indicator. This is an electronic indexing feature that eliminates the need for an index slip. Electronic tones are recorded on the medium to indicate messages or end of letter instructions.

3. Automatic Backspace Dial. This feature adjusts the amount of dictation the user will hear repeated when he or she starts transcribing again after a pause. Such machines have an adjustment that can vary the length of backup or word recall according to the discretion of the individual transcriber.

4. High-Tech Microphones. Full-featured microphones are designed to give complete operating control in the palm of the dictator's hand. The unit can be automatically turned on by picking up the microphone. Some microphones have built-in speakers to give close-up replay of dictation. Audible cues can be inserted to signal the ends of letters and special instructions. When the operator places the mike back on its cradle, the system automatically shuts off.

5. Electronic Display. The index display gives the transcriber a clear, bright preview of the arrangement of dictation on the cassette. It shows where each letter begins and ends as well as the amount of tape left. It also pinpoints special instructions.

6. *Alphanumeric Readout.* The readout shows precisely the length of each letter. It allows the operator to set the page format before typing. This feature greatly reduces retyping.

7. *Automatic Search and Scan.* This feature automatically scans to set up the electronic display. It searches out and plays back special instructions automatically.

8. *Pulse-Touch Control Buttons.* Pressure-sensitive controls replace old-fashioned buttons and switches. When the appropriate control is touched the symbol lights up immediately to show that the unit is responding. When it is touched again, and the light goes out, it tells the operator that the function has stopped.

9. *Automatic Phone Message Recorder.* This is an advanced feature on some systems that provides for round-the-clock phone message service. By attaching the dictation system to the phone, it converts the unit to record telephone messages from any phone in the world—24 hours a day.

Dictation Management and Control

If large amounts of dictation are dispersed among several workers, a system is needed to monitor all work that flows in and out of the centralized area. Vendors have introduced word management computer systems that automatically log and report dictation in work groups and in centralized environments.

Managing the dictation function and keeping track of dictated material is essential to the smooth flow of work. In some instances, this is done by an individual whose job includes keeping a log of all incoming dictation. Automating the centralized dictation process can be accomplished through management computer systems.

A system, such as the one in Figure 7-7, keeps track of dictated material. It includes a desktop console with an alphanumeric light-emitting diode (LED) display, a report

Figure 7-7. The central recorder, right, uses microcassettes; the console, left, displays information about the dictated material.

Source: Courtesy of Harris/Lanier Corporation.

printer, and convenient keyboard, all of which are controlled by a microcomputer with associated memory.

As soon as an author completes dictation from a private-wire, touch-tone, or dial telephone, the system can automatically print out a log entry. It identifies (1) the author and type of work, (2) the recorder in which the job is contained, (3) the length of the dictation, and (4) the time when the author completed the dictation.

Used with a multiple-cassette recorder, the system identifies the cassette on which the job is recorded and its position on the cassette. When a completed cassette is ejected for transcription, the computer prints a self-adhesive label that includes detailed information on each job in the cassette. The label can be affixed to the cassette for simplified control of workflow. In this way, the transcriptionist has pertinent data about the job at all times, and a permanent log remains with the supervisor.

Off-line work is easily entered into the system through the computer. The LED display line "talks" to the supervisor. This prompts the proper keyboard entries in sequence. The display also alerts the supervisor automatically to certain unusual situations, such as an off-hook input station or a recorder requiring attention.

Special Features of Central Recording Management Systems

Central management systems tell managers everything they need to know about dictation and transcription work in their department. They do so more quickly and with greater ease than a manual log. Through computerized systems, they track documents, handle priorities, assign transcriptionists, and increase a department's productivity. A central recording management system contains the following features.

1. Multisearch. Locates work and handles priorities. The computer pinpoints a particular piece of work or searches for a special document with a simple command. This feature allows the user to search and retrieve documents by any variable information or key word within the document.

2. System Status. This feature provides up-to-the-minute information on the status of every piece of work in the system, as well as current backlog and other valuable management information. For example, the user can easily see at any time exactly where the jobs are and what must be done. This allows managers to make necessary staff assignments and take care of the oldest and most important jobs first.

3. Customized Data Control. The system can be customized according to the variables within a department or company. On a special page, the user can summarize the output of each transcriptionist from each recorder, and from the entire center. Just as you can search by any variable that is captured, so too can you summarize by any variable that is captured.

4. Expanded Memory. Through floppy disks, storage can be expanded to provide unlimited memory.

5. Instant Identification. Dictators and originators are recognized instantly. The file index can display identification numbers attached to the originator's name. This information provides the status of each person's work and the document can be searched. Files can be retrieved by name or by any component of the identification number.

6. Off-Premises Work. Transcription can be performed on both endless loop and microcassette systems from locations other than the department. Standard telephone lines can be used for direct access between the satellite information-processing workgroup and a worker working at home.

Whether the organization is large or small, work management can enhance the effectiveness of the dictation and transcription functions. Both functions can be automated using equipment that finds important information in seconds with simple commands. It allows managers to stay on top of what is going on while minimizing costly interruptions. Vendors are offering leading-edge systems with simple commands and direct benefits.

Digital Voice Processing

Today's computerized central dictation systems have offered managers greater improvement and control over the flow of work. But now, the industry is beginning to move toward digital recording technology. Digital recording allows voice storage on a hard disk, so that an author can insert and delete material anywhere on the disk verbally.

Using an ordinary telephone or an input station, voice is converted to digital impulses and stored on a hard magnetic disk. Here, it is instantly accessible for transcription or review.

Benefits

Use of digital technology offers much greater accessibility to dictated information.

1. It permits complete *random access*. Random access is a method of storage on external media. Information is stored in no specific order. Any document in the system may be instantly located. *Dictionary recovery* is a method of direct access of a list of specific words stored in a computer's memory. It is initiated directly from an index on the supervisor's console (central recording computer). Therefore, transcription scanning is eliminated. It facilitates priority and other out-of-sequence work. The following example will illustrate the way this feature works. In a hospital, the supervisor has just received a call from the floor requesting a particular presurgery assessment. Endless loop or cassette dictation systems would require an extensive physical search to find the proper tape or location of the document. The digital technology system allows any document to be located instantly at the touch of a button. The real benefit is added efficiency in handling these types of material.

2. It provides for step-by-step verbal prompts. This feature minimizes the training necessary to operate the system. Because it is programmable, experienced users may bypass these prompts by the touch of a button.

3. Because digital recording permits random access to the recording disk, designated work channels for specific types of dictation are unnecessary. This capacity eliminates confusion. Users do not have to remember which channel to use for different kinds of work.

4. At the touch of a button during input, the originator identifies the type of work being dictated. With dictation completed, the information is automatically routed to the disk and is available for immediate transcription.

5. There is a higher level of control and management of dictated material. The features described earlier dramatically reduce system misuse. Instant accessibility to dictation also permits a supervisor to establish separate *queues* (waiting sequences) for managing the flow of work to specific transcriptionists. Dictation may be handled in any manner desired—first-in, first-out; by the originator; by type of report; and so on.

6. In the past, special instructions were given when the originator thought of them. The instructions were not necessarily where the originator wanted them to be made in the text. Now, for the first

time, corrections, amendments, and special instructions can be placed exactly where they belong.
7. Security codes built into the central recorder assure that only individuals cleared to review sensitive dictation have access to the information.
8. Unlike magnetic tape, digital technology creates no voice distortion if a transcriptionist must slow the pace of dictation during playback. This enhances clarity and understanding during transcription and increases the transcriptionist's efficiency.

Digital voice storage is an exciting new technology that works hand in hand with central management consoles to enhance the management, control, and productivity of voice processing. The technology of voice disk storage has been a forerunner to other advanced voice technologies such as voice mail, voice store and forward, and variable speech compression.

Variable speech compression (VSC) lets the user play back dictation at speeds that are up to 100 percent faster than those at which they are recorded—and without "Mickey Mouse" sounds.

Voice Mail

A new dimension has emerged in voice processing. The ordinary telephone is turned into a multifaceted communication center at the touch of a button. This concept is called *voice mail*. Voice mail allows the user to send and receive natural voice messages from any telephone, anytime, anywhere in the world. Destined to become the electronic post office of the future, voice mail, or voice messaging systems, are a highly effective means to communication. This computerized voice storage and forward system provides convenient, cost-effective delivery of "voice telegrams" when direct telephone conversations are not possible or desired.

After entry of a message via a telephone handset, the sender's voice is converted for storage. Messages are forwarded to the recipient through a mailbox or callback system.

A mailbox system allows the recipient to call the system, enter an optional password, and play back the message. At the push of a single button on the telephone pad, one is offered a wide range of options, from receiving individual messages to sending group broadcasts, from message editing to call forwarding.

Because the systems are computer-based, the person receiving the voice message can review it then decide whether to replay it, save it, discard it, or forward it to another party. Voice messages can also be edited. Furthermore, a voice message can be recorded once and sent to several people at the same time.

Voice mail systems receive their commands in the form of tones. The tones are generated when the keys on a standard twelve-key telephone pad are pushed. Depressing the right key tells the system that you are ready to record a message or instructs the system to stop recording. The caller can skip backward or forward when replaying a message.

The key sequences are easy to learn. Some voice mail systems provide templates that fit neatly over the telephone dial face. No computers or peripherals are necessary. Voice mail systems can be used from the office or home and during travel. They provide convenient communication and preserve the natural tone and inflection of the speaker's voice.

Benefits

1. Cost Savings. Voice mail provides new opportunities for saving money and increases productivity. Local and long-distance telephone charges can be reduced by 50 to 70 percent. There is a reduction in "telephone tag," the frustrating cycle of placing phone calls and leaving messages to people who are not available. Otherwise, messages are left and the phone bills mount. Statistics show that the average business call reaches the intended recipient less than 30 percent of the time. And 6 percent of the calls that do get through are considered interruptions.

2. *Work Streamlining.* A second advantage inherent in such systems is the ability to eliminate some of the dictation and typing required for company memos and directives. The voice mail system with its automatic distribution and protected access, storage, and annotation facilities can provide this function via stored voice memos.

3. *Continuous Accessibility.* Using conventional phone systems, time zone differences meant that people located on opposite coasts were able to communicate with each other by phone only within a 4-hour time slot. Voice mail systems are always accessible. They allow users to accomplish tasks regardless of time or location.

Voice mail systems transform a user's phone into a powerful and productive voice message exchange. Voice processing technology helps workers to communicate and schedule their time better.

Advanced Voice Technologies

Voice processing may soon become the critical link for all the processing components. The tools and technologies that use the human voice are accomplishing tasks in the quickest, most efficient way. Since, by and large, executives and other information workers are talkers, their technological tools should be able to understand and respond to verbal instructions.

Many of our current methods of communicating and manipulating information, which traditionally depend on keyboard entry, may soon be replaced by voice-based procedures.

Processors that "talk" (speech synthesis) and "listen" (voice recognition) will speed the transfer of information (Figure 7-8). Digital recording and transcription machines will free office workers to perform more productive tasks. Cellular mobile phones will alter the definition of the office environment. Even the ubiquitous telephone will become vastly more sophisticated.

Advanced voice technology will thrive because some of the major computer and office automation vendors are actively pursuing ambitious development goals in the area.

The balance of the this chapter will discuss the advanced voice-based technologies that stress a more "human" worker-machine interface. This pertains to the ability to command a machine such as a computer verbally and the capacity of that machine to respond "verbally."

Voice Recognition

Voice recognition systems "acknowledge" spoken words, sending appropriate digital signals to an attached system or device (Figure 7-9). Instead of entering a command or request by a keyboard, voice recognition equipment transforms a vocal input into a format. The format can be used by a *digital* computer. Since speech generates sound waves, these must be converted into *digital* impulses. The system must then compare those impulses to the digitized sound patterns stored in its memory. Continuous speech—with its unlimited vocabularies, individual speech mannerisms, nuances, and accents—presents the computer with huge storage, processing, and interpretation tasks.

Voice Verification

Voice verification not only transfers vocal input into a digital format, but also has the ability to store an individual's unique voice print into a computer's memory. It is the equivalent of matching fingerprints to ascertain someone's identity.

One application of this technology is in office security. In this process, an authorized employee makes a verbal inquiry into a sys-

Speech Recognition

[Diagram: Source → Microphone → Analog Signal → Converter → Digital Signal → Storage → Analysis & Identification → Comparison to Reference Template → Output → Action, Display, Storage]

Speech Synthesis

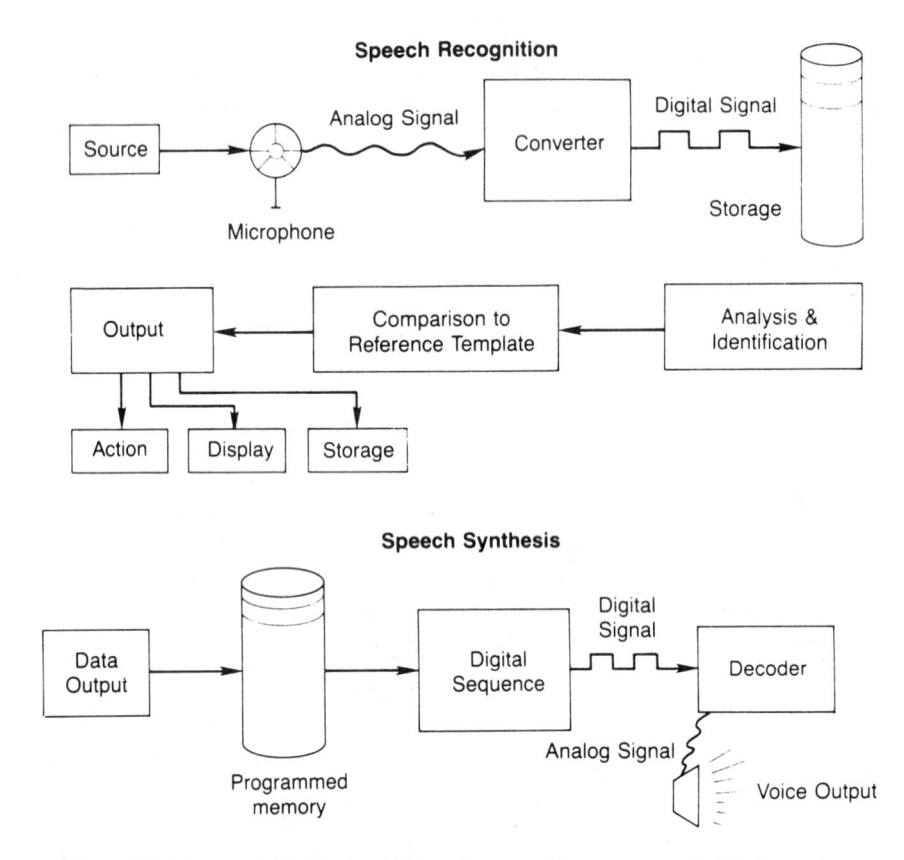

[Diagram: Data Output → Programmed memory → Digital Sequence → Digital Signal → Decoder → Analog Signal → Voice Output]

Figure 7-8. Advanced voice technology systems enable processors to talk (speech synthesis) and listen (speech recognition) with human users.

Source: Reprinted from the October issue of MODERN OFFICE TECHNOLOGY and copyrighted 1983 by Penton/IPC, subsidiary of Pittway Corporation.

tem, and the system determines whether the speaker is actually the person he or she claims to be or whether that individual is to have access to the information requested.

Applications of this technology are presently controlling access to buildings, security-area entrances, and confidential files. Some banks are using voice verifications for electronic funds transfer.

Speech Synthesis

Speech synthesis differs from voice recognition or verification in that the mahince, rather than the human user, "speaks." The computer reconstructs words from segments or phonemes, without the benefit of a real human voice imprint or pattern. The voice generated has a robotlike quality with little or no inflection, mood, or fluidity. Some auto manufacturers are using speech synthesis technology in cars. The voice responds to such potential problems as "low on fuel," "door ajar," "headlights on," and "check coolant." Supermarkets use speech synthesis at the checkout counter. The checkout clerk simply runs a product's Universal Product Code (UPC) symbol over the optical scanner. The speech synthesis system announces the price, and the visual display shows the product's name and price to the customer. Each item later appears on a receipt.

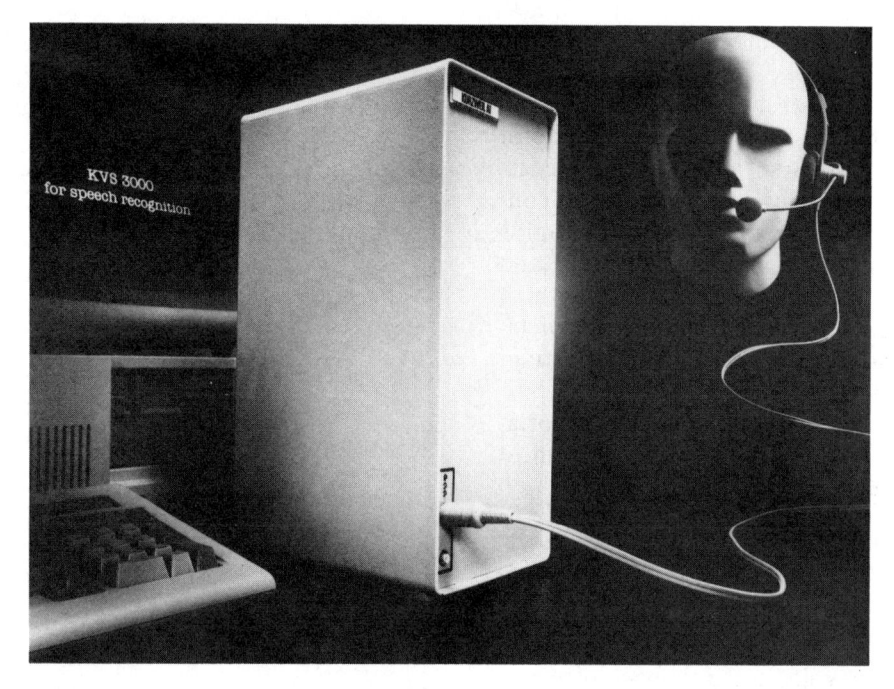

Figure 7-9. This voice system recognizes 1000 words and, so, expends virtually any host processor.
Source: Courtesy of Kurzweil Applied Intelligence, Inc.

Trends in Voice Technology

In all aspects of technological advances, automation is viewed as a positive force by some and a threat by others. Most industry experts agree that the future for dictating systems looks bright. Some, however, feel threatened by technology. In spite of the relative immaturity of the markets for voice products and the uncertainty regarding sales, large sums of money are currently being invested in voice technology developments. Present dictation/transcription systems will be challenged by advanced technology. Speech compression, voice recognition, verification, and speech synthesis will transform the dictation industry and alter the work habits of office workers.

In the future, voice recognition is expected to allow highly flexible direct voice interaction with the processors. It will be possible to convert the spoken word into the written page. This technology may do more than merely reduce the need for typing copy into a keyboard terminal; it could one day carry out commands spoken into it.

Each of these voice technologies will enrich the quality of life through the freedom resulting from more efficient work tools.

From today's smart phones through tomorrow's voice transcribers, voice technologies seem destined to help workers manage their time and communicate information more effectively.

Summary

Dictation evolved into voice processing. The common ancestor to all dictation devices dates back to 1877, when Thomas Edison invented the phonograph. Edison realized the worth of the phonograph as an office time saver. The phonograph permitted people to "write" with sound rather than pencil and paper.

One of the most recent advances in the evolution of dictation is the digital recording of voice. Digital dictation is a complete departure from all previous forms. Sound is converted to binary digits and the information is stored on magnetic media. Insertions and deletions can be made at the point at which they are logical, instead of elsewhere. Dictation can be accessed more quickly by random means, speeding up transcription.

Dictation systems include desktop, portable, and central dictation units. The most common means of dictation is the desktop dictation unit. Following the advances in voice recording, the desktop machine has evolved into today's unit using mini-, micro-, and picocassettes (the smallest form available).

Advances in voice technology—voice recognition, voice verification, and speech synthesis—may alter the workplace as much as computers have. Many of our current methods of communicating and manipulating information, which are traditionally dependent on keyboard entry, may soon be replaced by voice-based procedures.

Review Questions

1. Describe the role of voice processing in the overall information processing cycle. What are some of the technologies that constitute the concept?
2. What is input, and how does it differ from information?
3. Compare dictation with transcription.
4. List some of the benefits of machine dictation.
5. How did the introduction of magnetic recording improve voice processing?
6. Describe the way a desktop recorder records and transcribes dictation.
7. What is remote dictation?
8. Define and compare endless loop systems with multiple cassette systems.
9. List the various types of recording media. Which one is the most popular today?
10. What is the function of an index slip? Describe the automated features on dictation units that are replacing the index slip.
11. How can managers keep track of dictated material in a central recording environment?
12. What is the recording medium used in digital voice recording? What benefit does this have over other recording media?
13. Define random access. Describe dictionary recovery as it pertains to digital recording technology.
14. Briefly define the following advanced voice technologies:
    a. Voice recognition
    b. Voice verification
    c. Speech synthesis
15. Why is "telephone tag" a problem? How does voice mail remedy it?

Projects

1. Speaker verification systems not only transform vocal input into a digital format but match the "voice print" in the computer memory. They have been compared to matching someone's fingerprints to determine identification. What implications does this technology have for security and protection? You are to research this technology and describe an organization that uses voice verification as a security device. Write a brief report or case study of an organization in your community or a national company.
2. Study vendors in the dictation equipment industry. Compile a product comparison chart that details the dictation equipment available from a number of vendors. Design a matrix or chart and include the following headings:

Vendor
Model
Function
    Dictation

Transcription
Combination dictation/transcription unit
Portable
Desktop
Central System
Digital
Recording medium
Features
Options
Price (if available)

3. There are many applications for voice mail in a law or medical office. Describe three uses that an attorney or physician might have for a voice mail system.

4. Advanced voice processing systems, no matter how sophisticated they become, are still tools. As technology advances, designers of these tools must not neglect the human factor.

Write a position paper about the interrelation of humans and voice-based technologies. Take two positions: positive (acceptance) and negative (rejection). In what way might some of the designers of these talking machines be guilty of sexism when they assign male and female voices in various applications?

CHAPTER 8

The Computer

After reading this chapter, you will understand
1. The evolution of computers
2. The basic elements of a computer system
3. The input/process/output cycle
4. The basic differences between mainframes, minis, and micros
5. The rise in popularity of the personal computer
6. The processing and storage components and terminology of computers

Although strategic planning and organization design may occupy most of the time and attention of information managers, it is the hardware, the *computers,* that inspire the most awe. They are the ones with beeping sounds and blinking lights.

People can relate best to the hardware. Hands can depress keys, turn switches, insert disks, and feed printers. Computers are the engines that power the flow of information. The computer processes data into information. Remember that *data* are defined as raw facts that must be processed to produce information. *Information* is processed, structured, meaningful data.

The epitome of information processing, of course, is the computer. The computer is practically synonymous with information processing. Computers once filled rooms the size of basketball courts. Now they sit on laps. Computers, once named with numbers and acronyms like UNIVAC and 3270, are being replaced by the friendly "Macintosh" or "Office Assistant."

One would think with all the hoopla and media hype that these processors have been around a long time. But the history of computers is relatively short, beginning with the birth of the electronic digital computer in the 1940s.

The Evolution of Processors

When J. Presper Eckert and John Mauchly announced plans to build the Electronic Numerical Integrator and Computer (ENIAC) at the University of Pennsylvania in the 1940s, the idea seemed like folly. The new computer contained more than 18,000 vacuum tubes, running it required a huge amount of electric power, and it weighed some 30 tons. But, by 1946, when the project was completed, the machine not only worked, but worked well enough to see postwar service (Figure 8-1). It churned out calculations faster than any other device had before.

But even before digital computers became popular, there were other milestones in the evolution of processing machines. These events can help to explain what can and cannot be reasonably expected from technological advances. Knowing the way these various computational devices affected past societies will help us to understand the way future devices are likely to affect our society. It is appropriate then to pause and look back on a brief historical timeline of technology.

1896–1910

- In 1890, Herman Hollerith devised a punch-card tabulating system that counted and categorized the census. This was known as the Hollerith census counting machine.
- The term *white-collar worker* was first used in an American newspaper.
- The first successful electric typewriter, the Blickensderfer Electric, was introduced in 1902, but lack of electricity in most offices limited its popularity.

1910–1920

- James Powers invented an electric punching machine for use by the U.S. Census Bureau.
- Regular airmail service was established between New York City and Washington, D.C.

1920–1940

- IBM introduced a new electric typewriter that gained wide acceptance in U.S. offices.
- Bell Laboratories developed a two-way television communication system.
- Howard Aiken started developing a computer at Harvard University in 1937.

Figure 8-1. The ENIAC was the first all-electronic general purpose digital computer.
Source: Courtesy of Sperry Corporation.

1940-1950

- Construction of the first large electronic digital computer, ENIAC, was completed at the University of Pennsylvania in 1946.

1950-1960

- The first generation of computers (1951-1958) began with the introduction of the UNIVAC I. This computer used vacuum tubes to control operations (Figure 8-2).
- Texas Instruments developed a new generation of transistors that used silicon in place of germanium in 1954.

1960-1970

- IBM introduced a word processor that stored typed material in coded form on magnetic tape in 1964.
- The first computers that used silicon chips instead of transistors were introduced to the market.

1970-1980

- The microprocessor was introduced by Intel in 1971.
- Hewlett Packard introduced the first programmable pocket calculator.
- IBM develops ink-jet, nonimpact printing for addressing labels.
- The first Apple computer was built in a garage in Los Gatos, California.

1980-

- The U.S. market for word processing

1934   1942   1951   1959   1967   1975   1984

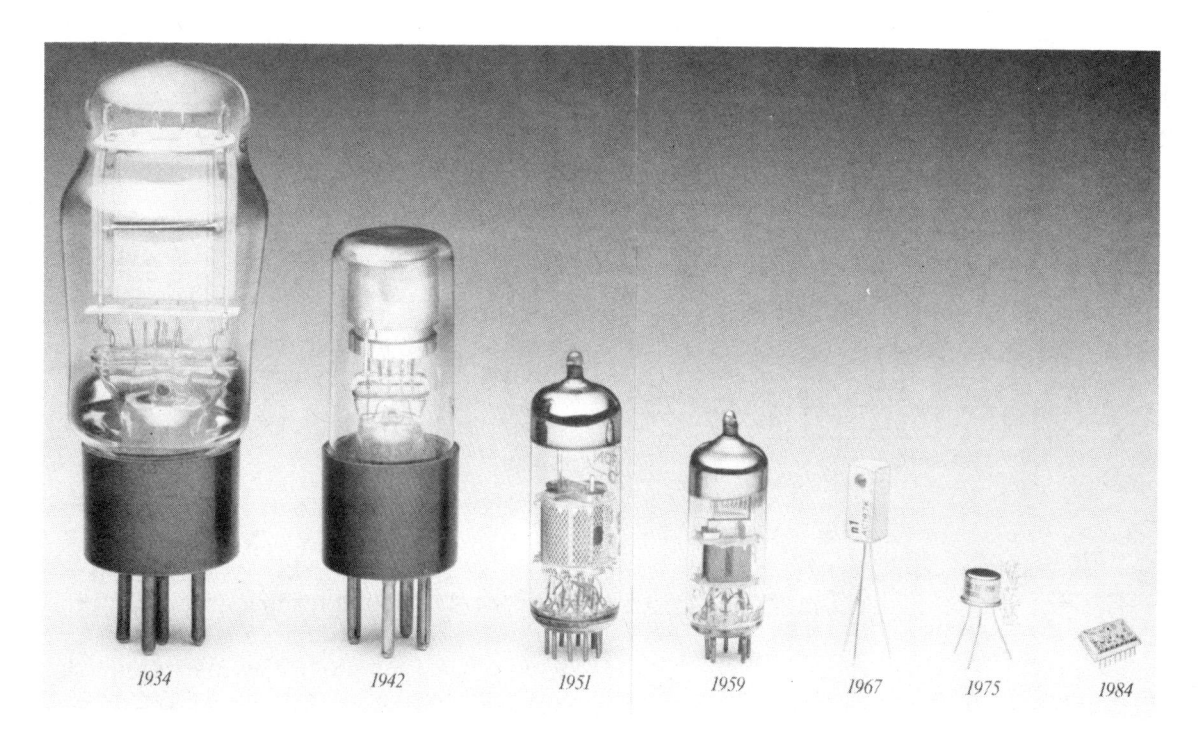

Figure 8-2. As vacuum tubes became smaller then gave way to ever-smaller transistors and, finally, to microchips, computers became smaller, more efficient, and less expensive.

Source: Schott Electronics.

equipment exceeded $100 million a year in 1980.

- *Time* magazine named the computer "Man of the Year" on January 3, 1983.
- Divestiture of AT&T occurred on January 1, 1984.

This timeline describes great leaps in technology. Present advances in technology make even greater strides. Imagine what $4000 could buy in personal computer power at the beginning of the decade of the 1980s. By 1986, the processing power and storage capacity of the main memory had increased twentyfold. And for the same $4000, the disk storage had increased five hundred times.

By understanding the growth stages inherent in this revolution, we will be able to use and apply new technology wisely. Now that we have briefly traced the evolution of the computer, let us examine the intricacies of these devices. The remainder of this chapter will discuss the array of processors and their function, features, and applications. By so doing, we can better understand why computers have so invaded our lives that we have become reliant, if not totally dependent, on them.

Computers and Computer Systems

What Is a Computer?

A *computer* is a machine designed to follow instructions. In doing so, it rapidly and tirelessly performs tasks by manipulating bits of information (see Figure 8-3). *Bits* are the most fundamental unit of information pertaining to

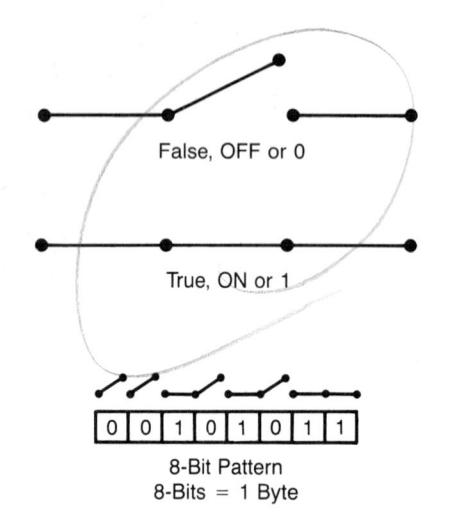

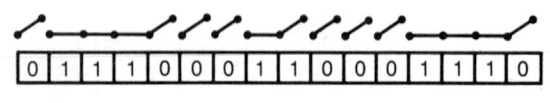

Figure 8-3. The way a computer represents a bit of information (top), and the way bits of information can be arranged in a pattern of 8 or 16 bits.

the computer. They are groups of binary digits (0s and 1s) generally called a *word*. You can compare it to an on-off switch. Either a bit is ON or it is OFF.

What Is a Computer System?

A *computer system* combines various components into a single, well-integrated system that includes the following.

Input Devices. Input devices convert data into a computer-compatible format. Input, therefore, consists of data that an input device coverts to a form the computer understands.

Output Devices. Output devices translate data into information.

The Central Processing Unit (CPU). This is the heart of the computer system that controls all its operations. It contains the primary storage and runs or executes the computer instructions.

Auxiliary or Secondary Storage Units. These provide additional storage for data and programs.

Figure 8-4 illustrates the basic elements of a computer system. It shows the flow of data through a computer system.

Processing Operations

The input-process-output cycle exemplifies the way various parts of a computer work together to complete tasks. Processing operations involve (1) the reading of input, (2) the processing of output data, and (3) the creation of output information. This operation may also be referred to as the *input/process/output cycle.*

Input devices allow the user to put information into the computer. A "brain," or central processing unit (CPU), assimilates this information. *Software* (also called programs) tells the computer how to interpret and act on this input information. Finally, output devices let the computer communicate with the user or the outside world. These processing devices are called *peripherals,* another term for accessories.

There are basically three classes of computers in use today, each with different, overlapping ranges of power. These are mainframe computers, minicomputers, and microcomputers.

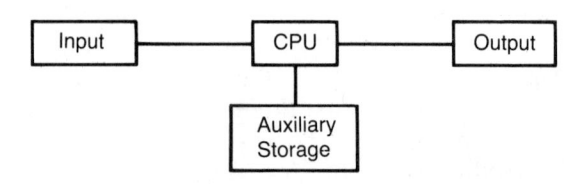

Figure 8-4. The basic elements of a computer system

Classification of Computers

Mainframe Computers

In the late 1950s, computer technology was in its infancy. Computers were large, delicate, extremely expensive machines. These early systems filled entire rooms and required whole departments to service and support them. The first computers were mainframes. They are the most powerful machines available and are capable of processing large amounts of information in a very short time (Figure 8-5). Today, almost all large organizations have mainframe computers that are used for the following:

- High-speed computation
- Data communication applications in which terminals at remote locations transmit data to a central processing unit
- Micro-mainframe links whereby terminals at remote locations retrieve and access information stored in the mainframe computer

Although mainframe computers are fast and have tremendous storage capacity, they are also very expensive. In addition, they must be installed in a special computer room free of dust and dirt and kept at a specific temperature and humidity level.

Minicomputers

In the 1960s, the first mass-produced *minicomputers* started a revolution in interactive computing for the everyday user. This computer was small enough to be installed in laboratories and offices, wherever people worked (Figure 8-6). It was accessible to anyone who cared to learn its relatively simple programming language and routines. Purchased for about $25,000, it put data processing within the reach of modest budgets for the first time. The minicomputers have become very popular in business, either to replace or supple-

Figure 8-5. The mainframe computer.
Source: Photo courtesy of Hewlett-Packard Company.

Figure 8-6. The minicomputer.

Source: Courtesy of Wang Laboratories.

ment existing computer power. These computers are usually used as general business computers. They handle accounting functions, sales, inventory, and payroll applications.

Microcomputers

The most striking trend in the hardware arena is the sudden ascendance of the *microcomputer* in general, and the *personal computer* (PC) in particular. Today's microcomputers are also known as personal computers because they are generally used by an individual worker to do his or her job better (Figure 8-7) rather than by a department in a central location. These revolutionary devices perform in essentially the same way as the mainframes and minicomputers. They all store, process,

Figure 8-7. The microcomputer.

Source: Photo courtesy of Hewlett-Packard Company.

retrieve, and communicate data. But microcomputers do so at the worker's desk.

As their name implies, *microcomputers* are the smallest computers. Although they have limited memory and speed, their cost makes them more attractive than mainframe computers and minicomputers.

It takes a large room to house a mainframe, whereas a micro may sit on a desktop or fit into a briefcase. To make mainframes cost-effective to operate, many people must use the large computers, frequently at the same time. If something goes wrong with the mainframe, the individual terminals tied to it cannot operate, so the mainframe must be repaired before the system is put back into service.

The Personal Computer

All of the classes of computers discussed have unique applications for offices. But the microprocessor (personal computer) is becoming so familiar a part of the information-processing landscape that it deserves special attention. There are many kinds of microcomputers, but the one that most people can relate to is the personal computer. It is a stand-alone device that places a wide range of capabilities in the hands of a user. Modern personal computers can be powerful, portable, versatile, inexpensive, and easy to use.

Hardware Components

Now, let us take a closer look at the equipment of the personal computer; it will help to understand how it works. The personal computer equipment is called *hardware*. The individual pieces, or *components,* can only operate when linked together by a variety of electronic wires.

As in all computers, data pass through the personal computer in four stages, entering as input, being organized by the processor and stored, and coming out in a new form as output. The input for a PC can be, for example, characters typed on a keyboard.

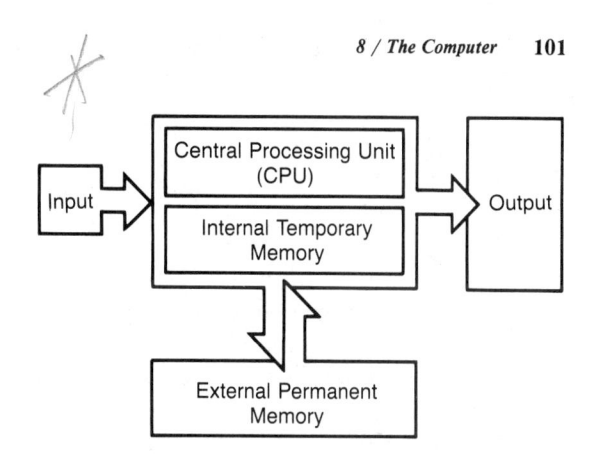

Figure 8-8. The way the components of a personal computer system are organized.

The components of a personal computer make up a system (Figure 8-8). Each personal computer configuration has several components, including input, process, storage, and output devices. Some personal computers have only one means of input and one of output—for example, a keyboard for input and a screen for output. Others may have a wide range of input and output devices. Some may be in the same box as the computer itself; or they may be separate, in which case they are called *peripherals.* Let us examine each component.

The Computer Terminal. The computer terminal, sometimes referred to as a *workstation,* has two main parts: the keyboard and the display screen. Sometimes these are combined into a single unit; at others, terminals have a detachable keyboard. *Ergonomists,* people who design machines so that people can use them easily, advocate using a detachable keyboard.

The Keyboard. The most common way of putting information into a microcomputer is through a keyboard (Figure 8-9). The keyboard is made up of letters and numbers arranged like those on a conventional typewriter. The difference is that special keys are added around the standard character keys.

Figure 8-9. The keyboard to a personal computer.

Source: Courtesy of Key Tronic.

These keys enable the user to create and revise text more easily. An optional numeric keypad helps in performing math calculations. Some numeric keypads are built right into the keyboard. Others (Figure 8-10) are separate and attach to the keyboard by a cable. Numeric data entry keyboards resemble a desk calculator and enable the user to process a column of numbers that may be contained in a document.

When you type information on the keyboard it passes directly into the computer. The computer then records these data and flashes them on the screen. It all happens so fast that you might think that you are typing directly on the screen. The display actually is an "echo" of what is going on in the computer. Therefore, when you see characters on the screen, you not only know that you have typed it, but also that the computer has received it.

Visual Displays. The most common way for computers to display information is on a video screen (Figure 8-11). The screen, a *cathode ray tube* (CRT), is located just above the keyboard. It displays information by projecting a beam at a screen coated with phosphor. *Phosphor* is a substance that gives out luminous light without sensible heat. The beam scans across the screen from side to side and top to bottom, fifty to sixty times a second, causing the phosphor to glow until the beam reaches it again; otherwise the screen would flicker.

Figure 8-10. A detached numeric keypad.

Source: Courtesy of Key Tronic.

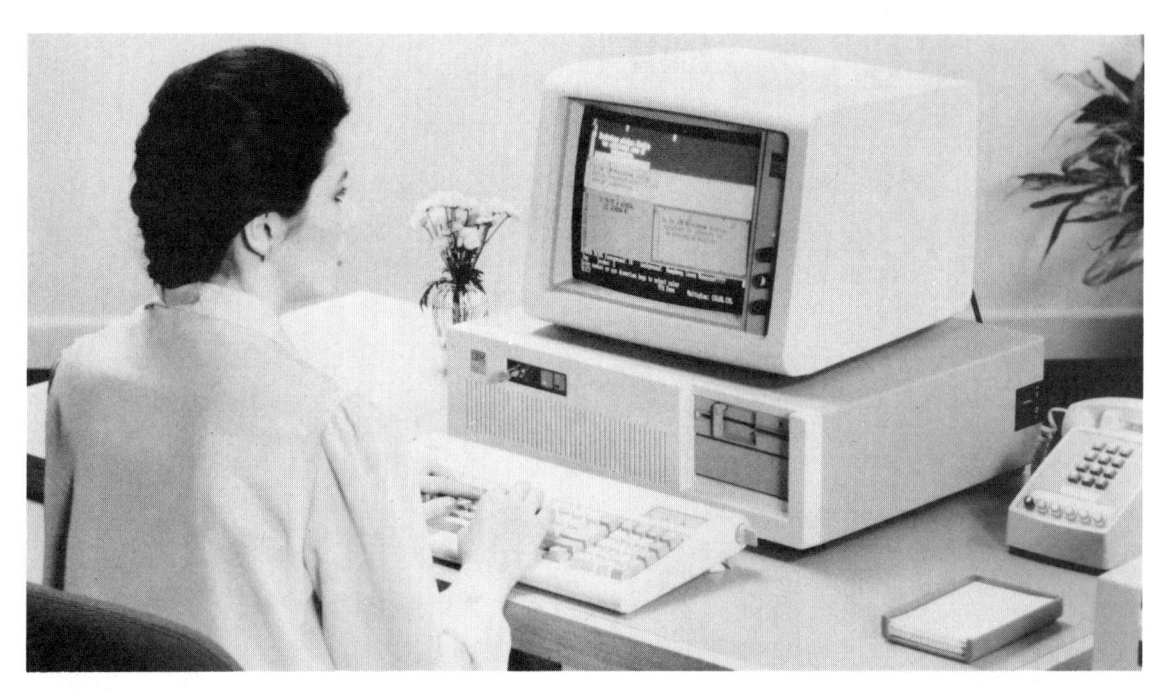

Figure 8-11. Information displayed on a video screen.
Source: Courtesy of International Business Machines Corporation.

The standard video screen displays 80 characters (or columns) across, and 24 or 25 lines in length. Some screens expand to 132 characters for electronic spreadsheets or accounting tasks. Twenty-four lines is about half a single-space typewritten page and is adequate for word/information processing applications. Some users feel that full-size (66-line) screens are actually a hindrance.

Screens can also be in color, but most business microcomputers, as opposed to home ones, use only one color for the characters. *Monochrome* is a one-color screen. The colors most often used in personal computers are green, white, and amber, against a black background. The most popular color is green.

The Microprocessor

The microprocessor, or the central processing unit (CPU), is the "heart" of the computer. It receives instructions and data, manipulates (processes) the data, and stores, displays, and/or prints them out as information. The unit (Figure 8-12) is sealed in a separate plastic or ceramic box or is built into the terminal. All the components of the personal computer are connected to the *CPU,* which consists of a microprocessor, the arithmetic/logic unit, and working memory.

There are two main types of primary storage, one that can store and access data and instructions and one that can only access prerecorded or preprogrammed instructions or functions.

Random Access Memory

Random access memory (RAM), also referred to as *working memory,* is the part of the memory used for storing programs and data. RAM can be accessed or altered as needed by each program.

Figure 8-12. The "heart" of a computer is the CPU; the heart of the CPU is often a printed circuit like this one.

Source: Courtesy of Intel Corporation.

Read-Only Memory

The CPU can not only access the contents of working memory but can also change them. *Read-only memory* (ROM) is the part of computer memory that contains prewired functions. Usually constructed by the manufacturer, it cannot be altered by programmed instructions. The type of information stored in ROM varies from computer to computer. For example, in a personal computer, ROM usually would contain the "system programs," such as those designed to handle operating-system functions. In small systems, ROM may be built into the system and thus not utilize any user memory.

Other memory chips, called *program read only memory* (PROM), can have their programs written on them after construction by "blowing" fuses with a pulse of electricity. Usually these are *erasable PROMs* (EPROMs), see Figure 8-13.

With EPROMs the memory patterns can be erased by shining ultraviolet light through a little window in the package. This connects the fuses again so that the chip can be programmed again and again.

Taken together, random access memory and read only memory comprise the computer's primary memory, and the forms of memory the CPU can access directly. Most computer systems have an auxiliary storage component as well.

Figure 8-13. An erasable programmable read only memory (EPROM).

Source: Courtesy of Intel Corporation.

Chips and Bits: The Measure of Computer Power

Within the central processing unit, working memories are stored on small *silicon chips.* The chip of silicon is often called the *microprocessor,* and it contains tens of thousands of electronic circuits. You can think of a chip as a kind of highway for information carried in digital form—numbers and letters represented as sequences of 1s and 0s. These electronic circuits can switch very rapidly indeed from one state (ON) to another state (OFF). The computer uses these two states to signify a *bit* of information. Each *bit* (short for *binary digit*) of information is the smallest amount of data the computer can handle.

The earliest computers had only 8-bit processor chips and 64K RAM. Sixteen-bit computers provide for faster operation and an easier route to greater accuracy in clerical computations. The major advantage, however, is the greatly increased RAM. With more (RAM) memory space, intricate programs and high-resolution graphics are possible, for example. Now computers and microprocessors deal with 16, 32, 64, 128, and 256 bits at once. In general, the more bits, the more powerful the computer. Using patterns of 8 bits, each one switched off as 0 or on as 1, the computer can represent all the letters of the alphabet.

In addition to bits and chips, another measure of CPU power is internal complexity. This includes the quality of the instruction set. The processor has many different functions. It has to "talk" to the rest of the system, so there is an input/output area on the chip assigned to that function. CPU chips process instructions at a rate set by the *clock speed.* There is also a *clock,* sometimes separate from the rest of the chip, that sets the overall computational speed. The clock regulates the speed so that all parts of the system can do their work in step with each other.

The final part of the CPU is the *arithmetic/logic unit* (ALU), which handles arithmetic functions and logical operations.

Auxiliary Storage

Information can be stored *off-line;* that is, out of the system, on compact storage devices. *Auxiliary storage,* or secondary storage, is necessary because the internal RAM and ROM of the system cannot meet most of a user's storage needs (Figure 8-14). You can put these devices in a drawer or on a shelf. The "back-up" file safely stores information in duplicate. A power failure could result in a loss of information contained *on-line,* or within your computer system.

There are several different types of storage devices for computer systems. Typically, aux-

Figure 8-14. Information is stored off-line, on a floppy disk, to expand the computer's memory.

Source: Courtesy of Xerox Corporation.

iliary storage for microcomputers is contained in the following devices and media:

Floppy Disks

The *floppy disk* is a flexible, portable piece of Mylar plastic that has been magnetically coated. It is the least expensive of all the auxiliary storage media for microcomputers. A protective covering encases the actual disk, which is inserted into the disk drive. The *disk drive* is a narrow opening that accepts the floppy disk. It may be a separate box or built into the terminal. The drive transfers information from the CPU to the diskette. A small mechanical device (like the needle on a record player) called a *read/write head* records the data on the disk. If the disk is designed to store files of text or data, the recording head can write to the disk. (You cannot write to the disks that contain special programs because the vendors have *write-protected,* or altered, the disk to prevent writing.)

After the data have been recorded, the disk can be removed from the drive and stored offline. Disk drives vary in size from 5¼ to 8 inches. The 5¼-inch disk drives have become the standard for personal computers. The newer 3½-inch drives are increasingly popular, whereas 8-inch drives have largely fallen into disuse. Storage capacity varies according to the density of recording, the number of tracks per inch, the use of one- or two-sided recording, and the size of the disk. The storage ranges from 140K (single-sided, single-density) to 650K and more.

Hard Disks

Hard disks are capable of storing large amounts of data (millions of characters per disk) and provide the computer with rapid access to the data stored. Compared to floppy disks, which make about 60 complete revolutions a minute, some hard disks make more than 50,000 revolutions. Hard disks are also referred to as *Winchester rigid disks.* Instead of thin flexible plastic, hard disks use machined aluminum platters coated with a magnetic surface. They not only provide greater storage capacity, but also superior protection from the environment. Hard disk technology seals the disk inside a dust-free environment. Winchesters designed for microcomputers can hold from 1 to 20 megabytes of information. There are 5,000 to 10,000 pages on a hard disk whose diameter is no larger than that of an 8-inch floppy.

Floppy disks provide inexpensive archival memory, they are portable, and they provide an easy means to switch from one computer to another. Unfortunately, the storage capacity of floppy disks is low, and accessing data on them is time-consuming. Hard disks, on the other hand, offer both quick access and large storage capacity. But they do so at the sacrifice of portability and the addition of considerable cost. They also introduce the factor of sensitivity to such outside influences as shock and dirt particles.

Now, there is a new method to store the software and data for each job on a separate disk. It is a hybrid technology that embodies the best of both features of floppy disks and hard disks. It is known as the *Bernoulli box* (Figure 8-15). The basic Bernoulli box package includes the drive unit, a controller card

Figure 8-15. The Bernoulli box combines the performance standards of a hard disk with the cost and convenience of a floppy disk.

Source: Courtesy of IOMEGA.

to be mounted in one of the computer's expansion slots, the cables, instructions, and a software diskette that allows the computer's operating system to recognize the addition of the unit. The disks (cartridges) can be treated as floppies, even though they have the speed and storage capacity of a hard disk. Using this system, the user makes backup copies of material using Bernoulli disks as a means of security. The ease and speed with which backups can be made by using this system make the technology worthwhile. It is a vast improvement, especially when compared to backing up the contents of a hard disk with countless floppies.

Printers

A printer is the major link between a computer and a user. It is the output device that produces a hard copy version of the information. *Hard copy* is the output printed on paper. Once you have made all necessary revisions on the text at your workstation, you can instruct the printer to print the final, error-free copy. The hard copy material is the lifeblood of office work, with applications ranging from word processing to statistics to graphics.

The quality and speed of printers vary, depending on users' needs (Figure 8-16). Microcomputer printers are for the most part *serial printers,* so named because, like typewriters, they print one character at a time. Serial printers can be separated into two general categories: impact and nonimpact printers. *Impact printers,* by far the more common, actually strike or have an impact on the paper through a ribbon to produce a printed image. *Nonimpact printers* use a higher form of technology that produces a character image on the paper without impact.

Dot Matrix Printer

Laser Printer

Daisywheel Printer

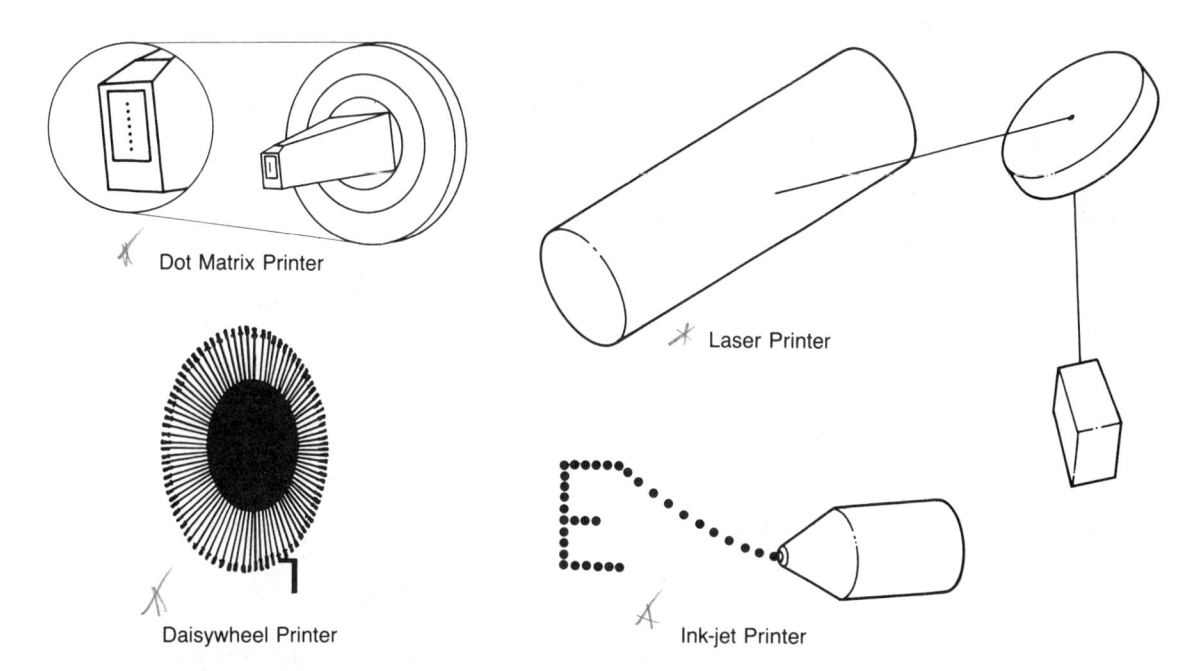

Ink-jet Printer

Figure 8-16. Daisy wheel, dot matrix, ink-jet, and laser printers.

Source: Copyright 1984 by Cahners Publishing Company, Division of Reed Holdings Inc. Reprint with permission from Business Computer Systems, July 1984.

Daisy Wheel Printers

Daisy wheel printers (also called *letter-quality printers*) are the most popular impact printers and produce characters of the highest legibility.

They are a direct descendant of the typewriter, but they print much faster. As the name *daisy wheel* indicates, various type fonts are available on small interchangeable wheels that resemble schematic mechanical daisies.

Daisy wheel printers use a ribbon and element (either a daisy wheel or thimble-type font) to make a solid impression of each letter. The letters are printed, one after another, just as in typing. It is difficult to tell the difference between the output of a good letter-quality printer and that of a conventional electric typewriter.

Dot-Matrix Printers

Dot-matrix printers form their letters by using dots, much like a scoreboard at a basketball game. A matrix printer has a battery of needles, one above the other. Each needle produces a dot on the paper as the print head moves across the paper. In a dot-matrix printer, the needles are moved sharply to mark the paper through a ribbon. The characters are communicated efficiently and generally much faster, but in most cases, not as precisely as in letter-quality printers.

The vast increase in microcomputer sales has contributed greatly to the popularity of dot-matrix printers. They represent a combination of versatility, print quality, speed, and low cost.

A new generation of dot-matrix printers produces what is called "correspondence quality" or "near-letter quality" hard copy, see Figure 8-17. By using two-pass printing, or more pins in the printer-head matrix, or both, these machines produce much better results than traditional dot-matrix devices. This feature is critical because it allows dot-matrix printers to compete with daisy wheels and other printers that produce letter-quality output.

Dot-matrix printers can also print graphic images, charts, or even complete pictures made up of individual dots.

Perhaps the most striking feature is color printing. Several vendors now offer color dot-matrix printers. With the increased emphasis on graphics in business and with color printing available to personal computer users, dot-matrix printers are growing in appeal.

The greatest advantages of dot-matrix printers are price and speed. They are the least expensive of all the printers. Dot-matrix devices can print at more than 150 characters per second, and more than a page a minute.

Nonimpact Printers

Nonimpact printers show the most promise of all the printer technologies. Led by laser, ink-jet, and thermal transfer technologies, they are becoming competitively priced alternatives for business printing needs. Nonimpact printers are fast, quiet, nonmechanical, and have the potential to produce color text and graphics of photographic quality. The following printing devices use nonimpact printing technologies:

Thermal Printers. A thermal printer passes an electric current through the print head to make a mark on heat-sensitive paper. This device is an inexpensive way to produce nonimpact images on paper. Consequently, it will become more popular in the low end of the printer market.

Electrosensitive Printers. In the electrosensitive printing method, a print head discharges electricity to aluminum-coated paper, producing an image. Although they are capable of very fast print speeds, because of the coating on the paper the resulting hard copy is inferior to that generated by other printing methods.

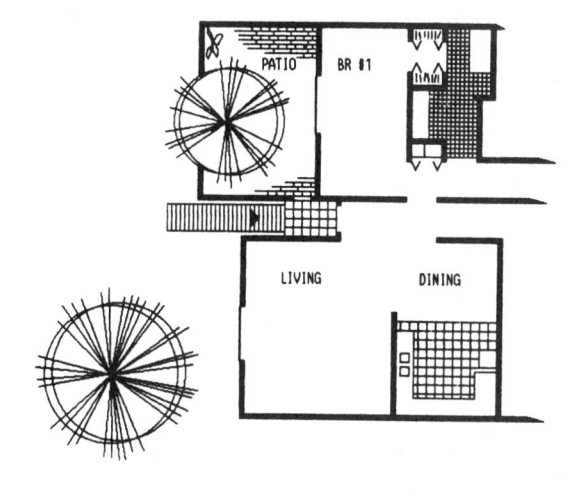

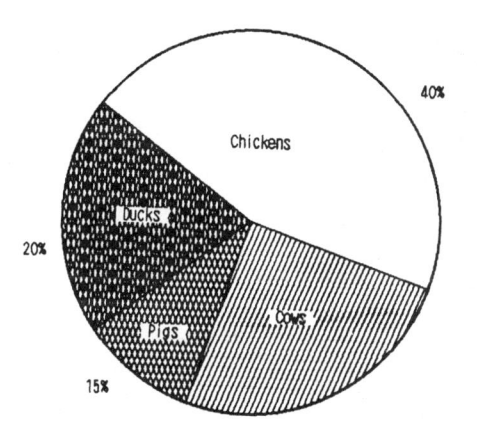

Ink-Jet Printers. Ink-jet printers are the most common nonimpact printers for microcomputers. They are similar to dot-matrix printers, however, instead of needles firing out to strike a ribbon, droplets of ink are sprayed onto the paper in specified arrays.

First-generation ink jets suffered from ink flow problems. Ink coagulated and clogged the nozzle, and only the thickness of the paper determined whether ink came out in blobs or smoothly. However, key advances in the methods used to spray ink on paper improved the quality of output. The ability to produce vibrant color quietly with low maintenance has improved ink-jet technology. Ink-jet printers require little maintenance. That and their quiet operation are primary advantages.

Laser Printers. In the vanguard of nonimpact printers are *laser printers* (Figure 8-18). The laser process is similar to that of producing documents from photocopy machines. Instead of working photographically from a printed page, the printer creates the image with a laser beam.

Laser printers fulfill the broadest range of business needs: speed, letter quality, quiet operation, and graphics (Figure 8-19). Lasers print at eight pages per minute and are faster than the fastest daisy wheel (80 characters per second) or dot-matrix printers (50–500 characters per second) depending on printing mode.

Although laser printers are expensive, advances in semiconductor laser technology are driving down their price and broadening their appeal.

Figure 8-17. Hard copy from a new generation of dot matrix printers.

Source: Courtesy of Toshiba.

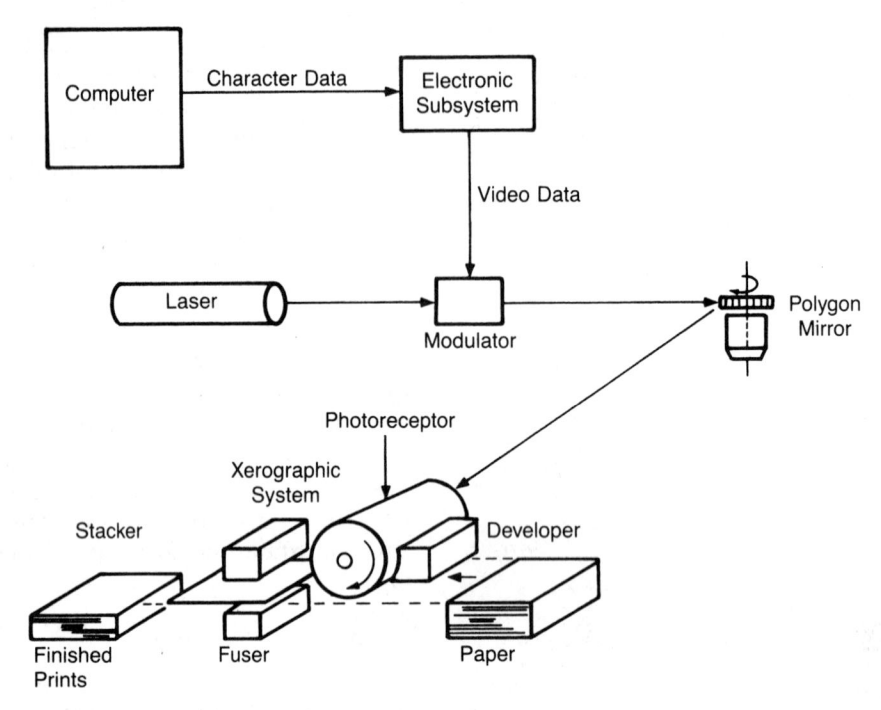

Figure 8-18. In a typical laser printer, a video signal rapidly switches a laser beam on and off as the beam is repeatedly scanned across a charged xerographic photoreceptor by a spinning mirror. As the photoreceptor drum rotates, an electrostatic image is formed. This image is developed by charged ink, then transferred to plain paper and fused by heat and pressure.

Source: Courtesy of Xerox Corporation.

Criteria for Selecting Printers

Selecting a printer can be confusing because there are many vendors, each offering a variety of models. Printers can print text at 400 characters per second (cps); they can produce characters of perfect quality; they can dazzle with color graphics; and they sometimes cost less than a portable typewriter. But unfortunately, no one printer can do all of these tasks well. Several guidelines should be considered in the selection of printers.

1. Print Quality. Decide how the printer will be used. If most of your applications are detailed transactions through which hundreds of pages are generated for in-house use, your choice may be limited to dot-matrix printers. On the other hand, if low-volume, letter-quality printing is required, a daisy wheel or thimble printer is the best choice.

2. Graphics. Graphics present complex numbers in an easy-to-understand format. The quality of graphics varies from printer to printer. Your specific application determines the choice in this category. A printer that would be adequate for in-house reporting work may not be appropriate for formal presentations to clients because of the chart density.

THE WATERMILL RESTAURANT
GRAND OPENING
COUPON

This Coupon entitles you to one free glass of wine or one slice of Chocolate Toffee Pie

- Cut Here

Our Newest Watermill Restaurant is located at 101 Savoy Ave.

The Watermill Restaurant is located between Olmstead St. and Taylor St. on Savoy Ave. Plenty of Free Parking. Open 11am-12pm Mon. thru Sun.

THE WATERMILL RESTAURANT

First Class Mail

ANNOUNCING THE OPENING OF
THE WATERMILL RESTAURANT
AT 101 SAVOY AVE.

G·R·A·N·D O·P·E·N·I·N·G

Figure 8-19. A laser printed advertisement.
Source: Courtesy of Apple Computer, Inc.

3. Page Size. If your applications call for long horizontal statistical typing, a printer must be selected to accommodate this requirement (Figure 8-20). If your work calls for nothing over 132-column pin-fed forms, the selection of models is broader.

4. Speed. Print speeds vary anywhere from 18 cps to 1000 lpm (lines per minute). Speed is a very good indication of printer price. For example, if you need a printer for general ledger applications in which hundreds of pages of detailed transactions are listed, your choice is limited to line and page printers.

5. Technical Interface. Printers must be compatible *(interface)* with the computer. Printers receive output information from the computer in one of two ways, either serial or parallel. *Serial* reception occurs with information that travels over wire located in the cable, one impulse after another. *Parallel* reception involves receiving information by several impulses (usually eight) traveling over separate wires simultaneously.

6. Vendor Service and Support. Reliability and maintenance are important considerations. Because printers are basically mechanical devices, they are subject to failure more often than the computer to which they are attached. It is important to take into consideration the reliability and promptness of the vendor in responding to service calls. Thus, one of the most important criteria in printer selection is service/reliability. Once the printer is purchased, how often must it be serviced? How expensive is maintenance/part replacement? And how much maintenance can the user perform unassisted? These questions provide useful "before-the-fact" information for prospective buyers.

7. Cost of Supplies. In Chapter 16, we will be discussing maintenance and supplies.

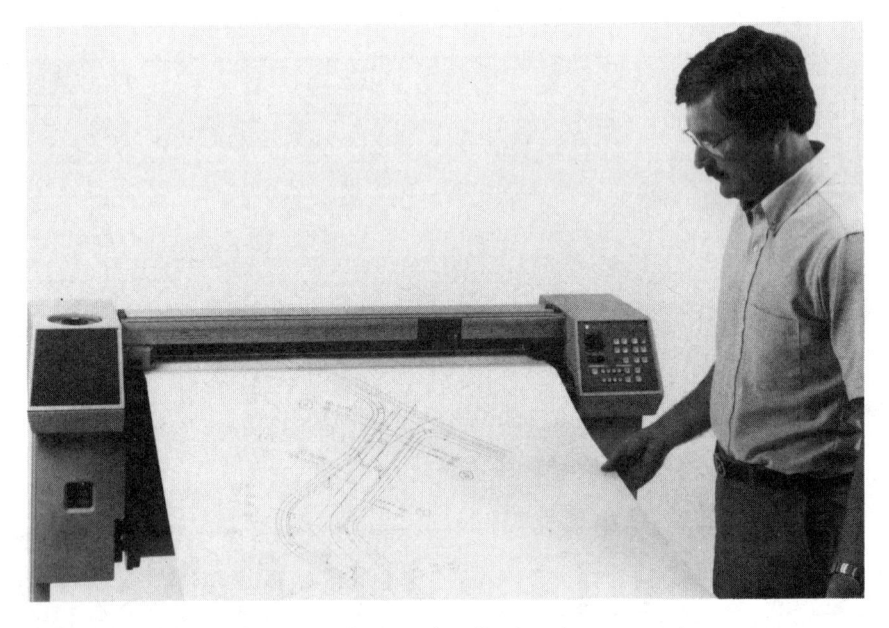

Figure 8-20. Today, there are printers to handle almost any page size.

Source: Photo courtesy of Hewlett-Packard Company.

There are many attachments and supplies that accompany the basic printer. It is pertinent to mention some of these factors in this section because they have a bearing on printer selection. For example, unless you have an unusual business application, it does not make sense to buy a printer that requires special electrostatic paper or special ribbons. Even if the initial purchase price is significantly lower, normal usage would probably cost more in the long run because of the cost of special supplies.

The microcomputer printer market provides an ever-growing number of ways to produce hard copy, text, graphics, and color. In addition, printers are available in models that offer a range of speed, quality, and cost. The choices also enable users to select the printers and technologies that best meet the needs of their office applications.

The key, then, is that the user define the application and compare the contenders by considering some of the points suggested in this section. Printer technology continues to evolve, and the proliferation of printer features is not about to stop. Refined impact and nonimpact technologies will continue to bring marked improvements to print quality and costs. No matter what changes are brought about in computer systems, printers will always be needed.

Personal Computers Move into the Business World

Personal computers continue to stream into the workplace. The chaotic invasion of these devices has compelled management information systems groups to

- Plan for new acquisition strategies
- Enforce data integrity
- Demand device compatibility
- Lay the groundwork for assimilating end-user workstations (personal computers) into an overall corporate information system

Many organizations are encouraging the acquisition and use of micros among professional staffs in the belief that these machines can aid the worker and ultimately the corporation.

Expanding Applications Promote Acceptance

Even with generally favorable acceptance of the microcomputer as a business tool, many users are still unsure of the exact role it should fill and the ways it can help them in their job. The reason for this confusion is that many corporations are undergoing a change in corporate structure and organization. Job titles and responsibilities within departments are changing almost as quickly as the technology of personal computers. The original intent of the personal computer was defined by the relatively limited processing capacity of early machines and the lack of suitable software programs.

New computer technology has changed all of this. The availability of new software and connectivity of microcomputers to corporate files have expanded the role of the PC, making it a powerful business tool.

When PCs are linked together in a corporate environment, the opportunities for information interchange and other new applications are virtually unlimited. Once a PC has been assigned to an information worker, it can be used in a variety of ways. How will the PCs be linked? Will the terminals be connected to larger systems, or integrated into stand-alone networks, or some combination of both?

The use of personal computers will also encourage their integration with existing systems—mainframes and minicomputers. The integration of PCs with centralized computers will meet the growing demand for data communication from central data processing departments to individual offices.

The personal computer will also connect

with other office devices through local area networks and PBX networks linked to host computers. What can personal computers do? The variety of abilities is too enormous to list, but the most popular uses are discussed next.

Data Bases. Data bases provide the management of information and its instant retrieval. Electronic data bases can be updated, cross-referenced, and used as a source of immediate decision support.

Graphics. Graphics is a useful feature that can generate a pie chart or bar graph to sell a product or communicate the way things are going.

Information Retrieval Service (Data Banks). If stock market information, research, news, or the many other features external data banks offer is important to your organization, then being "on-line" with a data bank will be a necessary application.

Networking. As soon as an organization requires two or more personal computers, the advantages of networking become obvious (Figure 8-21). As explained in the previous section, networking (or linking) capability lets one computer "talk" to and share information with another.

Forecasting. The forecasting feature helps answer questions such as, Should we lease or buy? What are the financial or personal weaknesses in our current operations? Should we take on a new project? Can we project our sales forecast; our cash flow; our profit and loss; and our purchasing plans? Do we have enough insurance? Should we change suppliers? Should we expand?

These are by no means all of the applications for a PC. Not only can computers give you all this information; they can select data according to any parameters you demand and list them in alphanumeric order. They can produce word processing, spreadsheets, ac-

Figure 8-21. Shared computer system in an office.

Source: Courtesy of Xerox Corporation.

counting, and calendar management. These functions will be discussed in more detail in Chapter 14, "Software."

Personal computers provide organizations a modest tool to start small and the opportunity to link additional devices in the most functional and cost-effective way. Thus, personal computers will have the additional ability to tie desktop workstations together with mainframes and other desktop workstations. The PC has emerged as a universal business tool. It will serve and satisfy the information worker as long as there is appropriate software and communication capability.

The growth in personal computers may be the most striking processor trend, but it is by no means the only one.

Chapter 9 continues to look at the computer market, exploring new configurations, enhancements, and features.

Summary

The personal computer has achieved universal acceptance in the business world. Managers consider it a competitive necessity. Management of information systems once meant the care and feeding of large systems; now it means the distribution and support of small computers.

The computer with office-oriented features did not even appear until 1964, with the advent of the IBM 360 computer and the IBM Mag Card memory typewriter. But within ten years, the technology had evolved to giant mainframes and sophisticated software systems on the one hand and cathode ray tube (CRT)–based word processors on the other. Within ten years, computers as powerful as those original mainframes were sitting on office desks.

The "computer on a chip" and low-cost random access memory (RAM) devices led to the development of the ubiquitous personal computer (PC). The PC has been a major catalyst in the convergence of communication and computing in the office. The benefits of the PC have eased the tedium of office work for America's growing army of information workers. Personal computing supplements an individual's job activity, provides more information for better decision making, facilitates communication, and is a tool for convenience.

Review Questions

1. What two key improvements in computers best characterize the evolution of computer technology?
2. Define a computer. Define a computer system. List the basic elements of a computer system.
3. Briefly describe the input-process-output cycle.
4. What are some of the basic differences between a mainframe and a minicomputer?
5. Why has the personal computer gained such popularity in the office?
6. How does the personal computer differ from the traditional typewriter keyboard?
7. Compare random access memory (RAM) with read-only memory (ROM).
8. What is a bit? What is the difference between an 8-bit and a 16-bit microcomputer?
9. Describe the way a computer stores information off-line.
10. What are the main advantages of a hard disk system?
11. Compare daisy wheel printers with dot matrix printers. List applications for each.
12. Briefly describe the benefits of linking a personal computer to a corporate network, and to a mainframe computer.

Projects

1. The last entry in the timeline technology table in this chapter recorded the divestiture of

AT&T. Expand this timeline by including at least ten significant events from the last date recorded to the present.

2. Visit the computer facility of your college or a large organization in your community. Interview the manager of information systems (or equivalent supervisor). Prepare a brief report describing their computer system. Include the following:

   Type of computers used
   Classification (mainframe, mini, micro)
   Applications
   Peripherals (printers, disk storage, network system, and so on)

   Draw an organization chart showing the structure of the company, the title of the department that handles computer and information processing, and its position on the chart.

3. Some of the computer ads in newspapers do not make sense to the average consumer, terms like *128K, calculator-style keyboard, monochrome display,* and *dot-matrix printer.* Cut out an advertisement for a personal computer in your local newspaper. To a person who is not familiar with technical terms, the ad may not make sense. Using the definitions and explanations in this chapter, rewrite the ad, explaining all of the terms so that the material is understandable to the average consumer.

Chapter 9

Processors: From Electronic Typewriters to Supercomputers

After reading this chapter, you will understand

1. The characteristics of electronic typewriters, word processors, and advanced computer devices
2. The evolution and future role of word processors as discrete types of equipment
3. The various components and peripherals of computers such as flat-panel displays, touch screens, mouse devices, and icons
4. Lap-size portables and their applications
5. The features of computer phones
6. The difference between a supermicro and a personal computer
7. Guidelines for selecting various processors

The discussion in Chapter 8 focused attention on the personal computer as a primary example of the microprocessor. Other processors also deserve attention and play unique roles in the information processing cycle. They vary in size, function, application, and configuration. This chapter will take an indepth look at some of the processing hardware systems and components that comprise the $60 billion worldwide computer market (Figure 9-1).

We will examine and classify each type of processor, discuss special features and applications, and finally offer some guidelines for selecting and acquiring computer systems.

Electronic Typewriters

The simplest category of processor is the *electronic typewriter* (ET), which is a cross between a standard electric typewriter and a computer (Figure 9-2). It looks and operates like a standard electric (electromechanical) typewriter, but the addition of a microprocessor gives the machine advanced features and capabilities.

Qyx, formally a division of Exxon Information Systems, was credited with introducing the first truly electronic typewriter in 1978. It had 1 kilobyte (1K) of memory and sold for approximately $1600. K is the symbol for 1000; thus, this device could store approximately 1000 characters. Other vendors soon entered the market. Their models ranged from low-cost home, student, or executive travel units to ETs that served as "advanced workstations." Although the prices and capabilities of the electronic typewriters on the market cover a wide range, they all offer ease of operation, compactness, and the ability to auto-

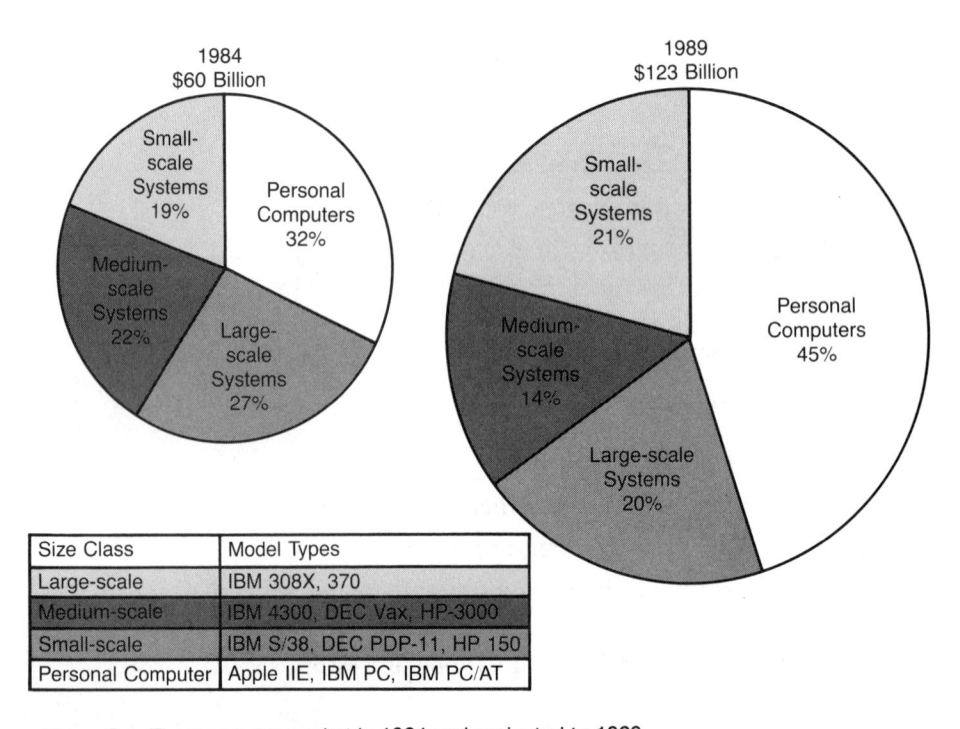

Size Class	Model Types
Large-scale	IBM 308X, 370
Medium-scale	IBM 4300, DEC Vax, HP-3000
Small-scale	IBM S/38, DEC PDP-11, HP 150
Personal Computer	Apple IIE, IBM PC, IBM PC/AT

Figure 9-1. The computer market in 1984 and projected to 1989.

Source: Courtesy of International Data Corporation.

Figure 9-2. The electronic typewriter.

Source: Courtesy of International Business Machines Corporation.

mate some of the repetitive manual functions of typing. The following are among the features and functions of electronic typewriters.

1. Memory. Memory can be internal and/or external. *Internal memory,* depending on the level of the ET, can store text for later recall or printout. Some machines have a direct type-to-printer mode, so they are more similar to the electric typewriters.

External memory is usually offered in the form of disks. Additional memory capacity ranging up to 300,000 characters can be provided by double-sided disks.

2. Expanded Memory. Most electronic typewriters include a *memory mode* feature that can store variable amounts of data until a printout is required. This type of memory is usually *volatile,* meaning that when the power is shut off, the stored text is erased. Battery protection ensures that even when the machine is turned off, the text is retained.

3. Keyboard and Daisy Wheels. Keyboards are electronic to the touch. Most units contain daisy wheel printing elements with 98 or more characters and a variety of available font styles and print speeds.

4. Built-in Features. Built-in features on most models include multistation calculators, easy-change daisy wheels, and lift-out platens.

5. Communication. Electronic typewriters can serve as part of a communication network. The optional equipment enables users to send and receive messages from other compatible ETs, such as word processors and computers by using regular telephone lines.

6. Display. Screens (Figure 9-3) that display a partial line or several lines of text can provide a level of "videotyping" for the user.

7. Automatic Word Processing Features. The following features are found on most electronic typewriter models:

Automatic erasure
Automatic phrase/format storage
Automatic centering
Automatic indenting
Automatic carrier return
Automatic right flush

Figure 9-3. Display screens allow the typist to catch errors before they are printed.

Source: Courtesy of Xerox Corporation.

Automatic boldface
Automatic line spacing
Automatic pitch selection
Automatic proportional spacing
Automatic underlining
Automatic column formats
Automatic statistical typing
Automatic superscript/subscript
Automatic hyphen mode
Automatic right justification
Automatic pagination
Automatic form layout
Global search and replace
Variable data insertion
Store/recall by document name
Programmable display prompts

These word processing features are explained in Chapter 11 "Word Processing." The terms can also be found in the Glossary.

Electronic typewriters offer many advantages. In spite of the number of functions, training is relatively simple. If instruction is required, operators are easily guided by factory or dealer sales personnel, training meetings, or training manuals. Another key advantage of the ET is its maintenance record. The electronic typewriter contains a greatly reduced number of parts subject to wear and tear. Its electronic components have withstood the test of time.

The electronic typewriter is becoming a processor that offers simplicity as a first step into an automated environment. It is a product that brings together automation, productivity, and easy use.

Word Processors

Word processors evolved from the *automatic typewriters* first introduced in the 1930s. Automatic typewriters were designed primarily for typing form letters over and over, using a punched paper roller that was similar to the tapes used in player pianos. The holes in the tapes instructed the typewriter to type certain characters.

IBM made the greatest impact on the development of the modern word processor when they introduced two new products in the early 1960s: the Selectric typewriter in 1961 and the Magnetic Tape/Selectric Typewriter (MT/ST) in 1964. The latter enabled the operator to accelerate production by "catching" (recording) on magnetic tape what was keyboarded and omitting words and phrases that were deleted by the operator during the editing phase, in automatic playback. The second magnetic media machine was introduced by IBM in 1969. The Magnetic Card/Selectric Typewriter (MC/ST) recorded keystrokes on magnetic cards and was designed to have a greater penetration into the typewriting market.

The next major development was the introduction of machines with display screens. Visual display screens differed from "blind" systems by allowing a typist to keyboard text onto a televisionlike screen, rather than onto paper. Errors are corrected and formats are changed by manipulating texts on a screen before they are printed on paper.

When it was first introduced, the *stand alone* word processor was marketed as a tool for improving the productivity of the secretary. Since then, it has matured as a single-function device. Many experts, however, feel that it will disappear.

Types of Word Processors

There are three types of word processors: stand-alone, clustered, and multifunction.

Stand-Alone Systems. A *stand-alone word processor* (sometimes referred to as a *dedicated word processor*) has its own CPU, functions independently, and is used primarily for word processing. The stand-alone system consists of a keyboard, display screen, CPU, storage unit, and printer (see Figure 9-4). The stand-alone system does not share programs or data with other terminals.

Figure 9-4. Stand-alone system.

Source: Courtesy of Xerox Corporation.

Multiuser Systems. The dedicated, or stand-alone, system did not address all of the varied office applications. As more tasks were automated, more systems had to be installed and the buyer had to spend more money. The answer to economical growth in systems is through the multiuser approach. *Multiuser systems* enable the users to arrange devices that can share storage space and peripherals. Also, workstations can share intelligence. There are two types of multiuser systems: shared-logic systems and shared-resource systems (Figure 9-5).

Shared-Logic Systems. Shared-logic systems comprise workstations that share a central processing unit, storage devices, and peripherals. Each workstation is linked to the others and to peripherals. The shared-logic system provides for easy and economical expansion. Each workstation runs off the same CPU. The workstations, however, cannot function independently, although some do contain limited amounts of intelligence. If the CPU goes down, the system is inoperable until the CPU is repaired.

Shared-Resource Systems. Shared resource systems also link individual workstations with a CPU, storage devices, and peripherals. However, each workstation within a shared-resource network has some intelligence and power of its own. If the CPU goes down, the shared-resource system retains the ability to perform tasks at the local workstation level.

Multifunction Systems. Multifunction or extended word processors offer additional applications such as math, personal computing, and communications. All extended systems have good to excellent word processors, as well as a range of other application packages. The vendors entering this market from the word processing area offer the best word processing (WP) software. The best applications packages in other areas are offered by the companies whose strength is in data processing. Multifunction word processors can be part of a shared network system or a stand-alone system.

The Executive Workstation

The electronic typewriter and dedicated word processor were originally used to increase the productivity of the secretary. Management quickly learned that it makes even more sense to increase the productivity of a high-priced executive, but this is not as easy to do. Organizations first encouraged executives to use personal computers as a tool to increase their own productivity, enhance decision support, and forecast trends. The personal computer has already made great inroads into the office environment and will eventually supplant the stand-alone word processor. Word processing

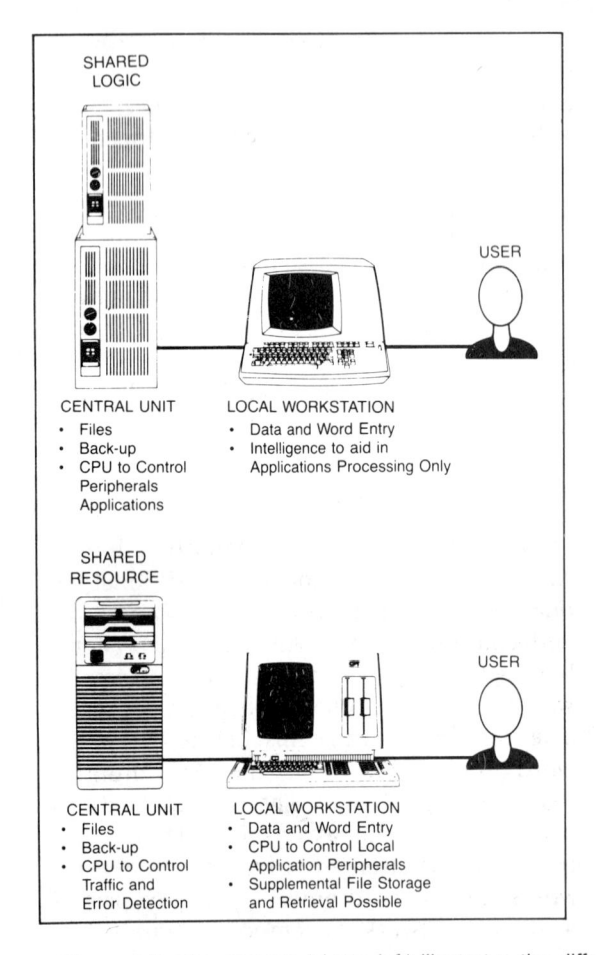

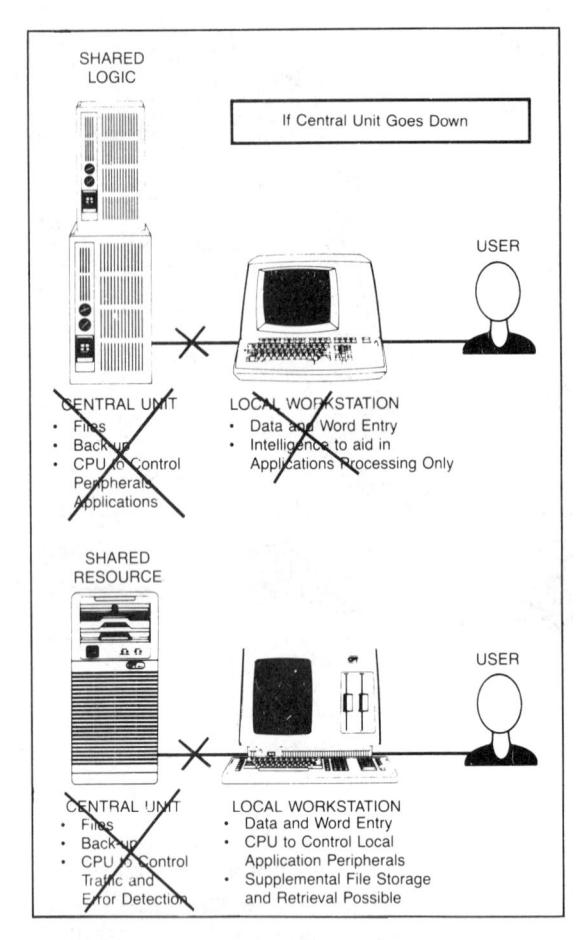

Figure 9-5. The diagram (above left) illustrates the differences between shared-logic and shared-resource office systems. The diagram (above right) depicts what occurs when the central processing unit goes down: the shared-logic system is inoperable until the situation is rectified, while the shared-resource system retains the ability to perform tasks at the local workstation level.

Source: Pam Jensen and Sharie Kimball, *Information & Word Processing Report*, "Multi-User Systems: What is Shared?" May 1, 1984. Reproduced courtesy of CPT Corporation.

will be a part of any new processing system. However, it will evolve more into an application rather than a specific piece of hardware (Figure 9-6).

Thus, the initial force to improve productivity in the office took two paths:

1. At the secretarial level, computers used were discrete, single-function systems. Word processors and personal computers were employed primarily in text editing.

2. At the professional managerial level personal computers were used for interactive computing and spreadsheet analysis.

Although both groups used their respective devices to increase their productivity, the en-

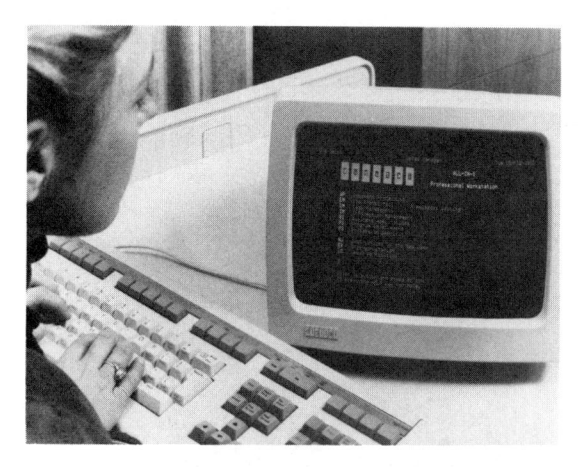

Figure 9-6. The user of this workstation has access to all the system's software: word processing, business data processing, electronic mail and filing. The computer can also be used as a desk calculator, phone directory, and calendar.

Source: Courtesy of Digital Equipment Corporation.

vironment lacked cohesiveness. A vast majority of the devices were stand-alones. An atmosphere of uncertainty about each one's role prevailed in the office. Each worker (secretary or manager) performed word processing, spreadsheet analysis, and project planning functions in blissful ignorance of what the neighbor in the next office was doing. This configuration did not lend itself to a team effort. Decentralization needed a new approach, that fostered teamwork in smaller work groups within an organization. A professional device that could function both as a stand-alone and in a network was also needed (Figure 9-7).

Invariably, technology transforms personnel—the more technology, the more sweeping the transformation. Word processors and personal computers set the stage for significant growth. They were new devices that combined the best features of both. But, instead of the single-function, piecemeal approach of the past, a system emerged that served the multiple needs of managers, professionals, and secretaries.

As vendors introduced devices that had more universal appeal, they stimulated the re-

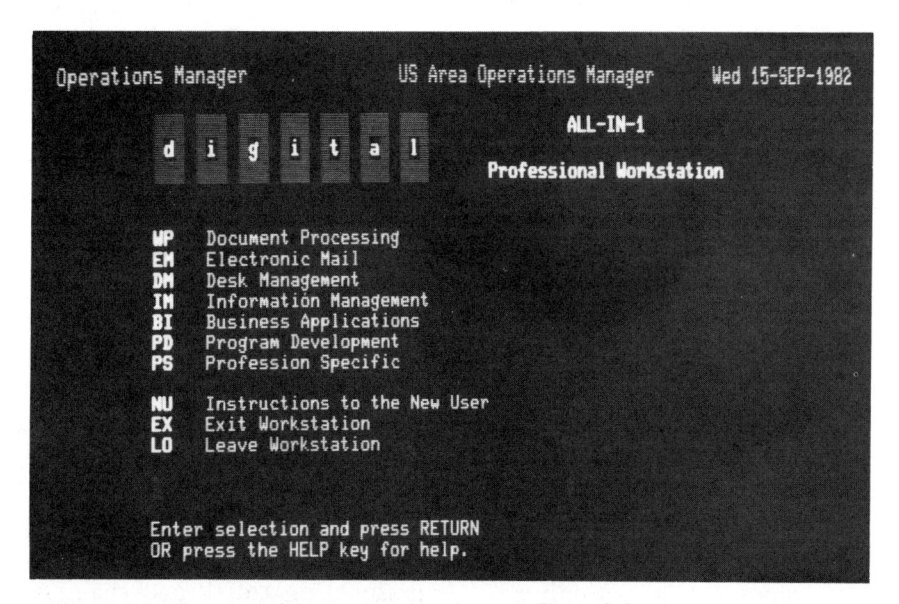

Figure 9-7. The menu of the workstation shown in Figure 9-6.

Source: Courtesy of Digital Equipment Corporation.

Figure 9-8. A universal workstation.

Source: Photo courtesy of Sun Microsystems, Inc.

organization of personnel working together in a smaller, team-oriented environment. The result was a new breed of terminals that began to move into more offices. The new devices were called *universal workstations.*

The Concept of a Universal Workstation

A *workstation* is a desktop device with local/personal computing power (Figure 9-8). It has communication capabilities for networking with other workstations and central systems. A *universal workstation* does all a workstation does, and more. As the word *universal* implies, it should be all things to all people. Such a workstation must serve executive, managerial, secretarial, and clerical staff with equal ease. Obviously no machine can be all things to all people, but vendors are designing machines to approach the ideal as closely as possible.

The ideal workstation should not only address the personal productivity needs of the individual user but also the growing business's needs. A truly integrated system should include specialized communication networks and services, data and word processing, image and audio processing, and multiple business applications software (Figure 9-9). If, indeed, a workstation is to be considered universal, it must replace separate, piecemeal tools—the telephone, calculator, word processor, dictation system, and whatever else presently fills the desktop as a regular workstation tool. The user should be able to switch from function to function with relative ease and to perform more than one activity at any given time.

Features of a Universal Workstation

The ideal universal workstation should have some or most of the following features.

User-Friendliness. The system should be easy to learn and easy to use. It should be as personal as a desk and chair. All functions should be integrated, and movement between the applications should be fluid. The user in-

Figure 9-9. Multiple-display windows and the ability to run most industry-standard personal computer programs enhance the performance of this workstation. The windows at the edges of the screen display text, graphics, and data from several programs or initiate office functions, such as electronic filing and printing.

Source: Courtesy of Xerox Corporation.

terface should be consistent through all applications.

Variety in Application Software. The system should be able to run the variety of available stand-alone applications such as word processing, spreadsheet, data base, and business graphics. In addition the system should be capable of using off-the-shelf software packages and specialized industry and professional application software.

Communications. The system should support a wide variety of local and remote communication options.

Storage. The system should have ample storage capacity on floppy, hard, and optical disks.

Future Expansion. The system's hardware and software should be fully upgradable in order to protect the corporation's investment. A user should be able to buy a professional computer with just the amount of functionality to do the job today.

The user needs full assurance that as job requirements dictate greater functionality, the workstation can be easily upgraded. *Upgrading* means adding functions, not replacing equipment.

Path to New Technologies

These features barely touch the surface for the ideal workstation. Most vendors presently offer pieces of this ideal package. The workstation of today is an office tool designed to be placed on a standard office desk. For this reason, it is generally smaller than a standard terminal. The newer workstations are ergonomically designed and have user-friendly aids such as flat-panel displays.

Flat-panel displays were first used in lightweight "lap" computers. Although they are ideal for portable computers, the major commercial market for flat panels will be offices and businesses. They include word processors, personal computers, and ultimately universal workstations. Flat-panel displays are slim, flat screens, less than three inches thick.

There are three types of flat panel displays: liquid crystal displays (LCDs), light-emitting diodes, (LEDs), and gas plasma displays.

The best known is *liquid crystal display.* Some special crystals change to a dark color when an electric current is passed through them behind special, polarized glass. Lines and dots of liquid crystal can be encased in plastic, wired up to a microprocessor, and used to display information. Liquid crystals also have the advantage of using very little power. That is why a digital watch can run for years on a very small battery. It displays information all the time. Scientists are refining this technology so that LCDs can produce smaller liquid crystal dots on larger screens.

Light-emitting diodes (LEDs) are like very small light bulbs. You can see them, unlike LCDs, in the dark. LEDs were used in watches, but are now mainly used in indicator lights, in calculators, and in some video games.

Gas plasma displays are made up of dots of fluorescent gas. They light up when electricity is passed through them. Their displays are clear, but they are bulky because of the control circuitry and the high-voltage electricity they need.

All the flat-screen technologies involve sandwiching a thin layer of chemicals between two or more layers of glass. They produce an image by electrically charging the chemicals in one way or another.

Early flat-panel screens held 25 lines of eighty characters. There are a variety of flat-panel screen sizes including full-size displays. The portable computer is hastening the evolution to flat screens. The bulkiness of the CRT screen may soon give way to the thin, flat screen in universal workstations.

User-Friendly Aids

Keyboards, screens, and printers are the usual ways of putting information into computers and getting it out. On most computer workstations, the user interacts with the computer by using a typewriter keyboard. These keyboards have a set of *cursor control keys* or directional arrows. A *cursor* is a flashing or bright dot, square, or underscore character. It shows the position on the screen where characters or spaces are being entered. Four or five keys are clustered together. Each is usually represented by a short arrow pointing in one of four directions: north, east, south, or west. If you touch a key, the cursor moves in the direction indicated.

There is a wide variety of other user-friendly ways to interact with the computer, too. The following are some exciting technological enhancements that allow more and more people to enjoy working on a computer. These aids make even the most complex devices incredibly easy to use.

Touch-Sensitive Screens

Touch-sensitive screen technology is a user-friendly aid (Figure 9-10). It is designed to make the operation of the computer simpler. Touch screen systems simply involve the user's touching a graphic symbol on the screen and receiving an immediate response. When you touch the screen with your finger, you

Figure 9-10. Touch makes this personal computer easy to learn and easy to use.

Source: Photo courtesy of Hewlett-Packard Company.

break one of the beams from top to bottom and another from side to side. Thus, the computer can figure out where you are touching. The system combines an *iconographic* touch screen and keyboard input. *Icons* are small pictorial representations of files, system services, and other resources that are pointed to or touched. They control a computer's operation.

Touch screen technology was first developed to help young children learn basic math and language skills on a computer-based learning system. In 1981, Hewlett Packard investigated the use of touch screens for industrial terminals. After experimentation, they designed a personal computer using a touch screen interface (Figure 9-11).

Today, touch systems have taken their place as the ideal input device in offices, factories, and schools. Although keyboards will continue to be necessary for high-volume data entry applications such as word processing and accounting, they can actually be a barrier in many applications.

Touch screens are gaining wider acceptance as they become more available. There are several reasons for their popularity.

Accurate and Efficient Input. Touch screens reduce errors and make computers accessible to everyone, regardless of training, experience, or typing skill.

When the user touches the screen, the computer acknowledges the hit and responds immediately. The complicated process beyond the screen is completely *transparent* to the user.

Ease of Use. With touch systems, there are no keyboard functions to learn, no programming languages to master, and no procedures to memorize. The user interacts with the system simply by touching the screen. To prompt the user to touch the screen when appropriate, application software creates menus, pictures, and other messages.

Speed. A user who presses the icon for file cabinet can leaf through pages in a given file at speeds far faster than that of fingers typing on a keyboard.

Immediate Interaction. Touch screens allow the user to interact with data bases immediately. This capacity eliminates the need for costly and time-consuming training.

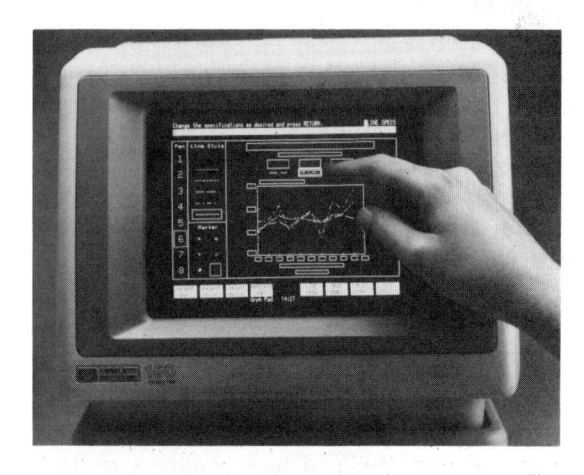

Figure 9-11. Touch makes applications programs like graphics easy to learn and use.

Source: Photo courtesy of Hewlett-Packard Corporation.

Figure 9-12. Using a light pen to enter data.
Source: Photo courtesy of Hewlett-Packard Corporation.

Variety of Applications. In addition to filing, main menu offerings include telephone and directory, electronic mail, graphics and slide capabilities, word processing, and spreadsheet.

For business professionals who use personal computers or universal workstations primarily as a decision-making tool, the touch screen is a superb feature. The absence of a keyboard means that there is less margin for input error and greater opportunity to concentrate on the task at hand. Users enjoy rapid, easy, accurate, responsive interaction with their computer systems.

Light Pens

Instead of keying into a terminal, an operator may use a *light pen* to enter data on a display screen (Figure 9-12). A light pen is a stylus-shaped device that contains light-sensing electronics at the pointer. The user touches the light pen to a particular area of the screen. The pen registers when the beam in the CRT scans past it, informing the user of exactly where on the screen the beam is printing. Light pens are used on computers as well as retail applications. They are able to read *bar codes* on packaged goods. Bar codes are a series of lines that identify items and prices for computers. They are read with a *bar code reader,* a scanning device that translates black and white bars of different widths into electrical impulses.

Mouse

A *mouse* is a device used mostly with personal computers to point to and select a specific function on the display screen (Figure 9-13). The mouse is a small rolling box. As the user moves the mouse around the desktop, the pointer on the screen moves in the direction indicated.

To tell the computer what you want, you

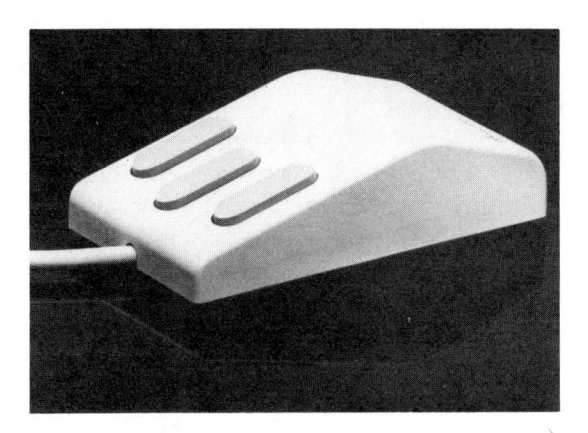

Figure 9-13. A mouse.
Source: Courtesy of Key tronic.

simply move the mouse until it is pointing to the object or function needed. When you get to the item you want to use, you click a button on the mouse to make the selection.

This can best be described as a "pick-and-select" sequence, which enables the user to make a selection. That selection may be a piece of data, an area of the screen, a token, or an icon.

Icons

Mouses interact with pictorial symbols *(icons)* to locate and retrieve information. Modern computer software is designed to display symbols that resemble familiar objects, such as file folders, clipboards, or desktops (Figure 9-14).

Users can relate to these icons; they are reassuring because people find themselves in a familiar setting. Much of the operation is intuitive.

Other desktop symbols are a calendar, phone, appointment book, and a tickler file. These can be added to existing icons.

The number and variety of applications using icons as graphic symbols are expanding. More software application packages include icons in their operating procedures.

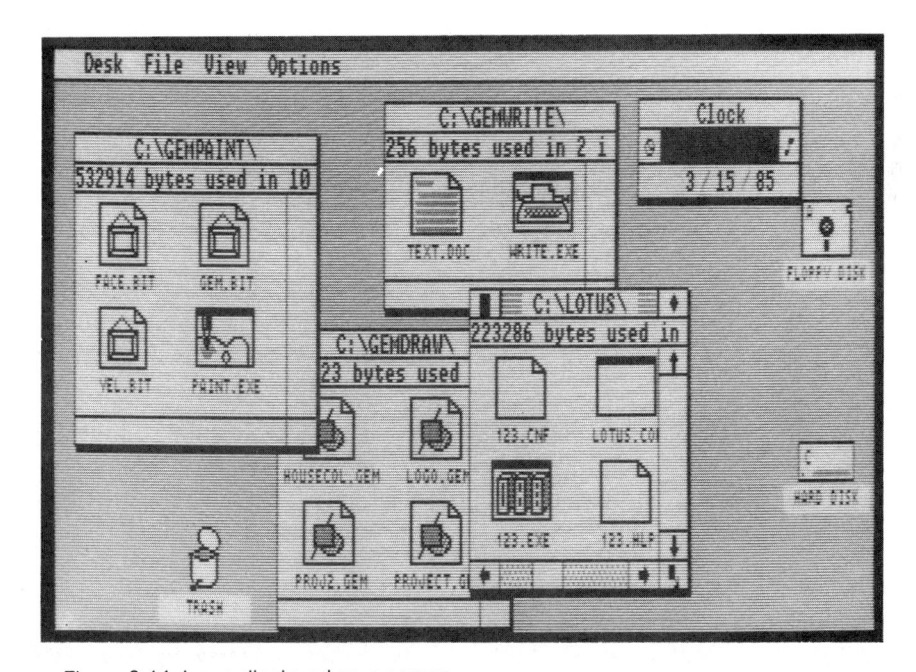

Figure 9-14. Icons displayed on a screen.
Source: Courtesy of Digital Research.

Mouses, touch screens, and similar devices have some disadvantages for experienced computer users. Some touch typists object to the need to lift an arm from the keyboard to touch the screen or the mouse. They find that this movement interrupts keyboard activity and becomes tiresome. Touch screen and mouse activity are used mostly to make selections among options. Most imput of data is through the keyboard, however. Generally, users spend the rest of the time reading the screen and thinking.

Actual use of the touch screen and mouse amounts to a small percentage of the total time a user spends at the workstation. Initially, the interruption of removing a hand from the keyboard can slow the data entry, but, with practice, the operator becomes skilled in using these aids. With skill development and mastery, mouses and touch screens can be at least as fast and psychologically more direct than traditional keyboards.

All of these computer aids are being incorporated into modern workstations. The applications are used in offices, homes, and schools. These devices make the operation of the computer simpler and reduce the degree of expertise needed to become a proficient user. They help to make computer operation unintimidating.

No one interface can be all things to all people. Nor can one workstation be considered *truly universal*. The thrust of future technology will continue to emphasize simplicity of operation. Once that is achieved, it will bring the power of the computer to more people. It will also bring a level of spontaneity that will let people concentrate on the problem at hand, without focusing attention on codes or specific keying sequences.

Portable Computers

The first portable computer was created and manufactured by Adam Osborne in 1981. It was called the Osborne I and cost $1795. Although Osborne's machine could be carried from place to place, it was not light: it weighed 23 pounds. Other vendors introduced similar machines in the 20-plus pounds category that were known as *transportables*.

A new category of micros has evolved, the *lap-size computers*. These machines generally weigh less than 10 pounds and can be carried in a standard-size briefcase. The difference between lap-size machines and first-generation portables is determined by what constitutes "portability." A portable is light enough to be carried and requires some form of direct-access mass storage. Portables are powered by batteries, attachable, or self-contained.

Devices that have been called *portable computers* actually fall into a variety of categories by size and weight (Figure 9-15).

Transportables. Transportables or *luggables* weigh more than 20 pounds and possess full-sized keyboard, large display, and built-in disk drives.

Notebook-size. The category of notebook-size computers covers devices that generally weigh less than a pound, have a one-line display from 12 to 40 characters long, and possess a miniature keyboard. They are referred to as *briefcase-size, notebook-size,* and *handheld.*

Lap-size. The category of lap-size computers represents the most popular among the true portables. They possess the following characteristics:

Ultratransportable, weighing 4 to 12 pounds
A flat-panel display offering a 25 line by 80-character display
A typewriterlike keyboard
A small but high-capacity disk drive
Extensive built-in software
Run up to ten hours on self-contained, rechargeable batteries
Fit easily inside a briefcase

Transportable Computer

Handheld Computer

Notebook-size Computer

Lap-size Computer

Figure 9-15. The varieties of portable computers.

Sources: *Transportable:* Courtesy of International Business Machines Corporation. *Notebook-size:* Courtesy of Epson America Inc. *Lap-size:* Courtesy of Kaypro Corporation. *Handheld:* Courtesy of Sharp Electronics Corporation.

The lap-size portable represents an important new tool for the mobile professional. It has the ability to offer computing performance that is at least comparable to that of desktops in a package that can be used practically anywhere.

Computer Phones

The evolution of the workstation has progressed from a single stand-alone device to one that can communicate electronically with the outside world. Advances in telecommunication have enabled PCs and workstations to add another dimension. It is a hybrid offering of the personal computer (PC) and the telephone. The *computer phone,* or *integrated voice/data workstation* (IVDW), combines voice and data communication in a single desktop unit (Figure 9-16). Typically, these devices are display workstations with telephone capabilities. They feature enough intelligence to compose and store messages and directories and have built-in modems for data transmission.

Among the applications they offer are the following:

1. Telephone number directory, which stores frequently used numbers (along with name and address data) for automatic dialing

Figure 9-16. Computer phone system.
Source: Courtesy of Data General.

2. Automatic redialing of the last number called
3. Multiple telephone line connection, so that voice and data transmissions can be made simultaneously
4. Hand-free dialing, so that number can be dialed without lifting the handset
5. Hand-free speaker phone, so that one or more persons can listen without lifting the handset
6. Directory search
7. Auto log-on to internal and public data bases
8. Calendar scheduling

Computer phones offer the user several benefits.

Efficiency. The user can do many things almost simultaneously. Voice and data information can be received simultaneously through one device, a capability that translates into a job done more quickly.

Economy. A single workstation performs the tasks previously handled by several devices and replaces the clutter of the phone, the desk calendar, the rolodex. The computer phone has one *footprint* that conserves desk space. (A *footprint* is the space a device takes up on a desktop.)

Employee Satisfaction. Computer phones enable the worker to perform multiple tasks. By doing jobs with devices that are self-contained and easy to use, they make the employee's job easier. When it is easier, it becomes more satisfying.

Input Terminals

Terminals are on-line devices, with little or no intelligence, used solely to communicate with a large computer system. *On-line* means that the terminal is linked directly to the computer. A terminal consists of a typewriter keyboard, which is used for data entry, and a cathode ray tube (CRT) screen (Figure 9-17). The demand for microcomputers has severely limited the use of these unsophisticated terminals. However, there is still a huge market for inexpensive terminals, especially among clerical workers who do not really need local processing capabilities.

There is a large price gap between input terminals and microcomputers. That gap is large enough to cause many companies to think twice about putting a micro on every desk. Many companies, such as banks and insurance firms, use terminals. They are used mainly as a means to give workers access to data stored in a host computer.

The main difference between terminals and microcomputers is that microcomputers contain intelligence. They have the ability to edit, manipulate, store, and process data. Terminals, sometimes referred to as *dumb terminals,* do not have intelligence or the capability of processing data.

The times are changing for terminal suppliers. The rapidly maturing end-user market is demanding more than a dumb box. The trend is definitely toward smartness. Users are looking for multifunctional, interactive desktop devices to fill their data processing and communication needs. In addition, modern terminals are beginning to look fashionable. They are sculptured and sleek, finely proportioned and well groomed. A goodlooking terminal adds status and class to an office. The new terminal is unlike first-generation terminals, which looked like cardboard boxes. Of course, the functionality of a terminal is paramount. But appearance clearly plays a vital role in a product's acceptance.

As a result, manufacturers are designing terminals with style, intelligence, and high-resolution graphics. These new devices are blurring the distinction between terminals and microcomputers.

Supercomputers

Supercomputers are the fastest, largest, costliest computers available. The Cray-1 was the world's fastest computer when it appeared in 1976. The company sold only three of the $8 million machines during its first two years. In 1985, this supercomputer was more powerful and sold for $20 million. Amdahl Corporation and IBM are other makers of supercomputers (Figure 9-18).

Analysts estimate that there are currently only 125 supercomputers in use worldwide. They expect that the market will expand to as many as 1000 by the end of the decade. Supercomputers are used for massive scientific calculations such as those in weapons design, seismic tracking in oil exploration, aircraft design, and weather prediction. Unlike conventional computers that add or multiply one equation at a time, supercomputers can process millions of equations per second. There will be other fields that will use supercomputer power in the coming years, such as pharmaceutical and biological engineering.

Companies that need supercomputing power currently have two options. They can either purchase a system at an average price of $10 million or purchase cycle time from a

Figure 9-17. An input terminal.

Source: Courtesy of Raytheon Data Corporation.

Figure 9-18. A supercomputer.

Source: Courtesy of Amdahl Corporation.

commercial time-sharing company. Both plans are extremely expensive. A new option that is more affordable will be micro-to-supercomputer links. Grants and subsidies from the U.S. government would make supercomputers available to qualified organizations. Companies at remote locations will be able to access supercomputers located at university centers. Supercomputer vendors are also supporting university centers to promote research and development projects.

Another category of super processors is the *supermicrocomputer*. Unlike the supercomputer described previously, this new machine is designed for the desktop. It fits in a new category between the personal computer and the low-end minicomputer. The supermicrocomputer is a powerful multiuser, multitask machine built around 16-bit and 32-bit architecture. Sixteen-bit machines give micros extreme processing power and speed. The performance levels of 32-bit machines rival those of large mainframe computers (Figure 9-19). Operating at levels around 1 million instructions per second (Mips), the supermicros are not quite as fast as traditional minicomputers, which are in the 1.5- to 2-Mips range.

The supermicro allows users in remote offices to run all the same applications as home office users. They are not dependent on a larger system. In addition, the supermicro can serve as a host and provide a more cost-effective solution for handling functions than giving a PC to everyone. The super (desktop) computer appears to be the evolutionary next step beyond the micro. It can provide power and services that a micro cannot provide.

Whether called *superminis* or *supermicros,* these desktop devices are offering more connectability, more power, and more features. Superminis and micros have greater numerical accuracy and faster response time, and they support larger data bases. Storage for some models is now reported in *gigabytes*

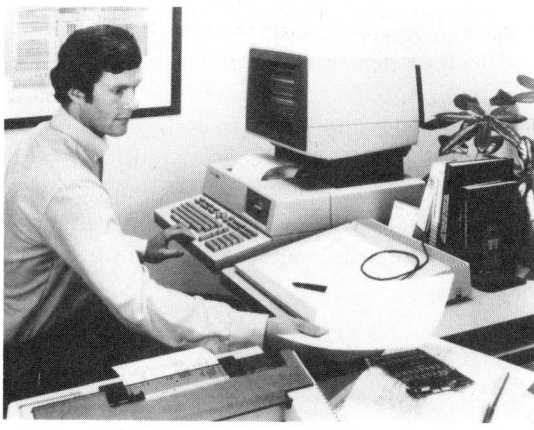

Figure 9-19. This 32-bit computer lets individual scientists and engineers have their own personal mainframes.

Source: Photo courtesy of Hewlett-Packard Company.

rather than *megabytes*. A *gigabyte* is 1024 megabytes, roughly a billion characters. A *megabyte (MB)* is 1024KB (kilobytes), or roughly a million characters. Whatever the future holds, these supermachines have aroused a lot of industry interest, particularly in the office. As an entry-level, multiuser, shared system, the supermicros and minis are attractive to both small businesses and large organizations.

Strategies for Acquiring Computer Systems

The debate over strategies and criteria for selection and purchase of computer hardware has gone on since the introduction of first-generation computers. New systems, peripherals, and components are introduced with increasing frequency. Hardware vendors are merged and bought out and simply go out of business. This turmoil makes reaching an intelligent purchase a risky exercise. As a result, a business computer shopper must not only evaluate products but also buy from a manufacturer with demonstrated financial and managerial stability.

Who Decides?

As the personal computers, supermicros, and universal workstations increase their market presence, it is clear that managers need to devote more serious attention to this influx. Data processing managers and office automation managers are likely candidates to act as corporate buyers for microcomputers.

Traditionally, data processing (DP) or management information systems (MIS) personnel controlled all computer hardware purchases. Organizational control of personal computers is as likely to be outside the area of DP as within it. Some organizations have formed purchasing committees made up of DP managers, office automation (OA) managers, and division heads. They are from key areas of the company and oversee the implementation of personal computers. In very large corporations, the responsibility for personal computers is often designated on the divisional level rather than corporatewide.

Regardless of who controls (or decides) the acquisition of computer systems, certain guidelines should be part of any decision.

Applications. The starting point for making a decision about hardware systems should be at the applications level. The fanciest workstation with the most streamlined functions will merely gather dust if it does not address existing needs.

Compatibility. It is not possible for one vendor's machine to be wholly compatible with another vendor's model. It is, however, possible to offer a degree of compatibility at various levels. The purpose for which the computer will be used will dictate the degree of compatibility necessary and desired. The term *media-* or *disk-compatible* refers to a machine's ability to read, and possibly to write, disks on compatible machines.

Processor-compatible refers to two computers that use the same microprocessors and can perform some similar operations. If files are to be used by other employees, care must be taken to ensure that they can be accessed on the other compatible devices.

Communications. Data communication capabilities on workstations are of prime importance. They should provide easy access to electronic mail, data banks, and micro-to-mainframe links.

Upgradability. The future needs of an organization must be considered. Therefore, the equipment selected must be upgradable.

Costs. Prices within the computer systems market fluctuate rapidly. In addition, a clear understanding of the amount the user is willing to pay must be present at the outset. Price does not always provide a sure indication of quality or the lack of it. Research should be done to verify present users' satisfaction with their equipment through consultation and reading buyer's reports.

In the midst of an ever-changing computer hardware market, leasing equipment may have attractive advantages over buying. *Leasing* is an agreement to make monthly payments over an extended period of time, ranging from two to five years. In most situations, the consumer can exercise an option to own the computer system at the end of the lease. The ability to lease a small computer system is still comparatively new, when compared to that for mainframe business systems.

Leasing, in the long run, is more costly than outright purchase. But, there are advantages:

> Less money required up front
> A hedge against obsolescence
> Substantial tax benefits

For a growing business, leasing is often the best decision. It allows use of a system that is otherwise beyond a company's reach.

The first decision the potential computer purchaser should make is whether it is worth purchasing a computer with cash or on credit. For some, outright purchase is still the best deal. But for a growing business that believes a computer is vital to its growth, leasing appears to be a viable alternative to immediate acquisition.

Selecting a Vendor

Once you have determined the applications, price range, and system features, you need to select a vendor. How can you make sure the next computer you buy comes from a vendor that will be around for a while? In considering which vendor to select, this and the following other questions should be answered.

Will the Vendor Be in Business Tomorrow? In judging a vendor's stability, one should consider the time the company has been in business and the extent that its sales and earnings have grown. The ideal, of course, would be to go see for yourself. If the supplier goes out of business, the user is left with a workable system for the present but will have difficulty expanding that system in the future.

Will the Vendor Continue to Represent a Product Line and a Strategy That Will Coincide with Your Company's Operation? The user must ask the vendor to reveal its plans for product development for the next five years. Knowing the vendor's product directions will help the user live with the products.

How Are the Quality and Responsiveness of Service and Support? Equipment that is not operating obviously cannot contribute to a smooth-flowing work operation. Therefore, it is essential that the prospective customer be assured that service is really available and that personnel are well trained. Will the vendor be willing and able to answer operational questions? Is there a hot line for immediate response?

Those who fail to develop and introduce a set of guidelines for computer purchases risk a host of problems, ranging from incompatible equipment, to poor employee morale, to uneconomical buying. There should be a reasonable set of standards based on one's particular needs. Otherwise, a company risks acquiring an array of expensive electronic equipment on the basis of sales pitches, advertising claims, and biased individual preferences.

In this fast-changing field, guidelines and criteria should be established, but they should not be permanent.

Managers involved with the selection process must possess the ability to evaluate the risks and opportunities involved in the final decisions.

When it comes to technology, there is no one-size-fits-all solution. To attain success, the organization must acquire equipment to suit specific needs. The technology must accommodate the way the organization works, not the other way around.

Once decisions are made, they should be justified and supported. Safe decisions are not necessarily sound decisions. Going with the number one vendor may not be the best strategy for the company's needs. Professionals do not have to cling to any manufacturer's coattails. Their future hinges on performance, not on the brand of equipment they select. They owe it to the company, the profession, and the industry to be independent thinkers.

Summary

The first modern office automation processor could be considered the typewriter. It was invented in 1714 and after that time functional improvements came slowly—electrification, increased mechanical efficiency, and the QWERTY keyboard. The computer revolution introduced processors with a dazzling array of features.

The stand-alone text processing workstation market basically consists of two major product areas: electronic typewriters (ETs) and stand-alone word processors (WPs). Although there is a trend toward sharing information among users and accessing more applications than word processing, a need for these stand-alone, single-function workstations still exists.

As technological change surfaces in products, it seems to insist on blurring product boundaries and confusing functional categories. Products that used to confine themselves to one function are now routinely covering two or more. Personal computers, executive workstations, integrated voice/data terminals, and supercomputers are processing devices that now provide multifunctional applications with increasing storage capacities. Computational power keeps increasing exponentially. Processing time and machine instructions are executed at blinding speeds.

Within twenty years, the marvels of processors have advanced from mag card typewriters and mainframe computers to running batch operations. Within the next twenty years, processors may use artificial intelligence and be a thousand times more powerful than they are now.

Review Questions

1. Briefly describe the types of memory found in electronic typewriters.
2. In the development of word processors, what was the chief advantage of using magnetic media over paper tape systems?
3. Which multiuser system can operate individual workstations when the central processor goes down? How is this possible?
4. Experts predict that the word processor as a piece of equipment will disappear. Why do they feel this way?
5. Is the universal workstation a tangible piece of hardware, or is it a myth? Explain.
6. Briefly describe flat-panel displays. How do they differ from traditional CRT displays?
7. Define liquid-crystal displays; light-emitting diodes; gas plasma displays.
8. When the user touches a touch-sensitive screen an immediate response results. This process is transparent to the user. What does *transparent* mean in this sense?
9. What similarities do mouses, touch screens, and icons have? What objections do users have to using these devices?
10. In describing lap-size portables, what is the main criterion that makes a machine portable? Explain.
11. Which occupations or professions would be suitable users of lap-size portables?
12. Briefly describe a computer phone. How does it help to conserve desk space? What is a "footprint"?
13. What is a dumb terminal? What function does it have in the automated office?
14. Which features differentiate a supermicro from a personal computer?
15. In selecting computer systems and workstations, guidelines and criteria should not be inflexible. Why?

Projects

1. Andrea Fastenberg hopes her parents do not read this. Last Christmas, the Fastenbergs impulsively bought a $2,000 computer system to help Andrea, a junior at Smith, crank out her term papers. But it proved inadequate for the job. Andrea retreated to her trustworthy typewriter. So far, she has not had the heart to tell her parents that they wasted their money.

   Why did the Fastenbergs fail the computer-buying test? For the same reason that students always flunk: they did not do their homework. Although the guidelines in this chapter apply to the business world, how do they apply to buying a computer for a college student? List guidelines and criteria that you would use in purchasing a computer for a college student. In addition, research and compile a list of colleges that require entering freshmen to purchase a computer.

2. The categories of microcomputers can be confusing and their applications can be overlapping. What does the future hold for the supermicrocomputer as described in this chapter? Will it maintain its separate and unique category or will the supermicro simply lose its identity and merge into the general micro market? Find an article that discusses this issue and write a short report. What do the experts think? What is your opinion?

3. Research vendors that offer electronic typewriters. Create a table that compares six vendors. Include the following headings within your chart:

   Vendor name and address
   Model name
   Features
      Memory
      Display screen
      Special word processing functions
      Communications
   Price range

   Be creative in your design of this chart.

4. Word processors have come a long way, technologically, in the last decade. The early stand-alone word processors have been replaced by smaller, less expensive, easier-to-use devices. Personal computers that run new word processing software are as good as dedicated word processors. Write a brief report describing the state of the word processor and its future. Will it absolutely disappear from the market, or do you see a segment of users who will continue to use it?

Include a pie chart or table from an office system consulting company that traces the trend of word processing sales and shipments for a period of five years.

CHAPTER 10

Image Processing

After reading this chapter, you will understand
1. Reprographic technology and devices
2. Copy technologies, devices, and applications
3. Advantages of using phototypesetting
4. The process of optical character recognition (OCR)
5. The concept of computer graphics, software, and hardware components
6. The emergence of desktop publishing—the computer's newest field

Image processing is another key technology in the information processing cycle. It includes photocopying, duplicating, phototypesetting, computer graphics, image printing, optical character recognition (OCR) and desktop publishing. Just as other processing procedures need planning, so too must an intelligent plan be established for image processing.

Reprographics refers to the reproduction of hard copies by photocopy, printing, and other office duplicating methods. Printed output is called *hard copy,* and output produced on a terminal screen other than a printed page or object is called *soft copy.* The major categories of reprographics tools are copiers, intelligent printers, and office duplicators.

Copiers

Many changes have taken place in output technology. These changes suggest alternate forms of electronic delivery systems and a "paperless" environment. Paper is still the most frequently used means of exchanging information in the business world. Billions of pages per year are created, filed, and copied. Modern technology has increased the use of paper by making it both faster and easier to produce professional-looking images.

The copier is perhaps the product most often taken for granted. It has gained recognition as a legitimate member of the automated office. However, the driving force in copier machines is the demand for copies, not for the machines themselves. Information workers want to see the results in hard copy. They want to pick up a piece of paper and read it. People are more comfortable with a tangible hard copy than with a soft copy on the screen.

Convenience Copiers

Convenience copiers are intended for low-volume applications. These machines are placed in scattered locations around a company and are near the people who will be using them. Convenience copiers are available in desk-top models or consoles. Desktop copiers are small enough to fit on the top of a desk or table; console copiers are floor models.

Personal copiers are a type of convenience copier intended for the low-end, low-user market. The significant feature of a personal copier is its disposable operating modules. This device incorporates a disposable cartridge that includes a photoconductor and a toner. The attraction for the user is that little or no need exists for outside servicing by trained technicians.

Convenience copiers are available in various speeds, ranging from three to fifty copies per minute. The slower machines cost less and are suitable for applications in which the volume of copies produced each month is less than 5000.

Full-Feature Copiers

The full-feature copiers offer users quality, speed, and limitless copying. Limitless copying implies nonstop operation, regardless of the capacity of any of the cassettes or trays. The machine will shift from a given cassette to the next, and so on, as each is emptied. Larger copiers are centralized and are capable of producing volume copies at great speeds. In addition they have excellent copy quality. Halftones are sharper (Figure 10-1), solids are denser, and texts are blacker. Service and maintenance procedures are dependable for these models. The user enjoys longer preventive maintenance periods and a straight-paper path, which is shorter and gives easier access to jams. There is less operator maintenance responsibility and minimum downtime.

Copier Technology

Copiers may be classified as either wet or dry. In the *wet process,* a negative is made of the

Figure 10-1. This has been reproduced from the *copy* of a photograph.
Source: Courtesy of Eastman Kodak Company.

original document by exposing it to a light-sensitive material. It must then be run through a developer bath in order to reproduce a copy. Because of the bother and mess of using chemicals, wet copiers are rare in today's market. There are five basic types of *dry copy systems*.

Coated Paper Copiers. This process uses coated, electrostatically charged paper. Thermography and electrostatic copying use coated papers. In the thermography process, material that is to be copied is placed under a heat-sensitive copy sheet. Infrared light generates heat on contact with the dark portions of the original document, producing black or white images. Coated paper copying initially gained popularity because it was the least expensive process. With a greater number of inexpensive plain paper models entering the market, coated copiers have declined in use.

Plain Paper Copiers. Plain paper copiers produce copies on plain bond paper. They are more versatile than coated copiers and produce a high-quality copy. Plain paper copiers usually copy faster than coated paper copiers. These machines have a drum or a belt inside to transfer the image of the original document onto the copy paper. The *toner* adheres to the image that is formed on the drum or belt, and then the image is transferred to the plain copy paper and fused to the paper with heat.

In addition to the professional appearance of the copies, users can make copies on different colored paper, special letterheads, and business forms.

Fiber Optic Copiers. Fiber optic copiers use hair-thin glass or plastic rods that transmit the light source. This technology replaces the lens and mirror arrangements used in conventional copiers. By simplifying the optical system in this way, these machines have been able to produce copiers that are smaller, lighter, and more reliable. They also operate with fewer maintenance problems and reduced energy consumption. These units give the user superior performance, especially in the low-volume machines, and at a competitive price.

Intelligent Copiers. Intelligent copiers use the optical systems of the copier and the digital technology of the computer. They are laser-driven systems. Intelligent copiers are linked to a computer to produce high-quality copies of the text that appears on the display screen. This eliminates the steps of printing out a copy of the text displayed on the screen and then reproducing that hard copy. With the intelligent copier, the user moves from the electronic information stored in a terminal or workstation to multiple copies without waiting for the system to print out the original document. Intelligent laser copiers combine advanced digital scanning and laser-beam printing technologies in one unit.

Color Copiers. Color copiers make full-color reproductions of a color original. The first-generation machines used toners to achieve the desired colors. These color toners, one for each color, were available in cartridge form. The user had to interchange cartridges in and out of the machine, using a different original for each color image. This was cumbersome. The latest innovation is keyboard selection from one or two color-toner cartridges in the machine. As technology improves, the ideal color copier will be commonplace. It will make true color copies in one pass and reproduce all the color tones of the original.

Copier Features

The most wanted features and options are available in every price range. Today's copier technology allows the user to select only those features that are currently needed or whose need is anticipated in the future. Office workers have become very astute shoppers. Whether selecting personal computers, dictation systems, or copiers, buyers are aware of their needs and the many competively priced, multifeatured copiers. Copiers now have many advanced features.

Duplexing. Duplexing is the ability to make two-sided copies (Figure 10-2). This feature saves on paper and cuts down on postage. A fully automatic duplexer handles both types of originals after the operator has pushed the correct button on the control panel. With a semiautomatic duplexer, the operator must reinsert each two-sided original for copying on the second side.

Two-sided copying means productivity plus economy.

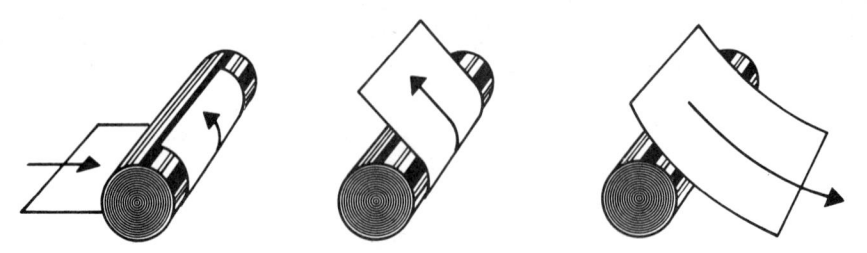

The secret behind high-speed single-pass duplexing is a unique drum that actually turns the paper over so an image can be put on the back at full operating speed. This, plus Kodak's famous short paper path, keeps it all running smoothly.

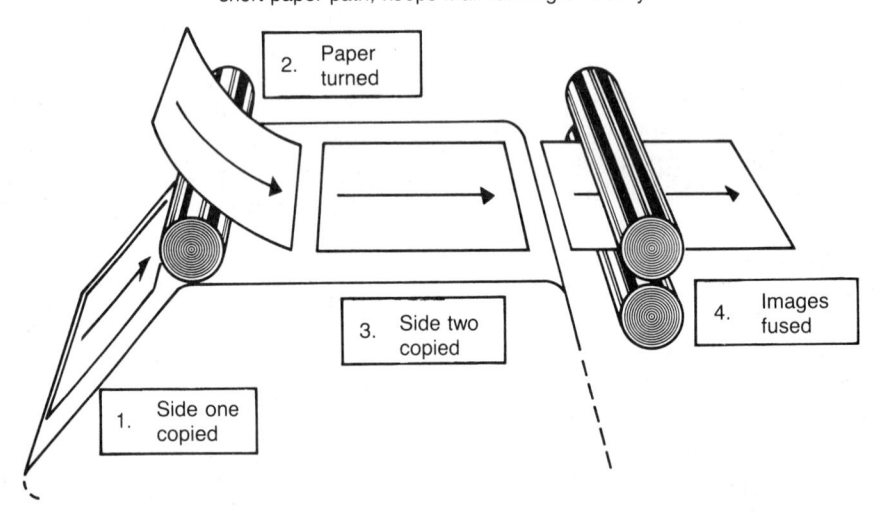

1. Side one copied
2. Paper turned
3. Side two copied
4. Images fused

Figure 10-2. Single-pass duplexing to copy on both sides of the paper.

Source: Courtesy of Eastman Kodak Company.

Zoom. *Zoom* is a photography term that relates to a lens system in which the focal distance can be changed by the camera user. A full zoom copier allows the operator to reduce or enlarge the size of the original copy. Models with this feature offer the user a wide range of reduction and enlargement sizes.

Automatic Document Feed. With the automatic document feed feature, the operator does not have to position each page on the copier surface: the documents are stacked into the feeder tray and the machine automatically feeds in one copy at a time (Figure 10-3).

Sorters. Sorters are output attachments that are essential for high-volume copying. Sorters or collaters are bins attached to the copier that provide automatic sorting into correct bins. Sorters vary in design and capacity for the number of sets that can be assembled and the number of pages per set. Compact ten-bin minisorters hang from low-speed units; faster, more sophisticated boxes have up to twenty-five bins.

Automatic Diagnostics. A display in the control panel tracks the paper path, stopping at any position that malfunctions.

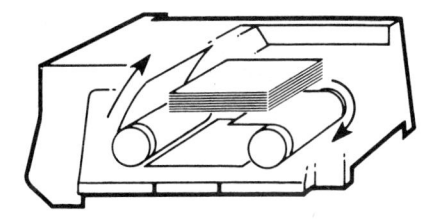

Automatic Feeder

The recirculating feeder automatically presents originals for copying in collated order. Takes up to 100 20-pound originals.

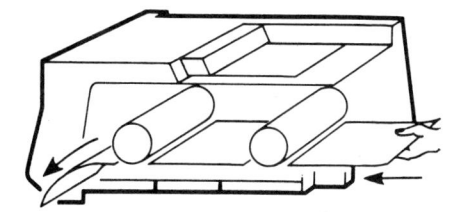

Versatile Positioner

Automatically positions originals for copying up to 11 × 17 inches, from 12 to 110 pounds.

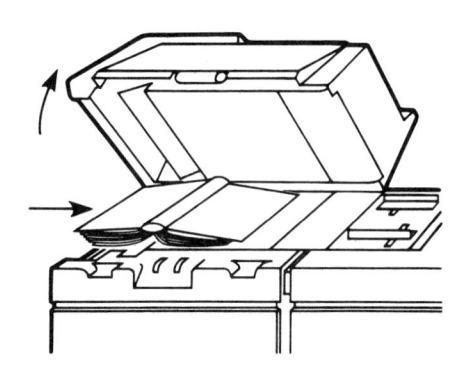

Easy Book Copying

The feeder-positioner easily opens for copying books of almost any size.

Figure 10-3. Some of the features that can be expected from a copier.
Source: Courtesy of Eastman Kodak Company.

Electronic Message Display. Using English-language words, a display screen flashes pertinent instructions to the operator.

Automatic Chapterization. With the chapterization feature, users can automatically create new chapters within sets of copies. Different paper stocks are used as slipsheets to create chapters within the sets.

Electronic Editing. An electronic editing device can be attached to copiers that can edit copies without altering the original. It can delete confidential information and move or center graphics and texts. It can even selectively reduce or enlarge key portions of the original.

Other attractive features available on today's copiers include automatic energy-saving mode, paper selection, image shift, advanced control panel, automatic exposure, and tray switching. Because of the availability of such features, it is important that a copier is easy for even the casual operator to use. Copiers will continue to play an important role in office operations. The demand for copies is growing, and decentralized machines are quick and convenient to use.

Computer-Interfaced Copiers

Copiers are now being tied into the total information systems network. Computer-interfaced copiers will be a key component as output devices. In addition to copying, they will have the capabilities of electronic typewriters, word processors, personal computers, facsimile machines, digital scanners, and laser printers. These machines will convert images into digital information. That information can then be stored in a computer's memory, transmitted over communication lines, and used to produce exact copies. By linking copiers with networks, speed, efficiency, and quality of output are achieved. Computer-linked copiers allow the retrieving and editing of documents without the waiting time for a copier to become available.

Image-Based Systems

Electronic imaging systems combine micrographics and information processing technologies to meet high-volume output needs. The Kodak Image Management System (KIMS) concept merges existing film media with the electronic transmission of scanned film images.

Using 16-mm microfilm as a foundation, KIMS scans an image on microfilm. It converts it into a digital bit stream and transmits the bit stream to a remote location for viewing on a high-resolution CRT workstation. Images can be transmitted electronically to other workstations on the network for additional processing. One benefit of the KIM system is that it reduces costs associated with organizing and accessing information. The system also reduces manual routing of paper, improves information security and integrity, and can be upgraded to meet current and future needs.

Scanning Terminals

Another computer-interface image processing device is the scanning terminal. This computer terminal scans a document's image for transmission within an office systems network. Images of graphs, printed forms, typed memos, and handwritten notes are digitized. Once digitized, the images can be displayed, stored, retrieved, altered, merged with text, and communicated. The IBM Scanmaster I and the Wang Professional Image Computer (PIC) are two leading products that offer scanning and computer-based image processing.

Computer-based copiers are fully integrated, compatible tools. They provide users easy-to-use, interactive data processing, word processing, and communication technology.

Phototypesetting

Phototypesetting is a method of preparing text and graphic material so that it is ready to be reproduced for newspapers, magazines, brochures, and books. (This book was phototypeset.) Phototypesetting is a computerized photographic printing process that uses film to produce an image *(camera-ready copy)*. The printing process can be compared to a photo process using a camera to take pictures of type characters.

Phototypesetting equipment allows the text to be condensed while improving its legibility and appearance. Whether the document is a promotional story, a multipage policy, or a lengthy report, phototypesetting enhances visual communication through a wide range of graphic effects.

Phototypesetting Process

Copy to be typeset can be input into the system in two basic ways: directly onto special phototypesetting terminals or through an intermediate, or peripheral, device. With the direct entry process, the operator must perform the following steps.

Figure 10-4. The typesetting keyboard unit displays the page composition for better page makeup.

Source: Courtesy of Alphatype Corporation.

1. Keyboarding. Copy must be keyboarded into the system by an operator. The operator keys the characters and machine instructions such as size, line length, and line spacing directly into the phototypesetter via the machine's keyboard. Certain words will need to be hyphenated at line endings, and, in certain cases, lines must be justified; that is, fitted into the proper line width. One can edit and merge text with graphics. The screen acts as a fluid window for scrolling and making changes and corrections. This permits the operator to view and correct the copy before it is typeset (Figure 10-4).

2. Typesetting. Typesetting is the actual process of putting characters on paper or film.

3. Processing. The exposed photosensitive paper must be developed in some manner in order to see the typeset copy.

4. Proofreading. Typeset copy is checked against originals for errors and omissions.

5. Correcting and Final Editing. If errors are found or changes made, the preceding steps are repeated.

When an intermediate device, such as a word processor or personal computer, is used, the copy and special codes are keyboarded. They specify the size and style of type to use, the spacing between lines, and whether the copy is set flush left, flush right, or flush left and right. The keystrokes are recorded on a disk or other storage device. This disk is then inserted into the phototypesetter, if necessary through a special interface device that translates the text and instructions for the phototypesetter.

Not all of these steps need be performed in-house. For example, by only keyboarding in-house, and sending a disk of the work to a commercial typesetter, organizations can save thousands of dollars each year, depending on their volume of work. Nearly 90 percent of all labor required for phototypesetting is in the preparation of copy for output, and in manual adjustments made after output; if they can be done in-house, the savings are substantial.

Benefits of In-House Typesetting

Savings. As indicated previously, keyboarding, proofreading, changes, corrections, quality control, and paste-ups are all labor-intensive. If a typesetting system is to save time and money, it must be automated at every step.

Quality. Typographic quality is not a frill; quality is a necessity in a competitive business environment. The variety, the "color," and the feeling of substance that high-quality typography imparts to the typeset page mark your message as important.

Condensation. Typesetting provides a decrease in the number of printed pages through compaction. Over three typewritten, double-spaced pages of text will fit on a single page if typeset. This is due to the clear, compact design of type and its improved hyphenation and justification. In this way, costs are reduced.

In addition, an organization enhances its prestige and maintains a corporate graphic identification through typesetting.

Desktop Publishing

A new and burgeoning field is emerging in personal computing: desktop publishing. It is the computer's newest application and is doing for publishing what word processing did for words. The benefits of desktop publishing are astonishing. Users can combine illustrations, graphics, and text in minutes instead of hours or days. From concept to camera-ready, the computer operator is in complete control every step of the way.

Basically, desktop publishing links an office computer with two new components—page layout software and a laser printer to print out a master that is close to professional typeset quality. An additional link is something the user may not even see: new software called "printer drivers" that allow various type styles, sizes, and graphic images to be printed in very high quality and be scaled up or down to a desired size.

With these tools, a user can bring in copy from a word processing program; set it in various sizes and faces of display and text type; set different line spacing, column widths and placement, coordinate the layout, and print the finished pages in a range of quality that can equal that of commerical publications.

Desktop publishing has all the obvious advantages of in-house typesetting without the disadvantages of expense and the requirement of a highly-trained staff. The newest image processing application, desktop publishing allows organizations to create camera-ready typeset layouts easily and economically, using the regular office staff.

Optical Character Recognition (OCR)

Optical character recognition (OCR) is a process that scans text images and stores the scanned characters in digital form. OCR devices can work as an extension of the phototypesetting process just described. The OCR device reads the typewritten copy and converts it into codes that can be understood by the phototypesetter. It then produces a magnetic medium that can be input into the phototypesetting system. This eliminates the need to rekeyboard the copy. Examples of OCR technology are the post office ZIP code scanners that sort the mail, the bank scanners that read the alphanumeric numbers at the bottom of checks, and the supermarket devices that can read the costs of the items and enter them simultaneously into the cash register.

OCR: An Important Link in the Automated Office

Advances in this technology have made OCR an important link in the automated office. Modern readers can now read a wide variety of characters and documents. They are able to interpret typewritten pages, adding-machine tapes, computer printouts, and handwritten characters (Figure 10-5). OCR readers can now read standard typestyles, including Courier, Letter Gothic, Prestige Elite and Pica, and Elite.

Readers can scan documents produced on paper from a variety of sources such as Selectric-like balls and daisy-wheel lasers or ink-jet printers.

Many of these devices are now the size of small desktop copiers and offer a variety of features. These units were designed for fast, accurate input of typewritten materials into a wide range of automated systems. Characters on typed pages are identified, converted to electronic signals, and transmitted at high speeds to word processors, computers, communications systems, and typesetters.

Figure 10-5. This page reader works with an information processor.

Source: Courtesy of CompuScan.

Benefits of OCR

OCR has many advantages.

1. *Increased productivity:* OCR eliminates redundant keyboarding. In one day, an OCR can input as many keystrokes as the average secretary produces in a week.
2. *Accuracy:* Accurate, error-free, and camera-ready copy is created.
3. *Less disruption:* When OCR devices are added to your work group, there is no need to reorganize your operations. They are simple to learn and simple to use.
4. *Flexible:* Many optical readers do not become obsolete if you change word processors or PC workstations. OCR devices can grow and change along with your work group.

When OCRs were first introduced, the chief benefit was their elimination of redundant keyboarding. The input typing function was transferred from a text editing machine costing $4000 to $10,000 to an office typewriter costing less than $1000. This benefit may not be as valid today because the price of workstations has decreased to the point where each person can have a terminal on his or her desk. Multifunction computers have replaced dedicated word processors, and in many cases electronic typewriters have replaced electromechanical typewriters.

Although OCR technology will not be as dynamic as other image technologies, it will continue to grow. As offices become more automated, there will not be a need to have one OCR unit per workstation. Instead, there will be a central unit that a number of workstations can use. The quantity of units will decrease, but the demand on the present units and the demand for the technology will increase. Improvements in OCR technology will take the following forms:

1. *Omni-front* technology is the ability of a system to recognize a greater variety of fonts and type styles.
2. *Electronic color scanners* combined with word processing technologies are leading to integrated computer image assembly systems. These units will provide for computer-to-plate systems for reproduction.
3. *Publishing applications* will evolve from technology that will give OCRs the capability to read both text and graphics. Machines that can blend scanning and graphics capabilities will be a boon to in-house publishing.

Furthermore, plans of OCR vendors include dramatic reductions in cost and size. There will be improved sensors to capture typed images more easily from a larger number of sources; and graphic arts capabilities for charts, graphs, and schematics. Graphics capabilities will become an expanding application of OCRs. People will want the equipment to read everything on the page—text and graphics. OCR will continue to be a strong image processing technology. The tremendous explosion in laser and nonimpact printing units may change the nature of the printed page, but not the fact that paper will continue to be used as a primary output source. As long as there is paper in the office, there will always be a need for technology that can enter data into a system in an easier, faster, and more accurate way.

Graphics

Graphics is a form of image processing. It presents information in the form of charts, lines, curves, and other visual displays (Figure 10-6). The process of translating business facts into graphic displays has changed from a manual, and painstaking, art to an automatic and instantaneous operation.

Prior to computer graphics, a person would have to use an art department to create any graphic presentation. The user would send a sheet of figures with instructions about the kind of chart needed to the department. A finished chart would be completed a day or two later. Maybe the title was not correct or the colors were not exactly what the user had in mind. If an organization did not have an art department, the user would probably have to design a crude sketch using a compass, ruler, and graph paper. The graphic would be sketched, shaded, typed, cut, and pasted onto the typewritten document. It was not perfect, but it was all that was available before computer graphics became an important part of image processing.

The blend of computer graphics hardware and software in the office has brought *presentation graphics* into the domain of managers and executives. Presentation graphics display information in the form of bar, pie, line, and scatter diagrams. The system presents information workers the option of not only seeing but also creating "action snapshots" from current operating data (Figure 10-7).

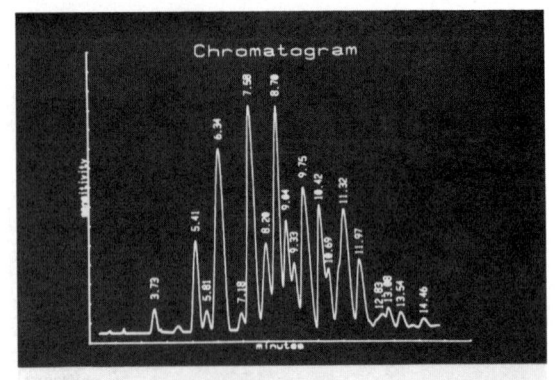

A chromatogram is displayed on an HP 1346A in a chemical-analysis system.

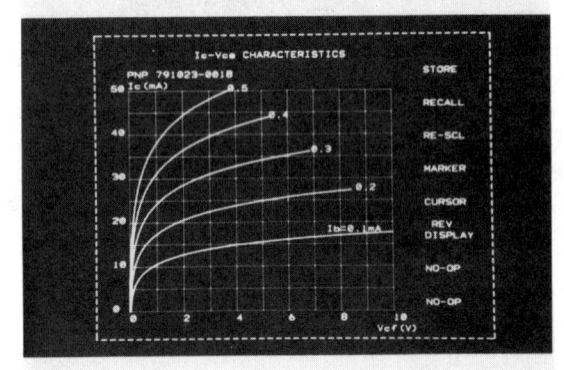

In instrumentation graphics applications, a plot of current-versus-voltage characteristics of a semiconductor component appears on the screen of an HP 1346A.

Figure 10-6. The painstaking work to plot the sensitivity of a chromatogram or the characteristics of a semiconductor component is quick and easy with computer graphics.

Source: Photo courtesy of Hewlett-Packard Company.

Today, it is no longer enough to see results only in printed form. In many cases, graphic output is now essential, and more firms are using the power of the computer to unearth information through graphic software and graphic hardware systems.

Graphic Software

Trends or patterns that may not be apparent in raw numbers on a worksheet emerge boldly and clearly in graphics. There is simply no better way to illustrate a point. The 2-day wait

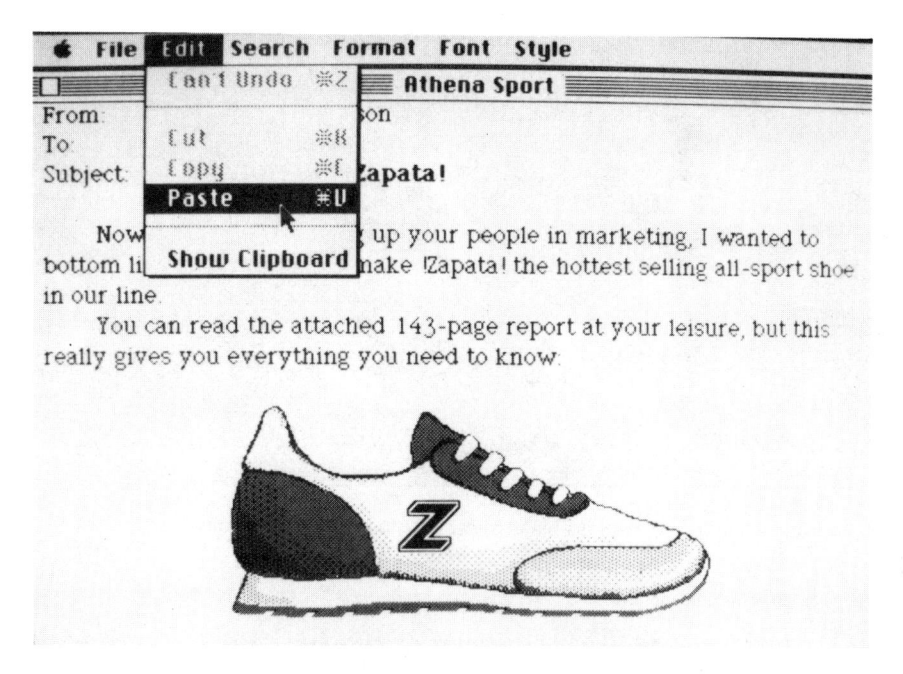

Figure 10-7. Memos take on a new look when illustrated with graphics. Portions of text or graphics can easily be moved around within one document or "cut" from one document and "pasted" into another one.

Source: Courtesy of Apple Computer, Inc.

to get a graph from the art department has been reduced to a 5-minute process, with a computer, plotter, or printer. The smudged, cut, pasted 5-hour labor of love has become a flawless, three-color, three-dimensional masterpiece, created with a few keystrokes. With modern software programs, a user can easily create a graph from information on a spreadsheet and display it on the screen in less than 2 seconds. Information can be changed and redisplayed in a single keystroke.

With the variety of software on the market, a user does not have to be an artist to create an effective presentation. The first generation of graphic software enabled a computer to plot data and insert them into a report (Figure 10-8). New graphic software can create shapes, colors, and animation. When this software is combined with peripherals such as windows, icons, and mouses the explosion in the use of graphics is easy to understand. With the software loaded into the computer and your hand on the mouse, you are free to let your imagination soar. Through artistic graphic software and a personal computer, a user can produce virtually any image the mind can create. The mouse allows the human hand to create it. If you want to draw a circle, you select the circle shape from a menu and draw it.

You click a mouse button to select the center of the circle and then move the mouse pointer away from the center. As you do, a perfect circle grows out radially from the center point. The farther you move the pointer, the bigger the circle. When it's the size you want, you click the button again and the circle solidifies.

Graphic software gives the user freedom to doodle, cross-hatch, spray paint, fill in, and erase. Even if a user does not have artistic tal-

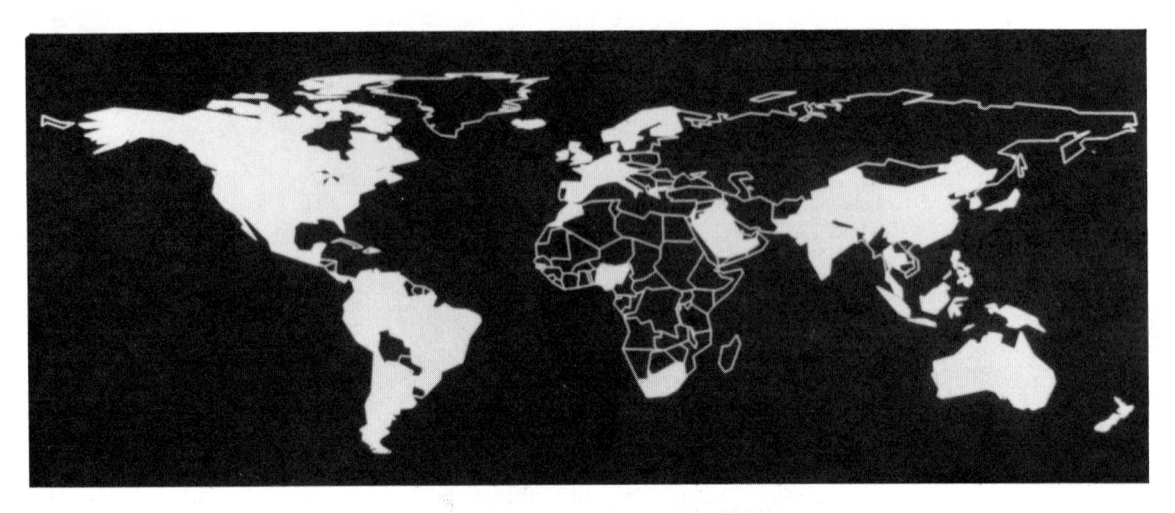

Figure 10-8. With its own software, SAS is able to show countries with SAS® software installations.
Source: Courtesy of the SAS Institute Inc. SAS is a registered trademark of SAS Institute Inc., Cary, NC, USA.

ent, menus, icons, and special tools for designing everything from office forms to technical illustrations are available to help. In addition, there are type styles to create captions, labels, and headlines.

Graphic Data Bases

In offices, electronic data bases usually are thought of as containing only text and numbers, even though image data bases exist in many forms; e.g., presentation slides, product photographs, and microfilm archives.

Data bases on a PC primarily contain text or numbers that can be searched to locate specific data files. In the past, any images associated with these files had to be kept separately, usually in paper form. Now image processing software allows images to be included in the data base files, so that they can be viewed on the screen along with the text or printed as an integrated text-image document. Applications, such as data bases of homes for sale, personnel files, or manufacturing parts lists, benefit from the inclusion of images.

Another type of data base is an archival system, such as those based on microfilm. Here permanent records are kept, to be reviewed periodically. Electronic archival systems are beginning to have a significant impact on film-based systems. Because they provide a fast, easy to use, accurate means of establishing large data bases of documents containing text, line art, and written annotations, they can be distributed to remote locations and retrieved on demand for viewing or printing.

Graphics Hardware

The hardware associated with a computer graphics system is an important consideration. Vendors offer numerous systems and features to provide an efficient link from the images on a computer screen to the production of hard copies.

Printers. Printing informal graphics can be done with most dot-matrix printers, ink-jets, and thermal-transfer printers. The better models of dot-matrix printers make images

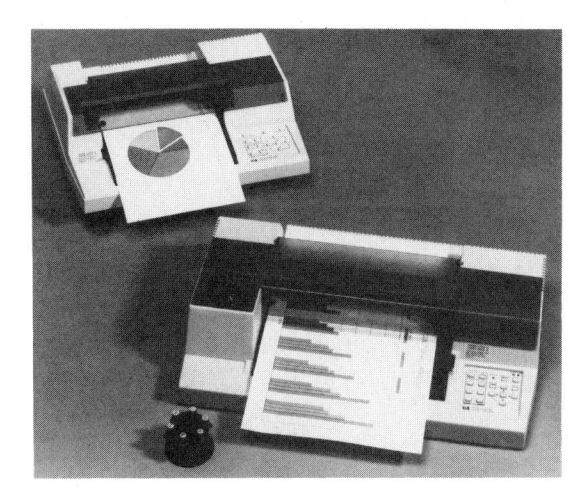

Figure 10-9. Hewlett-Packard's two-pen plotter and six-pen plotter both plot charts and graphs on paper or transparencies and can be hooked up to a variety of personal computers.

Source: Photo courtesy of Hewlett-Packard Company.

with greater dot densities. But the proponents of more recently developed technologies point to *plotters.* Plotters produce hard copy output by drawing graphs, maps, charts, and pictures on any type of paper.

Pen Plotters. Pen plotters are a form of plotter printers (Figure 10-9). They draw basically linear images with pens, rather than constructing images from dots. The pen comes into contact with paper. Either the pen moves, the paper moves (on a roll), or both paper and pen move. Pen plotters generally do report graphics very well, but they usually do not work for freehand-style graphics. However, it is very difficult to convert the dot-based graphics programs into the "vector" line-drawing mode plotters. Therefore, vendors provide a variety of plotters with prices ranging from $1000 to $4000. No other output device duplicates the full-color linear fidelity of a plotter. The cheapest models can produce images as handsome as the most expensive.

The only differences are speed and convenience.

Sophisticated plotters have features that include sheet feeders, eight pens, and drawing speeds that equal the dot-matrix graphics output.

Graphic Terminals. A graphic display terminal is a CRT with the ability to display a wide variety of graphic, pictorial, and even animated data on a screen (Figure 10-10). Graphic terminals may be monochromatic (one color) or multicolored terminals. The CRTs of a graphic terminal truly function in the way a television tube does, displaying almost any character or figure that can be created by the computer.

Color terminals can display eight to sixteen colors simultaneously, selecting from a sixty-four-color palette. Different models offer 13-

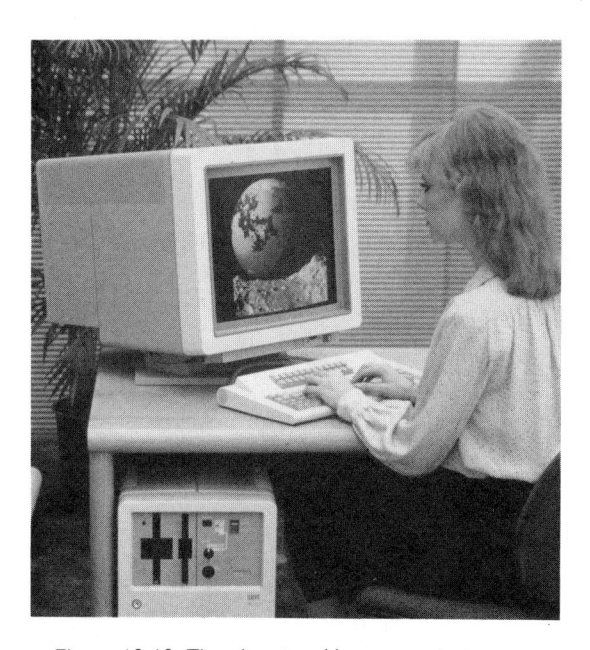

Figure 10-10. The planet and lunarscope were generated from a mathematical formula on this graphic terminal.

Source: Courtesy of International Business Machines Corporation.

inch or 19-inch display, *high resolution* (very clear figures on the CRT screen), zoom, scroll, two-dimensional transfers, and other features.

Developing Visual Literacy

Computer graphics are becoming a new visual expression among business executives and information workers. Information workers are becoming visually literate. *Visual literacy* is the ability to conceptualize and see images instead of facts and figures. It is the ability to substitute pictures for sheets of texts and columns of figures. Graphics technology is offering a new way to explore visual ideas.

Business executives will learn visual thinking by using computer graphics. The process will be similiar to the changes in thinking brought about by electronic spreadsheets and other software programs. At first, spreadsheets were used to replace the specific manual task of budgeting and forecasting. But as spreadsheets moved across the country, the number of uses skyrocketed.

Business graphics users can look forward to an abundance of technology. Vendors continue to upgrade their products within a variety of price ranges. The following are some of the trends and applications in graphics image processing:

- *Color:* There is a preference for amber phosphors terminals rather than the green so widely used in the past. More and more terminals will offer full color, if not quite "living color."
- *Slide-making applications:* Within graphic technology, the most progressive area may very well be computerized slide production. One of the main uses of business computer graphics is presentation, and the 35-mm slide is the preferred projectable medium. Systems currently available range from simple do-it-yourself methods of photographic on-screen computer displays to systems for mass producing slides overnight.
- *Integrating photos into documents:* Special software programs can merge real-life photographs into data base management files and other documents. Pictures of people, products, diagrams, maps, and company logos can be integrated with your data base. The applications can include security, signature verification, real estate, and electronic cataloging.

  Customers, distributors and sales personnel can quickly search data and view the resulting product/picture information on one screen. Files can be updated easily and quickly.
- *Animation.* Animation is a revolutionary feature of computer graphics that makes it possible to create and control the movement of up to 200 objects on the screen at once.

Electronic Presentation Boards

Blackboards have been an important tool for the presentation of charts, graphics, and other visual aids for meetings and lectures. The electronic *copyboard* or *whiteboard* offers a new electronic tool for visual aid presentation (Figure 10-11).

These electronic presentation boards take the ideas off the boards and put them onto paper. They make multiple crisp copies of notes, diagrams, and equations in seconds.

These boards work in much the same way as facsimile machines. The image on the board is scanned and the pattern of white and dark spots is recorded electronically. That information allows a printer to reproduce the image on a piece of paper.

When one begins to consider the potential use of graphics in business, it becomes obvious that the range of possibilities is very wide indeed.

Computer graphics is said to be one of the

Figure 10-11. The Fotoboard is the office blackboard of today.

Source: Courtesy of Eczel Corporation.

fastest-moving segments within image processing. Graphics should continue to grow as thousands of users put these new tools to use in countless ways. Electronic graphic systems can replace the architect's drafting table, the artist's easel, and the illustrator's sketch pad. Powerful graphics capabilities are here today. With the advent of highly integrated terminals and printers, dazzling images can be generated. They reflect the realities of the job, not the limitations of the computer.

Other Image Technologies

Several other areas of technology involve images. They are microfilm, facsimile, and videotex. Each of these could involve either original or nonoriginal input, depending on whether the source of the information has been internal or external to the organization. Most image technologies involve communication. They transfer their source over high-speed electronic circuits. Some of these technologies will be discussed in other chapters as they fall into other technology categories. Today's technologies are changing. They are capable of performing multiple tasks and applications. As office information systems continue to evolve, it sometimes becomes difficult to pinpoint where each segment belongs within this large umbrella. All of these technologies overlap into other phases of information processing. Whether they are classified as input, output, storage, retrieval, communication, or a combination of each, we must learn to identify them. Then we can make intelligent decisions regarding acquisition, purchasing, and applications.

Summary

There is a visual, as well as a literate, revolution taking place on computer screens. The graphic and image interface, typified by the use of pictures, usually referred to as icons, and pull down menus, windows, and mouse controls are part of a growing number of computer systems. Convenience copiers (plain-paper) include such features as automatic duplexing (printing the second side of a sheet), automatic collating, laser digitization, push-button color, and typewriter-size units.

Reprographics, thanks to lasers, microprocessors, and a raging price war, are contributing to the widespread use of this office automation sector. Advances in optical and microprocessor technologies have also increased the sales of optical character recognition (OCR) equipment. These machines essentially convert every typewriter into a word processing input terminal.

New OCR desktop scanners also permit the scanning of images as well as characters into a PC.

Other image processing technologies make it possible for users to create presentation graphics such as overhead transparencies and slides. With a desktop film recorder, it is possible to produce 35-mm slides of computer graphics. Electronic boards or whiteboards are a new image processing technology that creates paper copies of images as they appear on the board.

Review Questions

1. Define reprographics. What categories of image processing does it include?
2. What are disposable cartridges? Which type of copier uses this feature?
3. List some of the advantages of using plain-paper copiers.
4. Describe the latest color copy technology. What benefits has it brought to the user?
5. Briefly describe the following copier features: duplexing, zoom, automatic document feed, and automatic chapterization.
6. What is an image-based copier system?
7. How does a scanning terminal convert paper images to appear on a display screen?
8. What are the two primary advantages of using in-house phototypesetting?
9. How does desktop publishing differ from in-plant typesetting (printing)?
10. Briefly describe the process of optical character recognition (OCR). How does this method increase productivity and reduce cost?
11. Before computerized graphics were developed, how did a typical organization create graphics?
12. List some of the applications for computer graphics.
13. Describe pen plotter printers. Are pen plotters used for all types of graphics? Explain.

Projects

1. The office services department of a major broadcasting network prepares documents for all departments including the TV and radio networks, advertising and promotion, broadcast operations, and communications. Employee telephone directories are updated monthly by area, with a total update every four months. Manuals and brochures are prepared for both internal use and network affiliates. All of these documents are prepared in the central services department, using electronic typewriters and word processors. Materials that require graphics, charts, and logos are sent to the art department. When the materials are returned in 2 or 3 days, the entire job is sent to the central printing and publication department where it is rekeyed onto the typesetter. Turnaround time is usually 4 or 5 days.

    Describe how this operation can be improved. What changes that would speed up production and reduce costs can you suggest?

2. The chapter indicated that one of the newer trends of graphic technology is slide-making applications. Research this technique and describe the process. Summarize some of the major vendor packages in this area and list the special features that they offer.

3. *Creating Graphic Projects:* Create a pie chart from the data in Table 10-1. Create a line chart for the data in Table 10-2.

TABLE 10-1. Local Network Installed Value

1986	
Broadband	35%
Twisted pair	5%
Baseband	60%
Total $ Value of Installed Base $200M	
1990 (Projected)	
Broadband	50%
Twisted pair	13%
Baseband	37%
Total $ Value of Installed Base $110 M	

TABLE 10-2. Worldwide Telecommunications Expenditures

Year	Dollars (in Billions)
1986	$378
1987	$427
1988	$438
1989	$546
1990	$617
1991	$697
1992	$788
1993	$890
1994	$1006

CHAPTER 11

Word Processing

After reading this chapter, you will understand
1. The concept of word processing
2. The advantages of word processing
3. Some of the major features of word processing
4. Some of the major applications of word processing
5. The evolution and trend of word processing as a dedicated, discrete piece of hardware

Throughout this textbook we refer to the benefits of word processing. In every aspect of office information systems, word processing plays an important role. Word processing is part of the strategic planning and implementation discussed in Chapter 2. It is a building block toward the comprehensive integrated information system in Chapter 6. It is a key component of information processing in Chapter 3. It is identified as a separate hardware category in Chapter 9 and described as a popular software applications package in Chapter 14.

Word processing has undergone drastic changes over the past decade. It has been defined and redefined to mean different things to different people. The early meaning referred to word processing as a concept. In an early textbook on *Word Processing,* 2d edition, by Arnold Rosen and Rosemary Fielden (Prentice-Hall, 1982), word processing was defined as

> The fastest most efficient, and most economical method of expediting paper flow from its authorship to distribution to the printed word.

This definition implies the interaction of people, equipment, procedures, and environment.

And in another book, *Getting the Most Out of Your Word Processor,* the hardware aspect of word processing was stressed. However, one should not lose sight of the fact that word processing is part of a total concept approach. The following is an example of this philosophy:

> To begin to understand the boundless promise of your new machine and our new information society, we need to define some terms. Word processing is the use of electronic equipment to type, change, and permanently store information for final printing and future use. Word processing is really a process, not just a product; it is an attitude as much as it is an identifiable set of hardware and software. Implicit in this definition are several important ideas, all of which center on the people, procedures, equipment concept.*

Thus, word processing can be a concept, a piece of hardware, a software package, or a generic application. It can even be a job description or a person's title. We have touched upon all of these aspects throughout this book. For this chapter, however, we will discuss the essential generic functions and features of word processing and how word processing can be a powerful instrument in creating, editing, and distributing information.

Benefits of Word Processing

Whether one uses a dedicated word processor or a word processing software program, our emphasis in this chapter will be on describing the features, advantages, and power of word processing. Word processing has the following advantages:

1. It can help produce error-free documents.
2. It can eliminate repetitive typing.
3. It can enable easy input (keyboarding) of text.
4. It can store or memorize text for later revisions.
5. It has the power of fast and simple editing.

Word processing is a way to *simplify typing.* You type the text once. After that, the power of word processing takes over. You can change words, sentences, and paragraphs, effortlessly. With a traditional electric typewriter, the addition or deletion of a single word may mean retyping the entire page. With word processing, a change takes seconds, and printing a new page, less than a minute.

---

*Arnold Rosen, *Getting the Most Out of Your Word Processor,* Prentice-Hall, 1983, p. 5.

Better writing comes from editing and rewriting that which has been written. It produces clearer, more concise writing, and because change is so effortless, it encourages editing.

Features of Word Processing

Today's word processors and microcomputer-based word processing packages offer a wide variety of features, which can be classified into two levels: basic features and advanced features.

Basic Features

Basic feature word processing programs are inexpensive and easy to learn. Programs at this level can be used for small applications, where users are primarily interested in simplicity of operation. Some programs offer no frills, others offer all the features necessary to perform the following basic word processing tasks:

Word Wraparound. Word wraparound automatically moves words to the next line when the right hand margin is reached and the word does not fit within it.

Centering. The center function centers a horizontal line of text between the left and right margins of a document.

Underscoring. Text can be automatically underscored during typing or inserted after the typing is completed. Some systems also feature automatic deunderscore, which automatically erases the previous underscore. Text also can be double underscored for statistical charts or other applications.

Superscripts/Subscripts. Superscripts or subscripts may be required in some documents for mathematical equations, references, or other cases. A superscript raises a character above the regular text line, a subscript lowers text below the text line.

Scrolling. Your system may include vertical, horizontal, or page scrolling. Vertical scrolling is the ability of the system to move up and down through a page or more of text, one line at a time. Horizontal scrolling is the ability of the system to move horizontally along a line of text in order to access more characters than can be shown on the display screen at one time. Page scrolling allows the system to move backward and forward through the pages of a multipage document.

Display Highlighting. The display highlighting feature highlights (intensifies) the shading of certain portions of the text to emphasize it.

Automatic Error Correction. The following are simple error correction features:

Backspace Strikeover. While keyboarding, if you make a mistake, simply backspace and strikeover the character and then keyboard the correct character in the space.

Delete. To delete text, you simply move your cursor to the spot in the document that you wish deleted, and by a command you can delete a character, word, line, paragraph, or any remainder of the text.

Insert. Insert is a feature that allows a unit of text (spaces, characters, words, paragraphs, or pages) to be inserted in an existing document at a designated point.

Move. The move function allows you to move text from one location to another on the display screen.

Required Space. A required space is a space that connects words so they will not be separated during a text adjustment.

Page Breaks. There are two kinds of page breaks. A *temporary* page break is entered where you want the pages to end when you are typing a document. A *required* page break can be entered to end a page of text.

Automatic Number Alignment. The automatic number alignment feature allows you to type columns of numbers that are aligned from the right edge of the column, giving the columns a justified, uniform appearance.

Automatic Hyphenation. The hyphen feature allows you to even ragged right margins by filling empty space at the end of a line with characters from a word at the left margin on the line below. A temporary hyphen, which is eliminated if the page or paragraph is reformatted, is entered automatically.

Automatic Indent. Indent establishes a temporary margin on the left, right, or left and right sides of a page. Text, when ended, wraps at the indents.

Pagination. Pagination is the process of identifying the line where pages will end. This ensures that there will be the appropriate number of lines on a page during printout. During pagination, page breaks are inserted throughout the text to divide the text into individual pages. The system inserts temporary page breaks, which can be changed if the document length is altered and repagination is necessary. The *repagination* function is used to change the number of lines per page within a document automatically.

Headers and Footers. Headers and footers are segments of text that are repeated on each page of a document. A *header* is the text that is repeated at the top of each page, and a *footer* is the text that is repeated at the bottom of each page. A header may be a title, other descriptive text, a page number, or a combination of these. A footer usually includes a *footnote:* a note placed within a document that cites a reference for a designated part of the text or adds further information on a particular item in the text.

Bold Print. Bold print can be selected to emphasize a word or a portion of text. Bold is a form of thick printing that highlights specified text on the printout.

Overstrike. The overstrike or overprint function is used to cause two characters to print over the same point of a printed document. When using the overstrike function, the first character prints, the printer backspaces, and the second character overprints the first one.

List/Merge. The list/merge function retrieves information automatically, such as the name and address from a mailing list, and merges it with a specific letter format. The merged information conforms to the format on a screen, so there is no need to retype or reposition copy. Lists may consist of names and addresses, dates, signature blocks, technical terms, and frequently used phrases. Lists can also be merged automatically with a format to produce envelopes and address labels.

Search and Replace. The search and replace function will search through a page and/or entire document for a specific word or phrase. The word or phrase can then be automatically replaced, all without typing. Also, the words or phrases can be totally deleted or merely viewed on the screen. This can also be a background function, and it incorporates the following options.

Search and Selectively Replace. The search and selectively replace method allows you to search manually for a specified text string. You can then automatically substitute a particular replacement string if you desire or go on to the next occurrence of the word without making any changes.

Global Search and Replace. With global search and replace, all occurrences of a text string are replaced automatically throughout an entire document.

Filing and Disk Management. If you are not using a hard disk system, the number of floppy disks you accumulate will increase rapidly. To retrieve documents quickly, you will have to create a filing system that is efficient for your office and best meets the needs of your company. Most word processing systems have provisions for filing and disk management. Word processing systems allow you to create a variety of documents. Most systems have the following filing and disk management features:

Storing. The storage feature simply stores a page, transferring format setting and text on the screen to the data disk. The new page is automatically recorded in the file index.

The File Index. After you have created a document, it will be stored on the disk for editing and printout. When the document is recorded, it receives a name (or number) that is automatically logged on the disk index. The disk index is a listing of all the documents stored on the disk. It may provide specific data about each document as well as some overall disk information.

Once a document is stored on a file index, most systems provide the ability to do the following:

- *Recall*—"Recall" or "read" brings a page stored on a disk to the screen. The page displayed on the screen is actually a copy of the page on the disk.

- *Update document*—Updating allows you to record the changed version of a page on the screen in place of the original version on the disk. With this method, the new document is overstored over the original text.

- *Divide a document*—If a document gets too long, it might be a good idea to divide the document into two separate documents.

- *Rename a document*—A document can be renamed if you decide to change the name or number after it has been stored.

- *Copy a document*—A duplicate copy of a document may be needed for a certain application.

- *Delete a document*—To use the disk space more efficiently, it is a good idea to delete from the disks documents that are no longer being used.

The twenty basic features outlined in this section will usually be found in most high quality word processing systems. Obviously, some vendors will include more features; others will have less. Some basic features may be considered advanced features in some word processing packages.

Advanced Features

New generations of word processing packages have impressive lists of innovative features and applications. Advanced or enhanced features have taken word processing a step further from the first-generation word processing packages. The new applications software packages increase the user's productivity in a variety of tasks. In both dedicated or WP software packages, vendors are now offering records processing, math, forms, equations, and advanced stored keystrokes. These powerful capabilities work in conjunction with the basic word processing functions to let the user sort, analyze, and evaluate many different kinds of information—then easily combine the results in a single document. Advanced features support the integration of text and graphics so users can easily design and manipulate documents containing both. Advanced features also provide an array of helpful user aids such as spelling verification; revision marks; split-screen editing; an "undo" feature, to correct accidental mistakes; and on-line training packages.

Advanced word processing features address a variety of user needs. Vendors are now designing packages that are comprehensive in scope and, yet, easy to learn and easy to use. System menus and conversational screen prompts guide the user through most system operations and take the guesswork out of word processing.

The following are some of the advanced features found in modern word processing systems:

Glossary. The glossary feature provides storage and recall of keystrokes for frequently repeated typing tasks. This feature can be a real timesaver, especially with repetitive text strings or long sections of text that are used frequently. Glossary applications allow the user to do the following:

- To insert technical words that must be frequently looked up
- To insert frequently used paragraphs such as paragraphs used in contracts and letters
- To recall formats and tabs so that time-consuming formats, such as special columns, do not have to be set over and over again
- To create and utilize several different specialized glossaries such as legal and medical terms

Decision Processing. Decision processing enables the user to increase the sophistication of glossary entries and expand their capabilities. Users can display their own prompts and error messages, and temporarily halt the recall of a glossary to accept operator-entered keystrokes. Virtually any word processing function can be performed automatically through decision processing.

Spelling Checking Program. The spelling, or dictionary, program scans a document and identifies spelling errors in a variety of ways. Some WP packages scan a document and display spelling errors within the text on the screen, giving a list of probable alternatives for the user's selection. A *scanning* feature on some systems checks a document for spelling errors. The cursor scans every word and stops and beeps at a word that is spelled incorrectly or is not contained in this dictionary.

If the word is misspelled, the user can correct the word and continue scanning the document or can look up the proper spelling in the dictionary and correct it automatically. On some systems, when words are identified as misspelled and do not appear in the dictionary the user has two options: the user can correct the word or add the word to the dictionary.

List Processing. With list processing, users can efficiently set up and maintain customized lists. They can retrieve selected information from these lists and print the results in a variety of formats—all without writing or relying on computer programs. List processing eliminates the manual work commonly associated with record-keeping activities such as employee files, telephone directories, and customer correspondence. You can define and create each list, accommodating thousands of pieces of information and providing customized records that reflect your specific needs.

After all records have been input into a list,

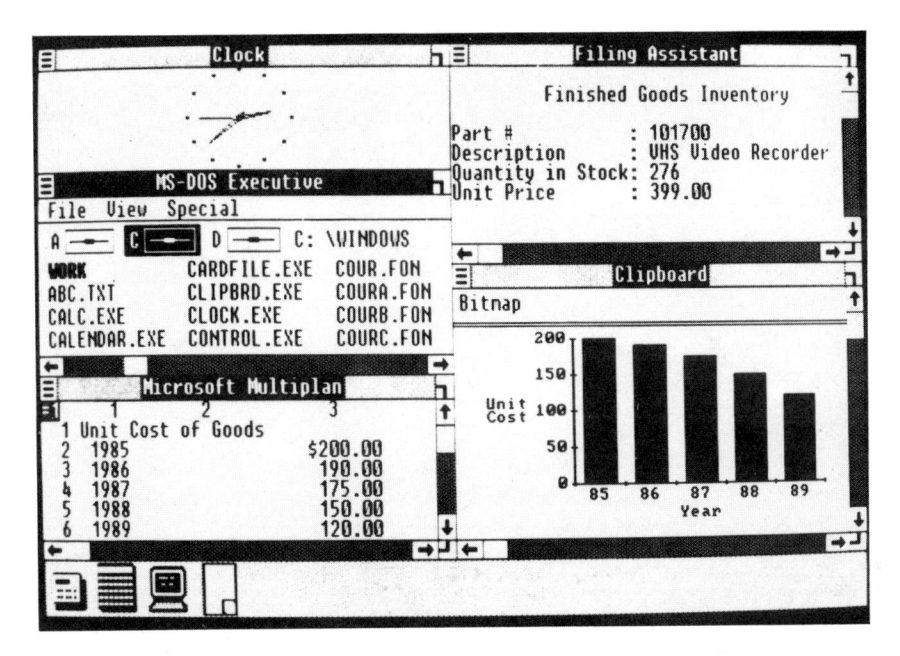

Figure 11-1. Windows and split screen editing.
Source: Courtesy of Microsoft.

specific records can be retrieved in a designated order. For example, from a list you may wish to generate letters only to the people who live in Vermont, or select only students enrolled in a certificate program. This is referred to as a *sort* feature. It is part of the list processing function. It allows the user to arrange tables, lists of information, and properly formatted sections of text in ascending, descending, alphabetical, and numerical order.

Math. A word processing system incorporates the standard four function math capabilities. Some include a ten-key pad for input. In addition to the four basic math functions of addition, subtraction, division, and multiplication, other options include:

Percentage
Column (vertical)
Horizontal
Numeric alignment

Calculator math (gives the user the ability to calculate any string of numbers at the top of the screen, using the four mathematical functions of addition, subtraction, multiplication, or division)

Split Screen Editing. The split screen editing feature allows the user to display different documents simultaneously (Figure 11-1). He or she can also move or copy between two windows and edit either one. For example, a user working on a new monthly sales report can compare it to last month's report.

Integration of Text and Graphics. Graphics can be mixed with text so that pages can be viewed on the screen before the document is printed (Figure 11-2). The user can enlarge, reduce, or edit the graph to design a new page. This feature results in the creation of professional-looking documents.

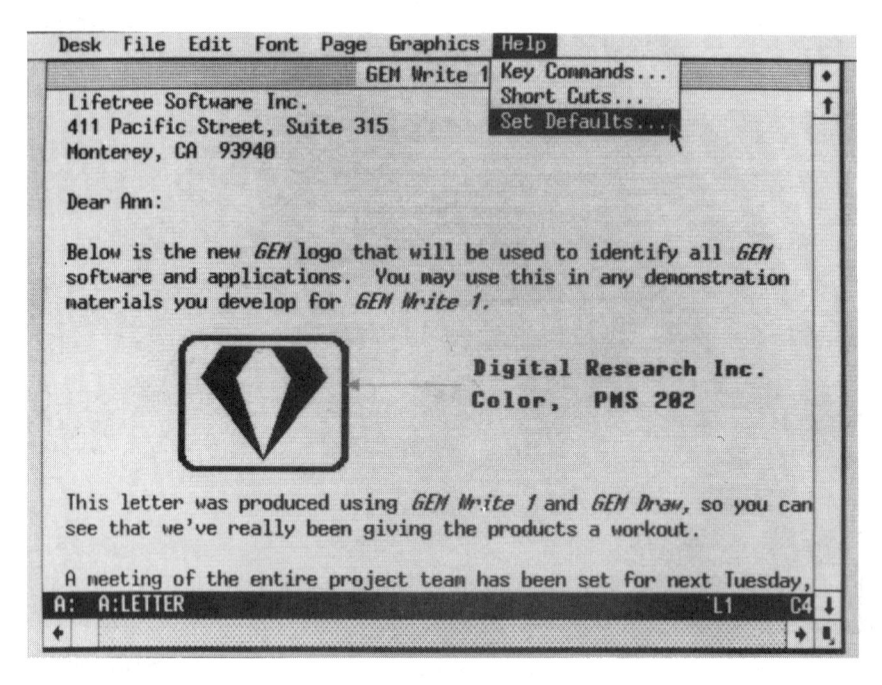

Figure 11-2. Integrating text and graphics on the screen.

Source: Reprinted with permission from Digital Research Inc.

Forms. Filling out forms can often be tedious. With a word processing system, a form can be easily stored only once and filled out each time you need it. Both preprinted and completely original forms can be created.

Widow and Orphan Lines. This feature is useful in preparing multipage documents. The system will automatically prevent the first line of a paragraph, title, heading, chart, table, and so on, from being the last line on a page. The operator simply selects this option from the print menu, and the system will automatically adjust during the pagination process.

Auto Outline. This feature allows the operator to create outlines in Roman numeral and alphabetical formats and renumber outline sections automatically. This is a tremendous asset, particularly for the legal and technical publishing fields.

Security. Protection against unauthorized access to documents stored in the file index is handled through a security feature. The user of any application would be granted access to the file only upon specification of the correct password.

Personal Computing. Personal computing has become an important tool in office automation. Multiple operating systems are now offered for functions on a dedicated word processor. The ability to run applications based on CP/M, MS-DOS, PC-DOS, UNIX, and XENIX allows the user to take advantage of the myriad of software packages offered under these operating systems. Programs such as Lotus 1-2-3, Symphony, and dBase II can be run with a variety of operating systems.

Index Generator. A word processing system that allows the user to generate a reference index is part of a records processing function. An index may contain thousands of key words. The system searches for all occurrences of the key words. In some systems, the index is printed with a command selection on the print menu. Others place numbers of the page on which the words appear in an index at the end of the document.

Library Services Feature. An advanced version of the index generator is the *library services feature.* In any large organization you have different classes of documents. Some have a very short lifetime, and some are very important to the corporate records. They may not be needed for a very long time and are in a data base or corporate file. The library system is one of the ways to get machine-readable text data into the computer. The problem we address is finding it again, once it is in there.

The library services feature solves the document retrieval problem. It is designed to locate office documents created through word processing or captured in word processing format. Therefore, it retrieves the documents by their own contents.

The system searches documents for words or phrases that are of interest to the user. And the user need only specify the search words at the time he or she wishes to find the associated documents. In other words, the user does not have to be concerned with how to retrieve the document at the time the document is filed.

Mailbox. The mailbox feature is an internal mail network that provides for sending and receiving short messages to word processing terminals hooked into the network. Information, such as short memos, messages, and reminders, is distributed through the mailbox feature.

To send a message or document, a send menu is displayed. The user fills in the blanks, including the send to, from, and so on, and keyboards the text of the message. The recipient can display a list of received messages in a menu (mailbox) and read the messages to a screen.

Industry and Professional Applications

In many environments, high-quality word processing is only part of the solution. All of the basic and advanced features described in the previous section can be applied to a variety of personal and professional applications. They can enhance creativity, improve business decisions, and increase productivity (Figure 11-3). Word processing hardware and software combined effectively have the power and flexibility to provide an unparalleled range of

Figure 11-3. Engineers and technicians combine word processing with financial applications and management control programs.

Source: Courtesy of the Sperry Corporation.

solutions. How can the variety of special software and advanced features be used in professional and personal applications? The following present a few examples of using word processing to create, edit, sort, analyze, and evaluate information in a variety of environments.

Engineering, Architecture, and Manufacturing

Word processing systems provide engineers, manufacturers, and architects with a wide variety of two-dimensional drafting capabilities. This enables users to apply word processing software by off-loading some work currently being done on CAD/CAM systems. Electronic template files can be created on the system, then used in conjunction with the line drawing and graphic features right on the screen. This feature provides complete, annotated mechanical drawings to facilitate a technical project.

Publishing

Word processing that interacts with typesetting can extend the limits of traditional word processing. Small in-house *desktop publishing* enables businesspeople to set type at their desks using word processors, personal computers, and laser printers (see Chapter 8). By using the power of laser technology in high speed computer printing systems, superb documents of excellent quality can be produced quickly, quietly, and cost-effectively. Laser printers can now allow smaller offices and work groups to enjoy desktop publishing. This small scale publishing can create newsletters, price lists, and other frequently issued and frequently revised material.

High-end desktop publishing is turning technical writers and illustrators into technical publishers. In-house newsletters, product specifications, contracts, and shareholder annual reports are excellent applications for these systems. Any printed work that needs fast turnaround, frequent revision, and short-run quantity is an excellent application for desktop publishing.

Law

The field of law was one of the first professions to recognize the importance of word processing as a powerful and useful tool to streamline the creation of legal documents. Without computers and word processors, too much time was spent on administrative details and too little was spent on practicing law. Word processing helps attorneys produce the best results for their clients.

Law firms of all sizes are using computer and word processing technology to achieve greater efficiency, cost-effectiveness, and performance. Through word processing they can do the following:

- Maintain accurate records by updating client records as time, expense, payment, and account adjustment entries are made
- Rapidly access account information with master reference and cross-reference lists and account inquiries
- Ensure accounting accuracy with audit lists for tracking fees, expenses, payments, and account adjustments
- Network with other word processors and mainframes; communicate with other branches, offices, and clients; and access legal data bases

Medicine

Medical and health services provide numerous applications for word processing. In a one-doctor office, a large hospital, or a group practice, accurate patient records and medical documentation are essential.

Malpractice and other forms of insurance have created an additional paper burden for the medical profession. However, a large part of a hospital's budget is utilized for administrative expenses. From the moment a patient is admitted, the paperwork cycle begins. Each medical department within the hospital—neurology, psychiatry, ob/gyn, and so forth—operates with a certain amount of autonomy. However, each depends on an administrative staff for support operations. A record must be kept of every test administered, every medication given, and every visit by a physician. Medical documents such as charts, operating reports, medical histories, and discharge summaries must be created, edited, and recorded.

The private practice physician has discovered the wonders of word processing. The doctor, in the daily tasks of office examinations, must also record accurate, up-to-date patient information. The word processor, combined with advanced software, allows the doctor to process billing, store patient records, process insurance documents, write medical papers, and correspond with other doctors. Communication links can be established from the doctor's desktop terminal to the mainframe that resides in the hospital, medical center, or university. Specialized medical data banks can provide the physician with accurate and up-to-date medical information.

Science and Research

Science and research professionals need advanced technical typing capabilities from word processing programs. They also need the speed and efficiency that a powerful advanced word processor can provide.

Science notations, multilevel equations, chemical formulas, Greek characters, and multicolumn formats for tabular information are features on WP systems that accommodate technical applications.

Raw data can be easily translated into understandable charts and graphs by means of the graphics-text-mix feature.

Writing—The Primary Application

Of all the applications, it is writing that benefits most from word processing. The power of the written word has been enhanced by the software for creating, manipulating, and processing words. Writers are a large category, including authors of books, magazine articles, stage and screen plays, novels, poems, and speeches; graduate students; and business or professional people.

The written word is used in business letters, memos, reports, contracts, catalogs, brochures, notices, and newsletters. The variety of documents and applications is limitless. The mastery of words can make the difference between a sale and no sale, between a raise and no raise, between a negotiated settlement and a court battle, between confusion and clarity.

For the businessperson, word processing leads to better correspondence. Small mistakes that might have passed because of time pressure can be corrected. Afterthoughts and clarifications can be added more freely. Secretaries, or the executives themselves, do not mind replacing a single word with a better word to improve a document.

Good writing comes from editing. Editing involves going over that which has been written and making changes and alterations. Word processing avoids the process of typing the same document twice.

Perhaps the most powerful editing shortcut is the *boilerplate*. Although it is called different names in different programs, it is a mechanism for building lists of words, phrases, or paragraphs that you then code and save to disk. Electronic boilerplating can cut down considerably on repetitive work at the keyboard. Form letters are probably the most common variety of boilerplated documents: "Thank you for your inquiry" "We are happy to enclose" "You are cordially invited to attend." If these letters are not personally typed, they tend to be ignored, so faithful secretaries type the same letters over and over.

Through word processing, the text of the letter can be stored on a disk, and only the names, addresses, and salutations are changed each time. Other applications for boilerplates are legal clauses found in petitions or contracts, standard clauses in insurance policies, and oft-used phrases in customer service correspondence.

Novelists, textbook authors, and professionals are switching to word processing and finding, to their pleasure, that it is revolutionizing the way they do their writing. Using the power of the computer and the capabilities of advanced programs, they are writing "smarter and faster." Writers are using these tools and technologies to improve the mechanics of writing. However, writing is still an intellectual and creative pursuit. No matter how sophisticated the tool, it will not help a person who does not have the creativity and talent to write. The combination of a good computer system and a comprehensive word processing program can aid an author considerably, but the craft of writing remains within the writer.

Computer Applications

Word processing	88%	Access info services	45%
Spreadsheets	81%	Networking	45%
Data base/file management	74%	Electronic mail	45%
Education/training	64%	Presentation graphics	44%
Entertainment	58%	Software development	34%
Personal finance	47%	Project management	28%
Business accounting	47%		

Top Three Applications over Time

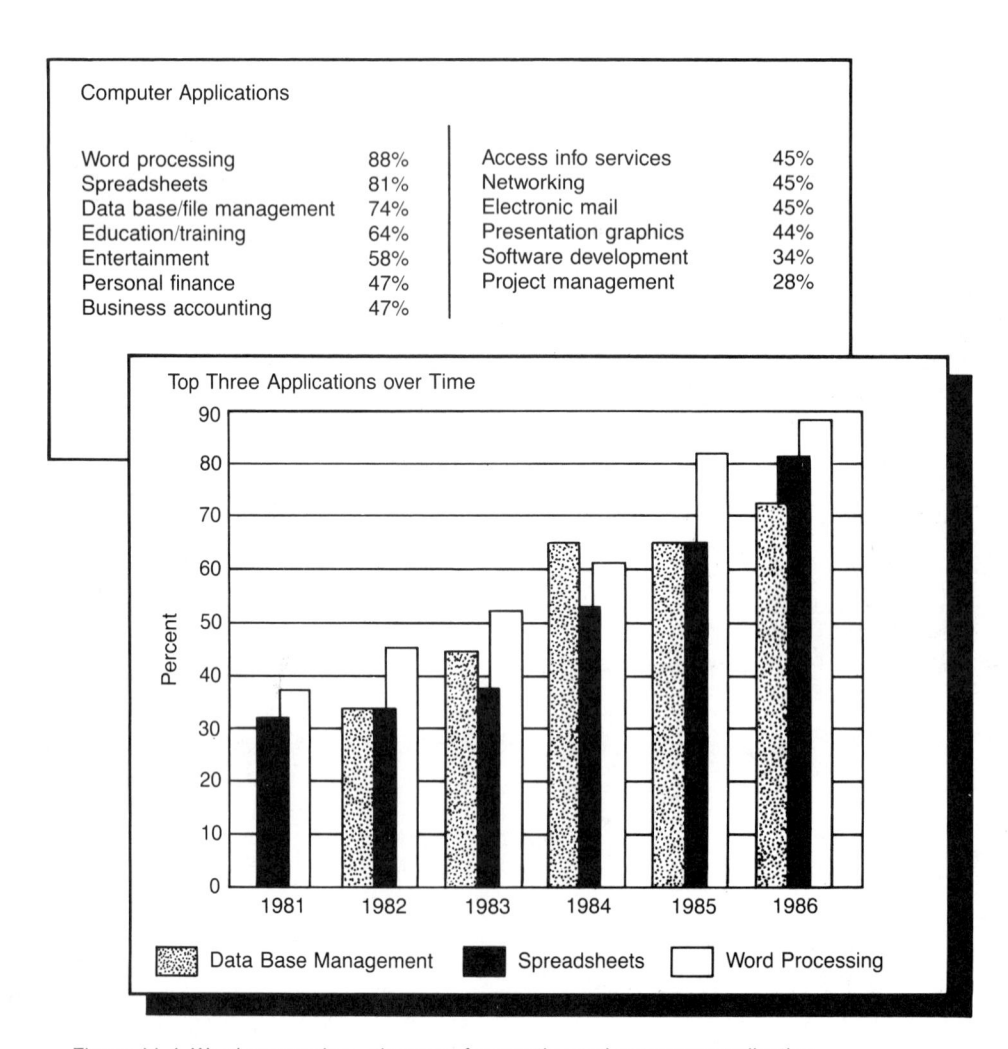

Figure 11-4. Word processing—the most frequently used computer application.

Source: Reprinted with permission from Personal Computing, May 1986. Copyright 1986, Hayden Publishing Company.

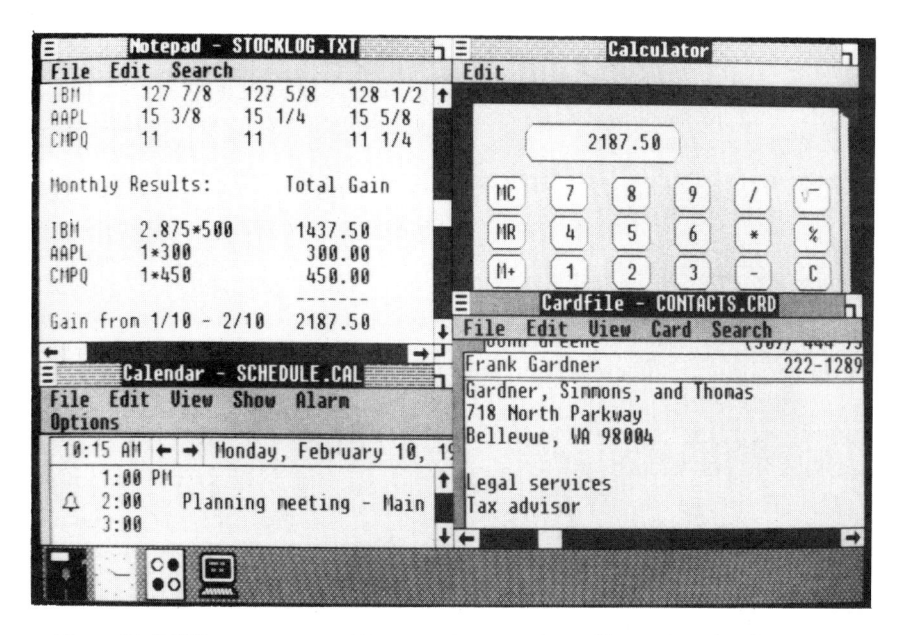

Figure 11-5. This program combines word processing with a notepad, calculator, calendar and cardfile.

Source: Courtesy of Microsoft.

Word Processing Is Still the Cornerstone of Office Automation

In spite of the tremendous changes taking place in information processing technology, word processing is still the nucleus of office automation (Figure 11-4). Although the dedicated word processing market has been steadily eroded by microcomputers, word processing software has grown to become a powerful office tool for a variety of applications.

Because of the popularity of micros, all levels of information workers are experiencing the benefits of doing their own word processing. Professionals are beginning to overcome their prejudice toward "doing their own keyboarding." They are beginning to view WP software as yet another tool that can be used on personal computers, to supplement the use of spreadsheets and other decision support software packages (Figure 11-5).

Modern word processing software packages offer more than just word processing. They offer a total integrated solution. This means that all the word processing capabilities one will ever want or need can be accessed quickly and easily. These advanced features open the door to new possibilities. First-generation WP software did not offer split screen editing, sophisticated graphics, and advanced records management. With the wealth of software available today, users are more likely to find packages tailored to their needs. As the market for this software becomes more competitive, even newer options and greater functions will be available.

Summary

Word processing uses a computer's power to store and manipulate information. Rather than writing or typing directly onto a piece of paper, the user enters the words, through the keyboard, directly into the computer. They are displayed on the video screen and can be moved about, changed, deleted, and added to at will.

But word processing goes beyond a way to input text. It should aid in the creation, organization, and expression of ideas. Its applications include proposals, reports, memos, legal documents, books, company newsletters, scientific papers, letters, form letters, and business forms.

Today, the number of features found in word processing programs is increasing. No longer constrained by the memory limitations of 64K machines, programs can do more and provide extensive help screens as well. In addition, most word processing software programs have powerful editing, filing, and reformatting features that are easy to use.

The stand-alone, dedicated word processor was a highly refined, single-purpose tool built to do one thing and one thing only. Today, the personal computer, a multifunctioned machine capable of doing many things well, including word processing, has taken over the word processor's market.

Finally, word processing is still the cornerstone of office automation. Recent surveys have shown that it is one of the most frequently used applications performed on personal computers.

Review Questions

1. Define word processing. How does this definition compare with an earlier version?
2. List some of the major advantages of word processing.
3. Describe scrolling; automatic hyphenation.
4. What is a footnote?
5. Differentiate between "search and selectively replace" and "global replace."
6. Describe glossary; widow and orphan lines.
7. What feature protects against unauthorized access to documents stored in a word processing system? How does the user access information from a "protected" system?
8. Can a dedicated word processor perform personal computing functions? Explain.
9. Describe how the graphics feature in a WP program helps engineering applications.
10. List some of the WP applications for a one-doctor office.
11. The text states that writers are using these (WP) tools and technologies to improve the mechanics of writing. What is meant by *mechanics* in this statement?

Projects

1. You are asked to select a full-featured word processing software program for your executives using personal computers. You have narrowed your search to two packages: menu-driven and command-driven. Explain the difference between the two. What are the advantages and disadvantages of each system?
2. Spelling checking software are programs that compare every word in your document against their own lists of correctly spelled words. Then they display all words from the document that do not match. What are the limitations in using a spelling checking program? Can it produce an accurate document? Describe other related programs that check for incorrect placement or usage of commas, periods, quotation marks, capitalization, and so on.
3. Interview a professional (doctor, lawyer, dentist, accountant) in your community who does not use a word processing or computer system. Ask him or her about his or her applications. Present a paper that points out ways a word processing system can improve the professional's efficiency and productivity.
4. Compare three popular word processing software programs. Create a table that lists and compares ten features. Include a summary at the bottom of your table that contains the following:

    Name and address of each manufacturer
    Name and price of WP software
    Operating system compatibility
    Outstanding feature or special requirements

CHAPTER 12

Telecommunications

After reading this chapter, you will understand
1. The historical account of events leading up to AT&T's divestiture
2. The broad aspects of telecommunications and the segments that comprise smaller technologies within telecommunications
3. Telephone systems and long-distance service
4. The concept of data transmission, transmission lines, and devices
5. Other technologies such as electronic mail, teleconferencing, cellular phone service, intelligent buildings, and satellite communications

A telephone rings in the middle of the night. A person in a remote corner of New Zealand sits listening, waiting for a voice from beyond the solar system. A mainframe computer in Allentown, Pennsylvania, dials up a point-of-sale terminal in Biloxi, Mississippi, and receives a full day's worth of sales and inventory data.

People, machines, and devices strive to communicate through voices, sounds, gestures, and simple ideas. "What hath God wrought?" was Samuel Morse's first commercial telegraph message. "Mr. Watson, come here, I want you," were the first words spoken over a new instrument—the telephone. "That's one small step for a man, one giant leap for mankind," said Neil Armstrong as he set foot on the moon. For the first time the human voice was heard from a body beyond the planet Earth. Crude transmissions of the past have given way to newer forms of electronic movement of information. Yesterday's conversation between the mainframe in Allentown to the point-of-sale terminal in Biloxi consisted of digital bleeps, which are transparent to the user. They lack the drama of the historic messages of Morse, Bell, and Armstrong, but computers, mainframes, and satellites communicate on a vastly different scale than a century, or even a decade, ago.

A global telecommunication revolution is in process and is doing more to reshape the information processing industry than any other single development. This revolution is changing the lives of anyone who picks up a telephone, switches on a television set, or logs on to a computer. The great proliferation of personal computers has made the transmission of data from one location to another virtually inseparable from its processing. IBM, for example, is convinced that transporting data between computers will eventually be nearly as important as manipulating data in a company's mainframe or an executive's personal computer.[1]

Those who purchase new equipment consider a computer's communications capabilities. Office equipment manufacturers now offer products that not only have power-processing capabilities but also offer a means to transmit information between devices. The vendors are now responding to the end users. Information workers, who have such an insatiable need to link up their rapidly multiplying personal computers and workstations, are likely to buy only from vendors that can link their equipment together.

Divestiture: Reorganization of an Institution

The seeds of the breakup of AT&T were sown when the U.S. Supreme Court handed down the Carterfone ruling in 1968, which ultimately revolutionized this country's telephone habit. The gist of this decision was that non-AT&T telephone equipment could be installed in residences and connected to AT&T lines. For the first time, Americans could purchase telephones instead of only renting or leasing them from their local telephone company.

Following divestiture of AT&T in 1984, the communication user faced turmoil, new product announcements, and dazzling technologies in the midst of the dismantling and redesigning of an institution they had known all their working lives.

Telecommunications: A Large "Umbrella" of Technologies

Telecommunication is a technology that provides the electronic movement of information between two locations. This traditional meaning is usually identified in the narrow sense as

---

[1] In a forceful expansion into telecommunications, IBM purchased Rolm Corporation, one of the leading makers of telephone switching equipment. The acquisition, in September 1984, was IBM's largest and its first in twenty-two years.

being synonymous with only telephone or voice communication. In a broader sense, telecommunications is a large umbrella. The technology embraces an assortment of devices, disciplines, and communication modes such as data, voice, video, and facsimile.

New generations of equipment descend upon users before the last have been evaluated. New transmission options make old networking and configurations obsolete. How can the information manager plan when whole product lines change so rapidly? This chapter will discuss the array of technologies available and the impact of those technologies on those who must know and skillfully use them to manage information.

Telephones and Telephone Systems

The dial phone was introduced in 1896 (Figure 12-1). That technology stood alone for sixty-seven years. Then the Touch Tone phone was released in 1963. Though the standard, plain phones have proved most popular in terms of overall acceptance, in the 1980s we have seen the advent of display phones, "intelligent" phones, cordless phones, and cellular mobile phones.

The two main types of business telephone systems are the key system and the private automatic branch exchange. (The abbreviation for the latter system is PABX or PBX. Some vendors use CBX, for computerized branch exchanges. For the most part these terms can be used interchangeably.)

PBX Technologies

The first automatic telephone switching device was introduced in 1891. This switch allowed callers to bypass telephone operators in placing and receiving calls. If this early mechanical switch represented the first generation of what we call private branch exchanges (PBXs), then the first generation lasted until the mid-1970s, or eighty-five years. The second-generation switches marked the end of mechanical switches. They replaced relays with transistors. At this point, computerized PBXs with all-electronic switching were introduced. Third-generation switches, which operate under microprocessor control and can

Figure 12-1. Telephones for Park Ridge, New York, 1909.

Source: Reproduced with permission of AT&T Corporate Archive.

Figure 12-2. A fourth-generation PBX.
Source: Courtesy of Fujitsu Business Communications.

be programmed, became operational at the beginning of the 1980s.

The fourth generation uses digital technology to provide integrated voice and data switching (Figure 12-2). This type of system transmits both data (digital) and voices (analog). The ability to handle digital transmissions allows personal computers, word processors, and other desktop terminals to transmit information over the phone lines. This means that all text and data transmitted to and from these devices go through the integrated PBX, which is also used for standard voice phone calls.

The PBX offers many capabilities including call forwarding, which automatically relays incoming calls to another extension, located elsewhere. PBX also has ability to automatically select the most economical circuit for a toll call. The PBX functions are activated by dialing a specific digit or digits on a standard telephone or on a telephone with additional function keys.

The PBX voice switching provides for major developments in voice messaging systems and multimedia conferencing. The enhanced digital PBX will evolve into the communciations hub of the office. It will bring enhanced multimedia communications to users. The integration of media applications and the interconnection of various systems are evolving realities. We will see more and more high technology vendors delivering multipurpose systems for the office as users demand more technology. Fourth-generation digital PBXs (Figure 12-2) will allow the information worker to access, process, and manage information efficiently.

Features of Telephone Systems

Modern telephone systems provide the following features:

Direct Inward Dialing (DID). With a DID system, each person in the company has his or her own telephone number. An outside caller can dial directly to an individual telephone on the PBX system without going through the company operator.

Direct Outward Dialing (DOD). The DOD system allows the businessperson to pick up the phone and dial an access number to obtain an outside line. The caller can then dial the desired number. All this is done without going through the switchboard or company operator.

Least-Cost Routing. Outgoing calls can be automatically placed over the telephone circuits that are least expensive for that particular call.

Call Forwarding. A user can program the phone system to forward calls to another phone or to the operator. This can be done if the user is out of the office, or calls come in when the user's phone is busy, or for calls that are not answered after a certain number of rings.

Call Restriction. Certain telephone users can be automatically prevented from placing long-distance calls.

Call Pickup. A user can answer any ringing phone in the system. One simply picks up his or her phone and dials a certain number and the extension number of the ringing phone. For example, if you hear a phone ringing in another area, you can pick up your phone, dial "6" (or some other predesignated number) and the call will be switched to your phone automatically.

Conference Calling. A user can set up a conference call among several people, both inside and outside the company, without the assistance of the operator.

Camp-On. If an extension is busy, the caller can *camp-on* the line until the previous call is completed.

Dialing Codes. After the user has entered two or three digits, the system can automatically dial frequently called seven- or ten-digit telephone numbers.

Call Waiting. A tone notifies a user engaged in a conversation that another party is attempting to reach him or her.

Long-Distance Service

The Birth of Alternate Long-Distance Service

In 1986, telephone users across the country were asked to pick one of several competing companies as their primary long-distance carrier. If they could not make up their minds, customers (both private and corporate) were automatically assigned a long-distance carrier.

These new developments in alternate long-distance services are a result of the divestiture decision of the Bell System, and of dozens of regulatory and technical changes in U.S. telecommunication during the past fifteen years. In fact, the genesis of non-Bell long-distance service goes back to shortly after World War II, when companies began experimenting with microwave radio. The prospect of large, privately owned networks made the big monopoly carriers very nervous. To forestall any competition, the big carriers established bulk discounts for their high-traffic customers.

But alternate transmission services continued to develop, and many saw opportunity in this new venture. By the 1970s non-Bell vendors had convinced the FCC and the courts that telecommunications capacity was a commodity. They reasoned that there was no legitimate reason it couldn't be bought and sold by others. The court agreed. As a result, *alternate* service was born.

The next major development came in 1982. It was the consent decree between AT&T and the Department of Justice that split the Bell System into eight entities. The entities include a new AT&T for long-distance service and seven regional holding companies for local phone calls and short-haul long distance.

New Choices in Long-Distance Service

Today, although it accounts for a fraction of the overall market, even a tiny slice of AT&T's long-distance pie is substantial. AT&T's share of the market is 81 percent, with the new regional companies and non-Bell independents handling an additional 19 percent. Here is a summary of some services offered by three major alternate long-distance carriers:

1. *MCI:* MCI was the forerunner of alternate long-distance service and is AT&T's closest competitor. Having started out as a supplier of private-line service to business customers, MCI diversified into other aspects of telecommunications. The company now handles international telecommunications and electronics and computer-generated mail. In 1986 MCI acquired Satellite Business Systems, a high-tech carrier, from IBM. MCI gained, among other things, the capability to offer satellite-based videoconferencing and data transmission. The link with IBM will marry IBM's information processing expertise with the voice and data transmission capabilities of MCI and IBM's Rolm unit, which makes telephone exchanges, to win business from corporate customers. MCI is winning over 10 percent of the customers in the "equal access" markets—those areas where phone users are being asked to choose a long-distance company.
2. *U.S. Sprint Communications:* The nation's third and fourth largest services competing with AT&T, GTE Corporation's Sprint and U.S. Telecom, merged in 1986 to complete the largest consolidation in the long-distance telephone industry. The two companies also merged their data communications subsidiaries, GTE Telenet and U.S. Teleco Data Communications Company (also known as Uninet).
3. *Allnet Communications:* Allnet was founded in 1980 and has grown to the third largest reseller of telephone circuits. Reselling is a new practice within long-distance carrier companies. The reseller leases lines from AT&T and sells the use of the lines to other customers. Instead of buying lines from AT&T, Allnet bought circuits from MCI, U.S. Sprint, and other non-Bell carriers and resold them to their customers.

There are over 400 other independent phone companies nibbling away at AT&T's lead in the $66 billion long-distance market. Figure 12-3 illustrates how the long-distance market is divided.

Equal Access

As part of the divestiture agreement, the courts ruled that the local telephone companies had to make it just as easy for people to use one of the competing long-distance carriers as to use AT&T. This is known as the "Equal Access" provision. Prior to this ruling, customers had to dial as many as twenty-four digits to make a phone call. By contrast, AT&T's customers have only to dial a maximum of eleven (1, the area code, and the number). The fewer numbers obviously gave AT&T an advantage. Equal access has placed all long-distance carriers on an equal footing.

Bypass

As the rates for long-distance service continue to increase, users will look to further savings.

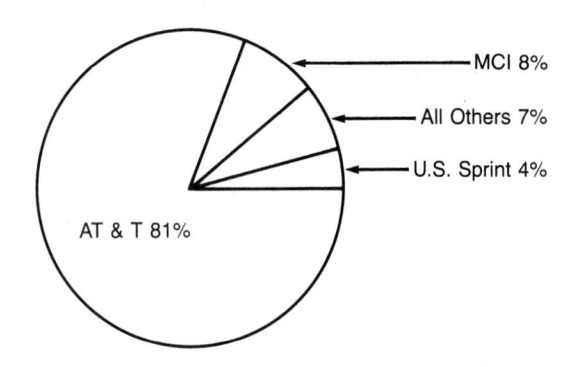

MCI 8%
All Others 7%
U.S. Sprint 4%
AT & T 81%

Figure 12-3. Shares of the long-distance phone market, in percentages, based on estimated 1986 sales.

A more drastic alternative is through *bypass.* By bypassing telephone companies—both local and long-distance carriers—users can take control of rates and realize other benefits. The benefits include implementation of wide-bandwidth facilities and control of installation schedules.

Bypassing can be achieved by three means: (1) Bypassing local telephone company circuits by linking buildings with microwave line-of-sight communications systems. (2) Bypassing the local telephone company by running a direct connection from an organization or corporation to a long-distance carrier's point-of-presence within a local telephone company's region. (3) Bypassing all telephone companies by operating a private point-to-point communications system.

Choosing Long-Distance Service

The communication user must ultimately choose a long-distance service that offers the best price, performance, and service. As telecommunications companies engage in a fierce struggle for position in the new era of telecommunications, the user is faced with a difficult choice in a climate of changing products, markets, and customer needs.

The following are factors to consider in selecting long-distance service:

1. Frequency of calls
2. Time of day
3. Coverage of cities
4. Billing
5. Service (operator assistance and credit card option)

Guidelines for Selecting a Telephone System

Is a basic telephone all your organization needs? Or do you also need a system to connect you with information? Most important, as you grow will your needs change? Your choice of telephone system is dependent upon your intelligent assessment of present telephone operations and your future communication strategies. The following guidelines should be helpful in selecting the right telephone system for a particular organization.

1. Reliability (of vendor)
    a. Size
    b. Track record
    c. Reputation
    d. Service
    e. Response time
    f. References
    g. Promise of on-time delivery
2. Financial considerations
    a. Pay-back period
    b. Buy versus rent
    c. Tax considerations
    d. Upgrading costs
3. System operation
    a. Integrated systems
    b. Upgradability
    c. Ease of operation
    d. Features that match your needs
4. Vendor support
    a. Training
    b. Instructional material

After a careful review of these guidelines, the information manager should assess the important aspects of his or her final decision. It is important to look for a company that offers a total service package and provides support right on the premises. The support should include computer-based instruction, instant-access hot lines, guaranteed response times, and software updates. Finally, one should select a company that has reputable people and comprehensive resources to deliver superior service across the board. The resources should include hardware, software, and training.

Electronic Mail

Imagine sending letters instantly—anytime, anywhere in the world—without touching a piece of paper, licking a stamp, or even leaving your home or office. Imagine getting responses within minutes after sending. Imagine doing all this automatically. Welcome to electronic mail!

Electronic mail can be defined as a means of transmitting text and data messages from one location to another. Electronic mail does for correspondence what word processing does for words and electronic spreadsheets do for numbers. It makes everything faster, more efficient, and more accurate.

Electronic mail, or E-Mail, covers a myriad of equipment and services. The services include Telex, facsimile, communicating word processors, personal computers, and a vast array of third-party services.

Facsimile transmission (Figure 12-4) is a fast-growing segment of business communications for sending and receiving documents, charts, graphs, drawings, letters, and photographs. It is different from other forms of electronic mail in that it transmits an exact duplicate of the original document. The facsimile is so accurate that it is accepted as such in a court of law.

In addition, facsimile eliminates the need for an operator to keyboard data. The person wishing to send a document simply inserts the paper copy into the facsimile (fax) unit, dials the phone number where the document is to be sent, and then places the phone receiver into the coupler on the fax machine. The coupler connects the telephone to the fax machine. The document is transmitted by means of a light source that scans the document line by line and converts light-density differences to electric current. These electronic signals are then sent over ordinary telephone wires. The facsimile receiver, at the other end, receives the electronic signals and converts them into a facsimile of the original document.

Some companies have a dedicated phone line for the fax machine. When the receiving unit answers, the signals are converted to a representation of the document, which is printed.

Compatibility between fax units can be a problem. Before transmitting, the user should be sure that the fax unit on the receiving end is compatible with the facsimile device in his or her office. Fax machines must be compatible if they are to communicate with each other. Vendors of fax units are moving in the direction of standardizing their machines so that a variety of brands will be able to communicate with one another.

Computer-Based Electronic Mail

The heart of electronic mail technology will ultimately involve computer-based message systems. Office information systems are moving toward a "terminal-oriented" environment. Here word processors, personal computers, computerphones, and multifunction information processors are the universal office tools. They will perform a variety of tasks. Computer-based mail is based on store-and-forward message systems. It uses the same technological underpinnings as computer-to-computer communications.

Process and Components of Computer Communications

Basically, computer-based message systems provide a means to transmit data from one terminal to another through a combination of devices. To communicate through a computer three components are necessary: a telephone line, a modem, and communications software.

Modem. A *modem* is a device that translates the stream of characters (the computer's signals) into analog signals (Figure 12-5). These analog signals then travel over the com-

Figure 12-4. This telecopier can send or receive text and graphics at a speed of 30 seconds per page.

Source: Courtesy of Xerox Corporation.

munication channel (your telephone line) to a modem at a remote location. This second modem translates (or demodulates) the analog signals back into data. The data are then transferred into signals that the receiving computer can understand. This is a process of modulation and demodulation, hence the modem's name (modulate/demodulate).

A modem is a key peripheral in computer communications and electronic mail. A number of vendors are currently working on developing faster modems by putting them on integrated circuit chips. Modems for personal computers are available in speeds of 300-, 1200-, or 2400-baud rates. One baud is 1 bit per second. A 300-baud rate would send information at 300 bits per second, or roughly 30 characters per second. In telecommunications, a blank space counts as a character. A 1200-baud modem operates four times as fast. It sends about 120 characters per second and costs four times as much. The speed not only cuts the time needed, it lowers "connect charges" from the phone company and from information utilities such as Dow Jones and The Source. A number of 2400-, 4800-, 9600-baud and higher speed modems already are available.

With the cost of using public telephone lines at a premium and sales for dial-up modems showing a 70 percent increase between 1982 and 1987,[2] the need for speedier, more efficient products is imminent.

The modem plugs into the *serial port* of a computer. A serial port is a socket. If a computer does not have a serial port, it may have *expansion slots.* An expansion slot is the portion of a computer that provides insertion of *communication cards* or "comm cards." These are small circuit boards containing microprocessor chips that provide processing power, memory, and intelligence to a computer.

Communication Software. Communication software is to hardware what gasoline is to an automobile. Without software, a computer is only an inarticulate mass of electronic parts. Then, all you can do with most modems is call or answer a number and perhaps view

---

[2]Dataquest Corporation.

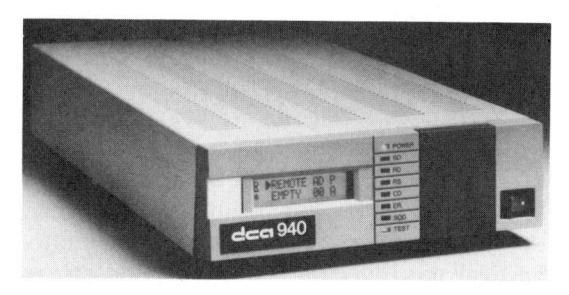

Figure 12-5. This modem operates over unconditioned phone lines.

Source: Courtesy of Digital Communications Associates.

data. You cannot capture and store information. The computer is basically dumb. Software, the programs that come on floppy vinyl disks, instruct the machine to carry out the commands given to it. Data communication software is therefore crucial and must be considered when you plan for a complete data communication system.

Telecommunication software and a modem allow you to connect your computer to the phone lines. You can then call distant data bases which are virtual libraries of information.

There is no shortage of selection when it comes to communications software. Over one hundred different packages are designed to link PCs with mainframe computers, local networks, electronic bulletin boards, and commercial data bases. Other software programs provide automatic electronic mail capabilities. Still others allow personal computers to combine several applications.

Communicating computers and word processors must "speak" to one another using a common set of rules. Modern transmission *protocol* must be used in establishing a computer-based link. A *protocol* is a set of rules that establishes how computers speak with one another. It spells out how and at what speed they will "talk," and most importantly, how they will ensure error-free transmission.

Transmission Lines. The selection of transmission lines and the mode of transmission are two important factors. Four basic transmission line groups service modems: narrowband, voice grade, wideband, and limited distance. There is no need to discuss the intricacies of each one; but it is important to know that the wider the bandwidth of the line, the faster the transmission speed. Therefore, narrowband lines are on the low end of the speed spectrum. They are limited to a top speed of 300 bits per second (bps). Wideband lines support speeds of 19,200 to 64,000 bps. The most common transmission line is voice grade. Voice grade lines support most modems used for business applications. They are ideal for public dial line data bank service. They accommodate speeds of 300 to 2400 bps.

Information flows through communication lines, which accommodate the direction in which information needs to be sent. The three types of communication lines are simplex, half-duplex, and full-duplex. On a *simplex* channel, the data stream moves in one direction only. In a *half-duplex* channel, information can flow in both directions, although not simultaneously. In a *full-duplex* channel, information can be transmitted in both directions at the same time.

There are two types of transmissions—synchronous and asynchronous. The difference between synchronous and asynchronous is in the manner of transmission. In the data stream, every individual character is made up of what are called *binary digits,* with each digit referred to as a *bit.* Each sequence of bits constitutes a *byte.* The most common kind of data communication used with PCs and communicating word processors is asynchronous. *Asynchronous* transmission is a mode that allows one character to be sent at a time with a single start/stop signal on either side of each character. A "start" and "stop" bit are sent down the line before and after each character of data, allowing the time interval between characters to vary. In *synchronous,* also referred to as *bisynchronous, mode,* data are sent down the line in a continuous stream of bits (blocks of 256 characters or more, with each block timed precisely before it is sent).

Third-Party Service. The acceptance of electronic mail as a useful business tool has enticed many leading companies into the business of third-party services, promising to deliver fast, cost-effective communications. Private ventures, such as MCI Mail, Western Union, CompuServe, The Source, and other public information utilities, are battling for a share of this lucrative market.

As office organizations grow, information overload and the lack of instantaneous communication become very real burdens. Once these conditions are apparent the need for electronic messaging is evident. An organization considering electronic mail may, if it has the resources, design its own system. It may buy and install a software package, or it may subscribe to a third-party service.

There are certain advantages to going with a third-party service. Few companies can justify the cost of building and operating the kind of worldwide network that some services offer. Companies with such resources generally prefer to put development costs into their own product lines. In addition, companies that design and install their own in-house systems must shoulder costs to stay abreast of the technology.

Office systems analysts believe that the future will see a melding of the pubic service agencies with in-house electronic mail systems. By providing access to mailboxes—within and/or outside the company—the technology can effectively create widespread coverage.

New Technologies—Expanding the Telecommunications Umbrella

Over the last several years, the communications industry has responded to business needs with a steady stream of high quality and reliable products and services. They spawned new technologies of their own. In addition to advanced telephone systems and electronic mail, a number of new communications technologies have been introduced. This section will summarize the many varied products and technologies that offer today's information manager excellent choices in fast and efficient communication.

Pagers

Radio paging is one of the most efficient means of sending and receiving information from person to person or to a group of people

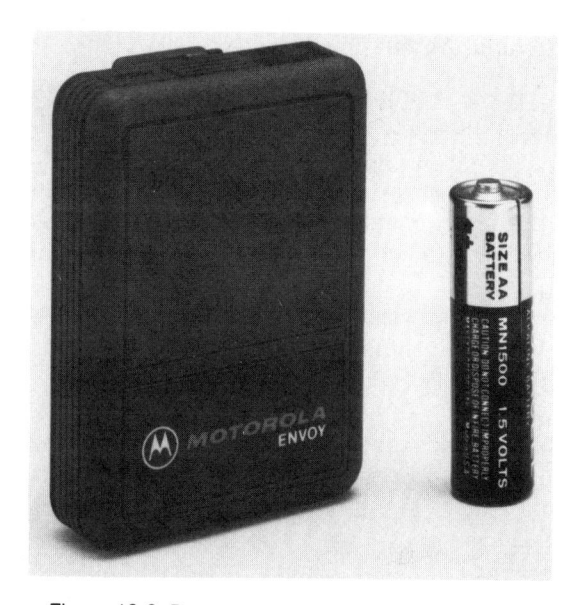

Figure 12-6. Pager.

Source: Courtesy of Motorola, Inc.

(Figure 12-6). Pagers, those little devices attached to a purse or a belt, are becoming a familiar sight. Doctors are not the only ones to use them: Company presidents and service technicians carry them, as do accountants and insurance adjustors. They can save steps, attract attention, and cause things to happen.

When you need to contact someone to give her information, you can send a signal directly to her pager. The pager will then alert the individual that she is needed. The alert tone or tones can have predetermined meanings such as "call your supervisor" or "call the operator." The alert tone can even be followed by a voice message. To avoid disturbing others, the alert tone could be replaced by a silent vibration.

Radio paging not only has the capability of sending a private message but is practically instantaneous. If you have personnel on call at home in the evenings or on the weekends, they don't have to stay at home to be near the telephone. They can take their pager with them wherever they go and still be contacted if necessary.

Cellular Mobile Phones

Cellular technology was developed by Bell Laboratories to meet the huge demand for efficient, high-quality mobile communication. The technology in mobile telephone communication is equal in quality to telephone calls placed in the home or office. A cellular system (Figure 12-7) represents one of the most exciting new services offered to business professionals. Eventually, it will be offered to the mass market.

Car phones have been available for years, but the old transmission severely limited the number of subscribers. Cellular radio uses computers to make better use of radio frequencies, thus opening up thousands of new mobile links. Through highly efficient use and reuse of the radio spectrum, cellular telephones permit thousands of simultaneous conversations to take place in a given geographic area, where earlier only hundreds could be handled.

In terms of operation, an area to be served by the new system is divided into many small geographical grid units (Figure 12-8). They are called "cells," each with a radius of about 8 miles. Within each cell a low-powered radio transmitter-receiver unit carries the calls over an antenna system for as many as seventy-two radio channels. A sophisticated computer-controlled call-switching system that is centrally located in the area controls the transmitter-receiver in each cell. It performs the switching task.

As the customer drives from one cell to another, sophisticated electronic equipment transfers or "hands off" the call to another cell site. This permits highly efficient use and reuse of the same radio spectrum.

As the demand for cellular services increases, the capacity of the system will be expanded through "cell splitting." This key feature of cellular technology adds new cell sites to existing ones and greatly increases the number of radio channels available to mobile telephone users.

Eventually, cellular mobile systems will be compatible with cellular systems in other regional Bell operating companies. When nationwide compatibility is achieved, customers will have access to mobile communications as they travel across the country.

Electronic Voice Messaging

The telephone is still a universal tool, but as a time waster, it has few equals. Trivial calls disrupt the workday. Group activities often involve separate calls to the individuals involved. There is the constant problem of "telephone tag," whereby two people return calls to each other repeatedly without making direct contact. In fact, many studies have proven that nearly 75 percent of business calls are not completed on the first attempt.

Help is on the way. A new breed of communications tool known as *electronic-voice messaging* (EVM), also called *voice mail* or *voice store-and-forward,* will bring practical solutions to these problems. Electronic voice messaging is similar to its text counterpart,

Figure 12-7. A cellular phone transforms an automobile into an office.

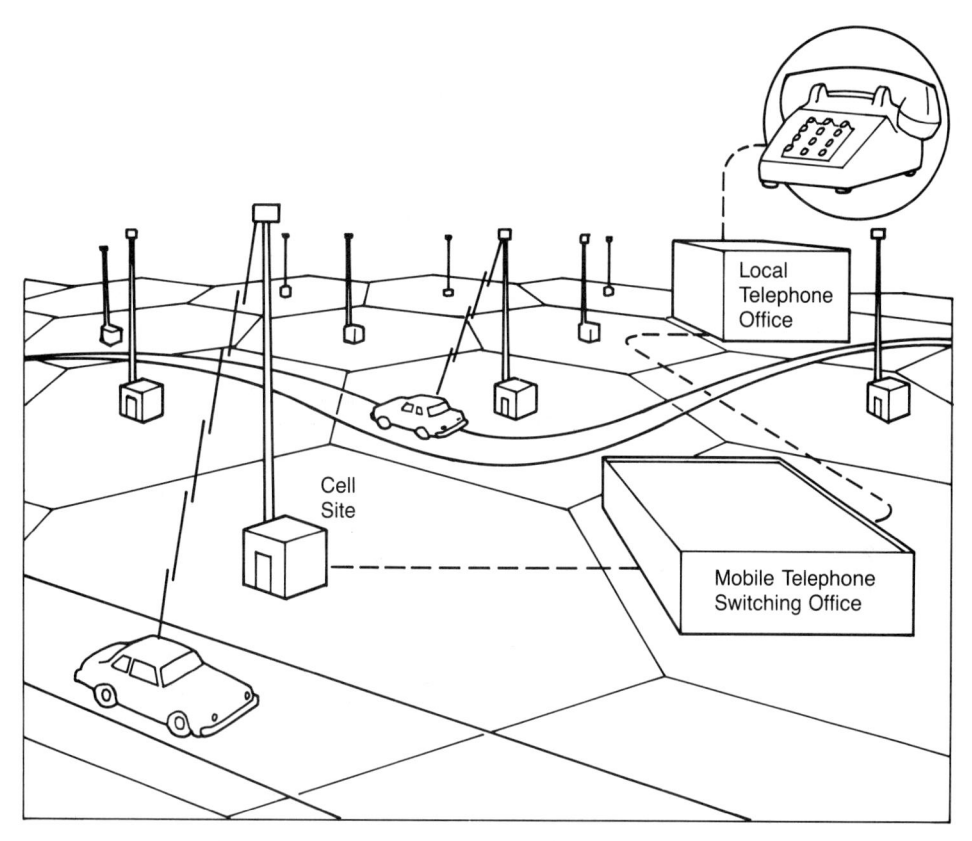

Figure 12-8. A transmitter in each cell relays telephone calls to cellular phones. As the phone travels from one hexagonal cell to another, it is "handed off" from one transmitter to another.

electronic mail, because the messages are stored in digital form for convenient delivery at a later time. However, with EVM the user simply dictates the message over the telephone instead of keyboarding it. The natural voice is then converted from analog to digital and stored on a disk or another storage device.

Each user of the system has a "mailbox" that stores voice messages from other users. To retrieve their messages, users simply call the system from any touch tone telephone. In addition, after hearing a message, a user can send his answer into the phone immediately, and the system automatically delivers the reply to the original caller. Unlike a telephone answering machine, a voice messaging system can take thousands of messages of unlimited length without garbling or losing them.

Electronic voice messaging is more than a mere cure for telephone tag or a fancy telephone-answering machine. It is, in fact, an exciting new way for business people to communicate dependably.

Local Area Networks

A *local area network* (LAN) is a system for moving information between devices located on the same premises (Figure 12-9). To expand this definition, *information* refers to data, voice, text, graphics, or image, and *devices* to large computers, personal computers,

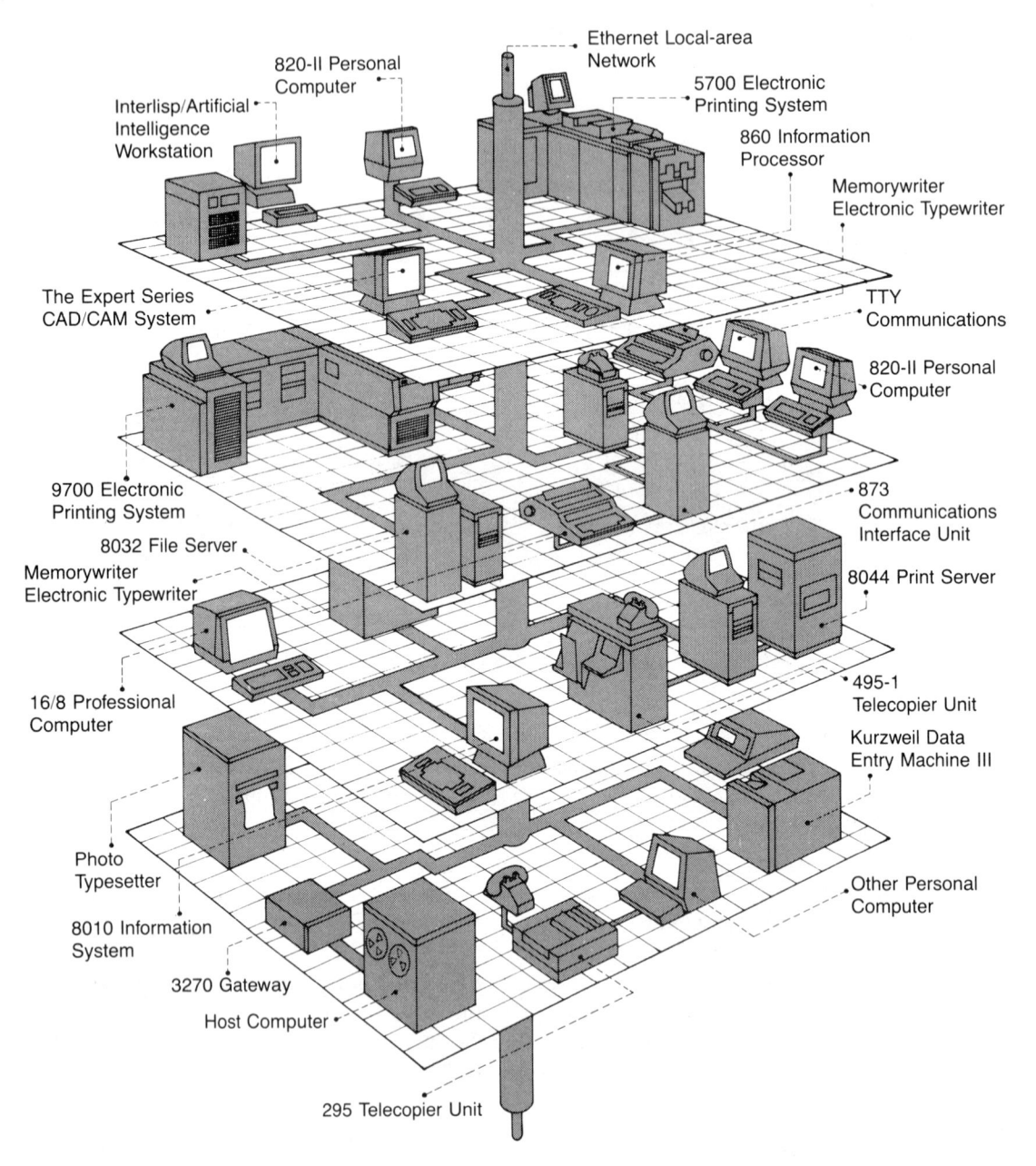

Figure 12-9. The stations in a building that make up one local area network.
Source: Courtesy of Xerox Corporation.

or other workstations, printers, telephones, and other peripherals. Finally, *same premises* includes the office building, plant, hospital, campus, or other geographically limited area. Local area networks are finding a place in organizations that are acquiring substantial numbers of personal computers. They serve as connective links and provide local information, which they exchange and store quickly and efficiently.

As more workstations and personal computers are added to improve productivity, the need to share information through the LAN link becomes even more crucial. In addition to linking personal computers, LANs can be set up to exchange information and share resources of large data bases. Thus, users connected to a LAN can have access to a large mainframe. At the same time they can enjoy a tremendous amount of flexibility.

Basically, a local area network consists of the following:

- *Hardware:* Made up of workstations and peripherals
- *Network interface:* An expansion card that plugs into the hardware
- *Central computer:* A central processing unit that houses the hardware, software, and operating system of the network (may be a microchip on the expansion card or a hard disk drive)
- *Network server:* Usually a hard disk drive that carries both the software and the programs available to network users
- *Cables:* Linking all the components

Network Configurations

Four main network configurations are used in LANs (see also Figure 12-10):

1. Bus. This is a network system in which stations are attached to a single transmission medium so that all stations hear all transmissions. This system has defined end points that can be added to. All devices are connected directly to the main bus line.

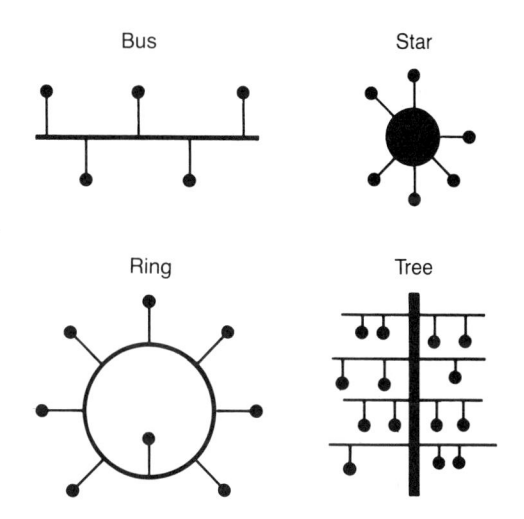

Figure 12-10. Network topologies.

2. Star. A star configuration connects end-users' workstations point-to-point to a master computer or controller. The end points can be intelligent or unintelligent (dumb) terminals.

3. Ring. The ring configuration forms a continuous path without defined ends. Signals pass around the ring and are pushed along by each device as they pass.

4. Tree. The tree configuration offers an expanded version of the bus configuration. By turning the bus to a vertical position and extending its branches, we now have a tree.

The intelligence of the network is in the central computer. This computer acts as a "traffic cop" to control the flow of information among these workstations. For example, the central computer may hold up data that one computer is sending to a printer because another computer on the network has already asked for printer time.

An efficient LAN configuration allows information to be exchanged among individual workstations directly, without going through the traffic cop (central computer) and thus eliminates the potential bottleneck.

As the number of personal computers grows, the ability to access information through corporate local area networks will enable workers to better analyze business problems. In addition, the LAN provides a means to distribute this information to other workers electronically so the business decision cycle can be shortened.

Teleconferencing

Today thousands of information executives are faced with the constant demand of trying to be in two places at the same time. If a corporate headquarters team wants to communicate with the sales staff of a branch office across town or across the country, the team can confer through the technology of teleconferencing (Figure 12-11). Teleconferencing is part of the telecommunication umbrella. The word has taken on a generic meaning, embracing a variety of technologies. The prefix *tele-,* from Greek, means "far." Teleconferencing refers to any electronic method used by people in at least two different locations to conduct a conference.

There are several different types of teleconferencing. *Audioconferencing* is limited to voice only. It is similar to a conference call except that the participants at each location are generally grouped together in a room equipped with a telephone, an audio speaker, and at least one microphone.

Audiographics is audioconferencing augmented by some kind of graphics capabilities, such as an electronic blackboard or a facsimile machine. It adds an extra dimension, but the participants have to wait for the graphics to be transmitted.

Slow-scan or freeze-frame is a type of conferencing that uses telephone lines to transmit

Figure 12-11. Teleconferencing makes possible meetings with people at widely scattered locations.

Source: Courtesy of CLI.

a single picture to a remote location. This system uses hardware programmed to send new pictures at regular intervals, whether the image has changed or not.

Computer conferencing is really a series of messages sent to computers or computer terminals. The participants take part when they are available. That is, the participants can be in a computer conference but not necessarily at the same time. This is an advantage when time zones create a problem.

Full-motion videoconferencing is essentially a two-way video/two-way audio (word-and-picture) teleconference. Participants in two separate locations can both see and hear each other. These telecasts can be in full color and offer full-motion, slow-scan, or freeze-frame video. This is the most expensive form of teleconferencing and responds to the primary human senses—sight and sound. It conveys the feeling that all the participants are actually present in the same room.

Teleconferencing can save time for executives, managers, and staffs. If implemented correctly, it can be used as a corporate asset, increasing productivity and saving precious travel dollars.

Videotex and Other Home Information Services

Videotex and teletext are rapidly emerging as important forces in communications. If predictions are correct, the technologies could drastically affect the way corporations and consumers communicate.

Videotex (without the final *t*) is a two-way, interactive communication service (Figure 12-12). It offers both information and transactions via text and graphics. The information is transmitted via cable, radio, or microwave to a display terminal. It can be a television set equipped with a decoder, a dedicated videotex terminal, or a personal computer.

Teletext (with the final *t*) is a limited, one-way service delivered to the user's standard

Figure 12-12. Videotex frame creation terminal.

Source: Reproduced with permission of AT&T Corporate Archive.

television set or cable television service. It is equipped with a simple decoder.

Videotex has turned generic, and this new label applies to a wide range of home information retrieval systems. The difference between videotex and teletext is that videotex is two-way. This means that users can carry on "conversations" with data bases and thus be able to make business transactions immediately. Thus, a user of Dow Jones News Retrieval who sees that a stock has reached the right price level can send his or her broker a buy or sell order.

Videotex services are designed for home and business use. Services include home shopping and bill paying, directory services, weather, sports, telebanking, airline reservations services, financial and legal services, electronic newspaper and magazine delivery, and even electronic education programs.

For the consumer at home, this "electronic cottage" technology may change the way people shop, bank, work, and communicate. It may permit them to do all of these things without leaving their living rooms. For business organizations, videotex can mean increased productivity resulting from speedy access to current information.

Telecommuting

The telecommuting concept was first presented by Alvin Toffler in *The Third Wave*. With the rapid advance of tools and computer technology, he pointed out, workers would be able to stay at home to complete their designated tasks (Figure 12-13). Though Toffler makes a good argument for "electronic home work," other advocates of telecommuting and remote work believe the future holds more than just working at home. They anticipate neighborhood office clusters or satellite work centers that will replace large inner city office complexes housing virtually all of a company's equipment and employees. Telecommuting will include not only secretaries but CEOs. Through local area networks, computer terminals with printers will move to the homes of managers and executives. Those information processing specialists who opt for a telecommuting environment will enjoy a sense of efficiency and a more flexible work schedule. Gone will be expenses incurred from commuter fares, day care of children, lunches, and clothing requirements. Travel to business conferences will be replaced by a walk across the hall to the teleconference room. Paper communication will be replaced by electronic mail.

There is, however, another school of thought that dispels the virtues of telecommuting. Telecommuting may not be for everyone. We may have so-called electronic cottages, but some people may not be willing to stay at home and tap out messages to the office. There are those who like to hear other voices, ringing phones, and the chatter of machines while they work. Many workers enjoy having other people around to bounce ideas off, to discuss plans with, and to console and be consoled. Telecommuters may miss the social and professional interaction with coworkers that improves job performance.

Figure 12-13. Working at home.

Source: Courtesy of Harris Corporation.

Teleports, Office Parks, and Intelligent Buildings

An important link is emerging between office building development and modern telecommunications (Figure 12-14). Soaring energy costs and more sophisticated demanding tenants are prompting municipal governments and private developers to integrate modern technologies in the construction of new buildings. Progressive builders are starting to incorporate advanced telecommunication services and equipment, such as sophisticated telephone switchboards, satellite antennas, and data transmission cables, into their buildings and office parks, just as they do electricity and air conditioning.

Teleports are clusters of commercial and residential buildings containing satellite antennas, networks of underground fiber optic cables and other advanced telecommunication facilities. They are springing up along major transportation routes across the country. Water, rail, road, and aircraft movements were once vital factors in maintaining a port city's preeminence. In the coming century, the ease of information movement using telecommunications will be a major factor in revitalizing and nurturing a region's economic health.

Office parks and *office campuses* are other projects that will incorporate advanced telecommunications services into their buildings.

Intelligent buildings will combine aesthetics with the realm of new technologies. In intelligent or "smart design" buildings, all operations will be linked through a central, or host, computer. All building systems for environmental control, energy, lighting, fire safety, and security are centrally controlled to the extent that they almost run themselves.

Tenants will be able to tap into the central computer to take advantage of word and data processing services, electronic mail, facsimile, data storage, and other communications facilities through *shared tenant services,* a plan in

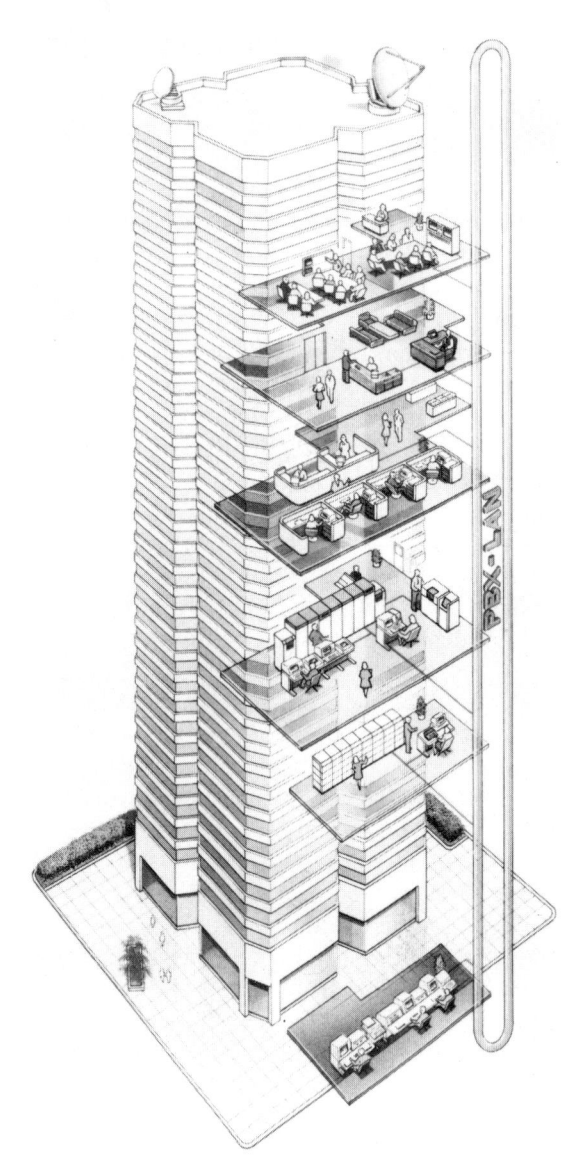

Figure 12-14. The intelligent building is a product of integrated computer and communications technology. With devices linked, information in all its forms—voice, data, text, image—is received and relayed automatically, not just between desks, departments, and divisions but also across town or across the seas.

Source: Courtesy of NEC Corporation.

which tenants pay for all of these services at substantially reduced prices.

Fiber Optics

Fiber optics is a remarkable technology that is reshaping long-distance telephone networks and will ultimately be the backbone of the U.S. communication system in the 1990s and beyond. Fiber optics are hair-thin filaments of transparent glass or plastic that use light instead of electricity to transmit voice, video, and data signals. Covered to prevent light loss along the line, the fibers are bundled together into a flexible cable.

These new fiber optic systems will be able to carry up to 400,000 light-borne calls simultaneously over a single cable less than an inch in diameter. This is ten times as many calls as today's densest microwave radio route.

No longer a laboratory curiosity with specialized telephone applications, fiber optics has burst into the real world. It offers unique capabilities at a good price. Companies are racing to install new fiber optic networks to meet the demand for applications such as video teleconferencing, facsimile, and videotex services.

Technical advances have been stunning from the standpoint of speed of transmission and capacity of data. For example, Bell Labs has demonstrated that it can transmit a gigabit (billion bytes) of information in one second—enough to connect three million home computers, 7000 two-way voice circuits, or 300 two-way video conferences—all from a single light source on a single fiber.

Satellite Communications

The most awesome network technology is satellite transmission (Figure 12-15). The recent and dramatic increase in the use of various kinds of communication systems has created a need for these sophisticated communication links. Since 1972, several vendors have launched satellites to compete with AT&T's

Figure 12-15. A satellite antenna (dish) picks up the communications relayed from space.

Source: Courtesy of Andrew Corporation.

terrestrial lines. Once launched, a satellite can stay in orbit for years, relaying phone calls, TV programs, or computer messages.

Basically the satellite functions as a microwave tower as it orbits 22,300 miles above the equator. It travels at exactly the same speed as the rotation of the earth. Satellites are being used more and more for business communications because they can handle voice, text, data, and video transmissions from all kinds of automated offices. Because of their very high altitudes, they can receive radio beams from any location in the country and reflect them back to a large portion of the world.

Many of the newer satellite systems will use new technology. They will have higher frequencies, which overcome obstacles that block line-of-sight transmission such as mountains and the curvature of the earth. As technology evolves, antennas (satellite dishes) are shrinking in diameter and price and becoming more common on rooftops and backyards. A Frost & Sullivan, Inc., study estimates sales of "earth stations" will increase to $3.7 billion in 1990 from $1.8 billion in 1985.

Summary

The theme of the integration of computers and communications has been interwoven into many chapters and sections of this book. This integration is changing the perception of office automation. Telecommunications has been thrust into the spotlight because of technology and deregulation. There is also a growing sophistication among top managers about the need to link groups of workers and provide them with devices and conduits for sending and sharing information.

This chapter examines the pattern of this revolution and highlights some of the opportunities the new environment will bring.

Much of this change is taking place within-forward-thinking corporations. Digital switches are replacing mechanical contraptions, millions of lines of software codes are programmed into computers, and copper wires are being dismantled in favor of microwave radio and fiber optics.

A proliferation of local area networks coupled with data bases will provide powerful tools for accessing vital information. Satellite communications offer a rich array of voice and data services to business users. Advanced digital PBXs and key telephone systems will be able to handle unprecedented numbers of calls, switch data, and talk directly to other remote switches. New data communications terminals are available for hooking computers and terminals to telephone lines.

Advances in local area networks and computer networks are now supported by a growing number of compatible devices. The ultimate network of the future, the Integrated Services Digital Networks (ISDN) will achieve a much sought-after goal. It will provide a standard interface into which computers, telephones, copiers, or any electronic device may be plugged.

Review Questions

1. How are office automation vendors responding to the increased demand for communications capabilities?
2. What single event accelerated the divestiture decision?
3. Describe the broad concept of telecommunications.
4. Briefly trace the growth of PBX technology.
5. What advantages does digital PBX bring to earlier PBX systems?
6. What is equal access? How did this provision help non-Bell long-distance carriers?
7. Why is reliability (stability) of vendors important in selecting a telephone system?
8. Define electronic mail.
9. Why is facsimile different from any of the other forms of electronic mail?
10. Briefly describe the function of a modem.
11. What is the unit of measure of data transmission? How does speed of transmission affect the cost of using public telephone lines?
12. Compare simplex, half, duplex, and full-duplex communication lines.
13. How can radio pagers alert users of a message without disturbing others in public places?
14. Describe the way cellular technology improves the quality of transmission compared to traditional mobile telephone service.
15. In addition to the applications for fiber optics cited in the text, can you think of other telecommunication uses for fiber optics?
16. If teleconferencing has the advantages of reduced travel time and expense, why has not this technology become widespread among American business corporations?

17. Which technology (videotex or teletext) is more appropriate for business applications? Explain your answer.
18. Why is telecommuting not for everyone?
19. What inducements do developers offer potential tenants to locate in intelligent buildings?
20. Describe satellite communications.

Projects

1. Research the regional Bell Operating Company that serves your area. Prepare a short report that includes its name, the names of its officers, services and products it offers, and organizational chart.
2. Describe the telephone system used by your college or employer.
3. Your neighbor has purchased a personal computer. He asks your help in setting up his computer for telecommunication capability. What should you tell him?
4. Your local board of education has jurisdiction over four senior high schools, three junior high schools, and seven elementary schools. Typical applications include documents sent to the community, and parents, as well as others concerning fund raising and social activities. Intraschool messages are typical school administrative documents. The board is considering installing an electronic mail system. What suggestions would you propose?
5. Find a current article on satellite communications. Write a summary of that article. Be prepared to give an oral report on the article in class.

CHAPTER 13

Computers and Communications: The Micro-Mainframe Link

After reading this chapter, you will understand

1. The concept of the micro-mainframe link
2. Applications and advantages of this link
3. Terminology such as *data, transactional,* and *operational integration*
4. Components and connecting links relating to networks and micro-mainframe connections

Expanding Information Resources

Personal computers have lived up to most of their claims. They are proficient tools to manipulate information. But creating and processing information at a single workstation does not utilize the full potential of interactive computing. Managers at all levels in the corporation are making business decisions in an increasingly competitive environment. They must gather the information required for their decision making. Personal computers provide these managers with an invaluable information analysis tool, but the computers are limited to the information stored within their memories. Managers need to tap other sources of information.

This need is the essence of information transfer. It is the ability to load corporate data into a personal computer, use the data for decision making, and then analyze the alternatives. Indeed, information stored on a corporate mainframe computer is essential to a manager's analysis, but it is often difficult or impossible to access. In addition to the information stored in the corporate mainframe computer, other information critical to corporate decision making is being generated on desktop computers scattered throughout the organization. A consistent and secure method is needed to channel this information when and where it is needed.

In most large organizations, corporate data reside in one or several large computer systems in a variety of formats. To obtain these data, information workers must ask mainframe programmers to extract information from one or more sources, including various corporate data bases. The process is expensive, time consuming, and hopelessly backlogged. Think of how a network of personal computers in the corporation, all having access to the corporate mainframe, could speed up interoffice communications.

The Micro-Mainframe Link

The need to expand computer applications and tap resources beyond the personal workstation is compelling to the corporate user. Desktop computers could work in isolation for just so long. A link had to be created between individual workstations and mainframe computers to share and distribute information quickly.

This concept, known as the *micro-mainframe link,* has captured the imagination of those who reasoned that if using a personal computer with 128K of main memory and 10 Mbytes of hard disk storage is good, why not get access to a mainframe? (*K* represents 1,024 bytes; 128K is approximately 128,000 characters or bytes of information. *M* represents a million; 10 Mbytes is approximately 10 million characters or bytes of information.)

Microcomputers and mainframes are two distinct ends of the communications link. Micros cannot take the place of mainframes, and mainframes cannot perform the tasks of microcomputers. Particularly in large organizations, the mainframe will continue to play a vital role simply because it is the system that controls the corporate data base. Mainframes have tremendous capacities to store information that can run into the *gigabytes* (one gigabyte or Gbyte is equal to 1 billion bytes of information). But mass storage may not be the only reason to link terminals to large computers.

Applications and Advantages of the Mainframe Connections

1. Electronic messages can be dispatched through the mainframe with a distribution list transmitted throughout the organization.

2. Vital information, such as corporate files, data bases, customer information, prod-

uct specifications, and sales figures, can be accessed instantly. Specific examples of their use include office workers' accessing and retrieving financial data to feed a spreadsheet application and customer lists from a corporate data base used to generate a mass mailing.

3. Standard application programs can be downloaded, copied, and run on the executive's micro. *Downloading* is the accessing or retrieving of a document or program from a large computer (mainframe) to a smaller computer (PC) and then copying it onto a disk. This procedure provides the end user with a program at his own disposal without accessing the mainframe each time.

4. Another important use is the need to upload data captured or revised on the micro to the mainframe for further analysis with mainframe tools. *Uploading* is a process of accessing or retrieving a document from the mainframe, revising it on the micro and sending it back to the mainframe as updated information. A growing need for uploading is in using the mainframe as a *storage* medium for the micro. This is useful for micro users, as they can use specialized mainframe peripherals, such as laser printers.

5. Moving data from micro to mainframe may not involve the corporate data base at all. It may be a function of using the most cost-effective storage technology.

Thus, the advantages of using a micro-mainframe connection include the following:

Information access
Information control
Information transfer
Information storage

Degree of Accessing Mainframe Files

Once the idea of a micro-mainframe link is accepted by corporate management-information systems (MIS) personnel, the amount of mobility to grant the microcomputer user within the files and programs of the mainframe must be decided. The desirability of this linking may be justified, but it is also fraught with peril. The keepers of the mainframe, the data processing people, have been accused of parochial thinking when it comes to their computers. That may be a justifiable accusation. There is a reason for the parochial thinking—a lot is at stake. Data bases, records, files, and confidential information have been created and filed at great expense and effort. Mistakes can be made, causing the deletion, change, or destruction of these records. The magnitude of such mistakes, even when there is no outside connection, can be catastrophic.

The data processing people are justified in their concern about "opening up" their computers to people who may know little or nothing about large computers. Unrestricted mobility within the mainframe is the PC user's goal. However, the mainframe people certainly cannot permit their satellite PCs to go roaming through their files and programs without controls and guidelines.

Controlling the Use of Mainframe Computers

In order to establish effective guidelines for the use of mainframe computers, the following concepts should be understood.

Accessing. Accessing allows microcomputer users to link up with the mainframe to read data only. A passive function, it can be a problem if you do not want to allow all information to be accessed (for example, personnel files, financial information, and product research).

Modification and Updating. Modification and updating mobility is a major problem. The personal computer is a potential source of constant surprise to the mainframe because it can perform significant processing outside the mainframe's scope of awareness. Access to modify or "write" into files should be restricted to authorized persons. *Writing* in files is the ability to record data on a storage medium. *Write-access* is the ability to access with a file, alter, record, or store new information in the file. These conditions imply the need for a strict level of communication security. A system should be set up in the mainframe for a user file area. This would allow personal computers access to those file areas only, with both read and write access.

Security. Protective measures should be established to prevent unauthorized personnel from accessing information. Passwords should be established for this purpose. Guidelines for security measures are discussed in Chapter 18.

Data Integration

The pursuit of true integration must involve a software system that links an organization's support system uniformly. It must be multidimensional and capable of changing as the business requirements change. True integration therefore must include data integration, transactional integration, and operational integration.

Data integration among different software application systems provides an organization with a common set of facts. All information about related business activities is stored in the corporate database. It is updated as events affecting these data occur. For example, sales, manufacturing, research, engineering, and accounting all use the same data to make decisions about their particular roles: selling, manufacturing, designing, or purchasing of any given product.

Without a uniform system of data integration, each department within an organization will have its own system or a separate data base to make decisions. The problem with this method is that the information is rarely coordinated. No one department really knows what the "facts" are. The only way to develop any reliable information is to integrate fully all the data about the organization.

Data integration requires the creation of a unified and common set of facts. The facts should be stored in subject data bases that contain all the information about a given subject. Customers, for example, could be a subject. The common data for customers could be stored in a subject data base and made available to any system or user requiring it. Customer data would then be accessed by age, geographical location, account, or other information. Any person in the organization interested in a particular customer would have access to the same data profile. Each department (sales, accounts receivable, customer relations) would be in constant communication, via the data base to make effective business decisions.

Transactional integration ties together all the events or transactions that occur in these systems. It is an on going update system that records events in the data base. It communicates the changes to other systems or users who, in turn, use the data to plan their business activities. The coordination and sharing of events make transactional integration a useful and necessary tool for managers and personnel in different departments.

Operational integration pertains to the sharing of information in one system by programs or screens in another. This is closely related to selecting the appropriate software to implement the desired integration.

Whether implementing data integration or transactional integration, a complete set of *complementary software,* designed to support the needs of every level of the organization, is the only practical way to achieve full integration.

Application software (discussed in more detail in Chapter 14) is changing to accommodate the levels of integrated information systems. A whole new generation of application software is emerging, along with a generation of software companies. Progressive software vendors are now creating major packaged application systems designed to support truly integrated systems. In addition, the software houses that built the leading data base management systems (the foundation for all applications) are now devoting their resources toward creating and marketing this new software for an integrated information environment.

The Convergence of Computing and Communicating Technologies

More and more devices are becoming available to improve the transfer of information from one location to another. Personal computers, multipurpose executive workstations, and high-speed output devices have improved the productivity of the modern information worker. The potential exists for greater productivity by working with tools that provide a significant amount of integrated functions. The tools should accept a common compatible information base and eliminate the need for manual conversion. These tools should harmonize, analyze, and synergize. The best resource for the automated office is people, not machines. People can analyze and think. They have the capacity to integrate their thought processes, solve problems, and communicate with one another. We are now trying to design our machines to perform in a similar manner.

The trends discussed in this chapter indicate that office systems managers are under pressure to move away from the stand-alone, mainframe-based computing environment and develop integrated network systems. A major stumbling block has been in finding a means to link together what often is a communications tangle. This can best be described as an installed base of personal computers, workstation terminals, and mainframes that require different methods, or protocols, to communicate with one other.

The hardware and software exists to make integration and the interconnection of office systems a reality. Linking micros to mainframes and installing local area networks are the key factors for achieving true integration.

At present the market is rich in plug-in adapter boards and related software (Figure 13-1). They help establish communication networks and solve the problem of *multiple protocol* (different machines communicating with one another). There is also a growing need for manufacturers of hardware and software to work closely with information systems managers to solve particular communication problems. Progressive vendors are finally offering communications products and tailored services aimed at this new market for network communications.

Electronic Connective Tissues: The Missing Link in Network Communications

Although office systems analysts predict that online terminals will soon be used by nearly every office worker, this link-up may not be as easy as it seems.

The personal computers and universal workstations now being installed usually do not "talk" to each other; nor do they communicate directly with the terminals and mainframe an organization already has installed.

The lack of standards has frustrated computer users, who want to connect all the computers in their offices and factories into networks to exchange information. Many analysts and executives think that such lack of

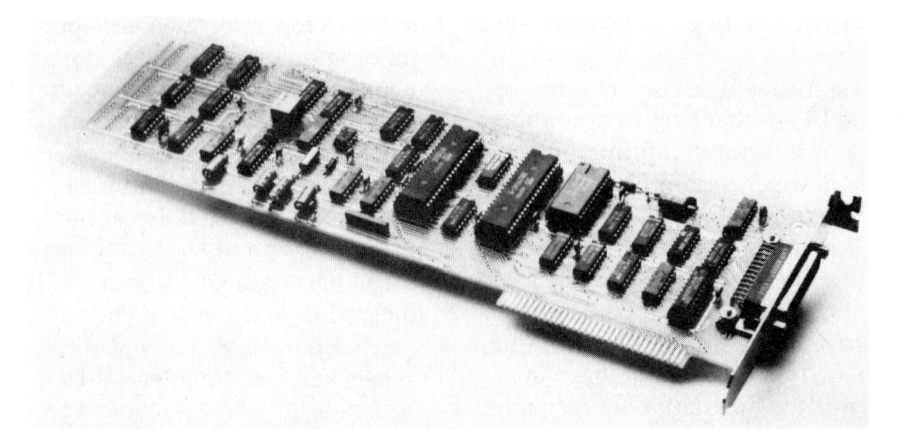

Figure 13-1. The IRMAcom plug-in adapter board allows the IBM PC and compatible PCs to attach to IBM 3270 networks.

Source: Courtesy of Digital Communications Associates.

connectivity has contributed to the slowdown in the growth of computer sales in the latter half of the 1980s.

Some organizations, such as General Motors, have designed their own standards to achieve compatibility for connecting computers, machine tools, robots, and other electronic gear in their factories. They have developed what they call *MAP,* manufacturing automation protocol.

Because the need to develop industrywide standards is crucial, the nation's leading computer companies are moving toward developing standards that will allow machines made by different manufacturers to communicate with one another and share information.

Major vendors are banding together to form a new nonprofit organization that will specify standards and test for compliance with them. The new group is called the *Corporation for Open Systems* (COS).

The greatest challenge to managers is integrating the new devices into the organization's existing communication environment at optimum cost.

Products are being designed for the most productive transmission networks. They will integrate components for voice, data, and image applications. Users can look forward to applying the hardware and software systems to their present and future business needs.

Office system designers plan to have a larger role in communications. It's no longer a luxury option for word processors to communicate with one another. Now, most systems can also emulate computer terminals and run computer software under standard operating systems. *Emulation* is a process by which one computer system is made to function like another, in order to accept the same kind of data, execute the same programs, and achieve the same kind of results. Today, more than 150 terminal emulation programs of software and hardware are on the market and still more are being planned. A terminal emulation program makes the micro act as if it were a "standard" or popular terminal. Even IBM has begun to emulate its massive office products line so devices can ship documents to one another.

Microcomputers dispersed throughout an organization are able to tie into larger computer networks. Through emulation, for example, the IBM PC and other models can be linked into mainframes. An example is the 5520 Administration System: It emulates the

IBM 3270 inquiry response terminal. Full-featured word processing systems have added the capability for its boxes to hook into IBM mainframes by emulating IBM data terminals.

Protocol converters are another method of connecting devices. Their primary function is to support communication among terminals and computers that use different character codes. Essentially, the converter changes the sending device's code.

Protocol conversion and emulation technology are similar. The two methods deal with changes in code and transmission protocol.

Gateways and Other Connecting Links

With these examples of new technologies, organizations are a step closer to meeting the information integration challenge. The next consideration is a means of linking all these discrete but cooperating elements across the barriers of multiple data types, file formats, and data locations. The following devices will accomplish this.

A *gateway* is a software tool that interconnects users (terminal operators, applications, and so on) (Figure 13-2). It functions as a gatekeeper and a translator for an information source. As a gatekeeper, the gateway is highly intelligent with respect to both the location and the data structures of its source. It understands the unique file structures, data types, access methods, and so forth.

Through gateways, companies will now be able to interconnect local area networks (LANs) with stations of different architectures from different vendors over multiple media. Gateways provide the solution for the LAN-to-LAN interconnection.

The device for transporting information to and from the gateway is the data bus. A system would include several gateways, one for each information source, all feeding into the data bus. The data bus is similar to a hardware bus (see Chapter 8).

Servers

As groups of individuals within departments realize the potential of office automation, they discover that by sharing resources such as printers, storage, and software, they are more efficient as a group.

Servers link workstations and printers together. This type of device merges the concepts of communication and computing into a single entity. Servers are used as a connect-

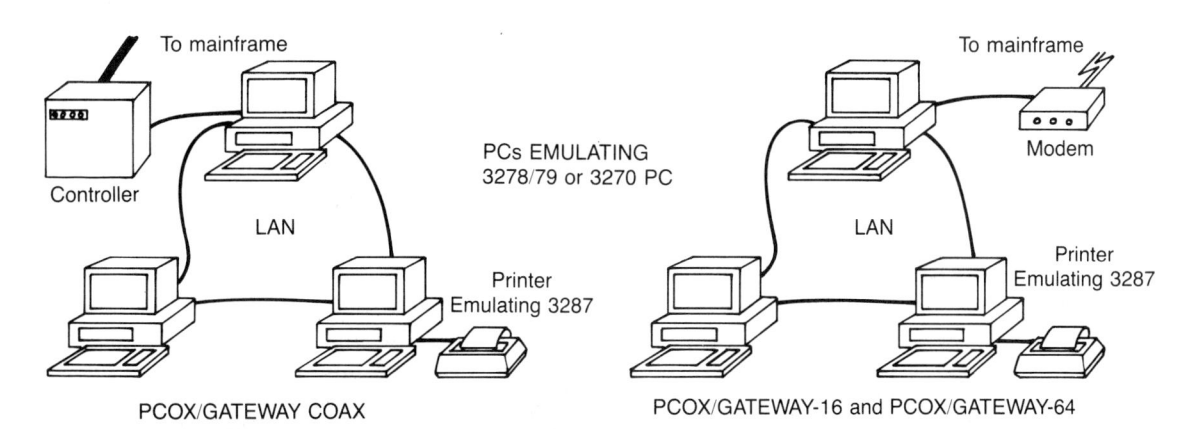

Figure 13-2. These gateways interconnect LANs and micros to mainframes.

Source: Used with permission from CXI, Inc.

ing link in local area networking. There are various forms of servers:

1. *Communication server:* In this type of device, the LAN may access public or global networks. Communication servers may also connect two LANs, similar to a gateway or bridge.
2. *Print server:* This type of server enables workstations to share printers.
3. *File server:* File servers provide a large amount of disk storage which can be made available to individual workstations. The file server contains a directory of files and a security system so that only authorized individuals linked to the network can access the files.

The *router* is another device to interconnect LANs by using an internetwork protocol. Like gateways, routers must be visible to the user stations. However, unlike gateways, routers are used to interconnect computers with similar architecture.

The X.400 Plan

Proprietary systems to connect one machine or mail service, such as MCI Mail, to another are beginning to appear on the marketplace. But the widespread interconnection of these electronic mail islands will only come with standardization. Soon office automation managers will be able to interconnect their messaging systems through X.400-compatible products. X.400 is the message interconnection standard supported by virtually every computer manufacturer. Once X.400 products begin appearing, the long-predicted global interconnection of private corporate electronic mail systems and public subscriber-based E-mail services will become a reality.

The X.400 series is the vehicle that will move information from one company to another. The standard provides the necessary protocols and technical specifications for the interconnection of separate computer-based message handling systems on a potentially global scale, according to its creators at the Consultative Committee on International Telephony and Telegraphy (CCITT), a United Nations–sponsored group that recommends telecommunications standards.

As a result of these efforts, during the 1990s one worker's computer system will be interconnected with virtually every other computer screen on the planet, making the task of sending an electronic letter, a telex, a facsimile, or a voice message as easy as international dialing is today.

The combination of these connecting devices makes possible the integration of data from many locations and in a variety of formats. The key to integrating devices from different vendors within a network is to achieve a degree of compatibility. Networks, gateways, and other links are a cost-effective way to expand a data base and share resources.

Summary

Microcomputers are fueling the demand for access to data stored in the corporate mainframe. In an ideal office environment, desktop terminals would provide corporate users with autonomous processing power as well as a friendly link to the resources of larger computers. One way to provide decision makers with this information is to give them direct access to the data base. Linking PCs and other desktop microcomputers to management-information system (MIS) mainframes is the micro-mainframe link.

Users could download information from the mainframe to their machines. *Downloading* refers to accessing data from a large computer to a smaller computer and then copying it onto a disk. In order to use a data base stored within a corporate mainframe computer effectively, data integration must be established. *Data integration* among different software applications will provide an organization with a uniform system of facts stored in a corporate data base.

This chapter describes some of methods and devices necessary to establish a micro-mainframe link. They include terminal emulation, protocol conversion, file transfer, integrated software, and network integration.

The perfect micro-to-mainframe link may not exist yet, but many of its elements currently do. Effective terminal emulation is available from many vendors. Building from that base, more complete integration is starting to occur with the advent of sophisticated file transfer methods.

Review Questions

1. Briefly describe the concept of the micro-mainframe link.
2. What is the difference between downloading and uploading?
3. Compare transactional integration with operational integration.
4. What steps have computer vendors taken to develop industrywide standards?
5. Define emulation; protocol.
6. List and briefly describe three types of servers.
7. In developing a micro-mainframe link that allows individual users to access information, what precautions should be observed?
8. Describe one method of linking incompatible workstations.
9. What is a gateway? A router?

Projects

1. A Health Maintenance Organization (HMO) was established in your community. As part of the committee to integrate office systems you have been assigned the selection of decision support tools to enable three end users (medical secretaries) to use their personal computers with maximum efficiency. The PCs are linked to a host computer in a micro-mainframe link. Typical applications include preparing patient medical records, histories, updating insurance forms, corresponding with other physicians, and accessing the GTE Medical Data Bank. The health center employs the services of five physicians, two X-ray technicians, and two registered nurses. What decision support software packages would you recommend? Justify your choice.

2. Research vendors that provide the hardware, software, and peripherals necessary to establish a micro-mainframe link. Create a matrix that compares the features of each vendor.

PART FOUR

Software and Storage: Components and Technology

CHAPTER 14

Software

After reading this chapter, you will understand

1. The variety and categories of software systems
2. The functions of systems software and utility software
3. Some of the applications for communication software
4. Expert systems and their applications
5. The variety of applications available within an integrated software package
6. Guidelines and precautions necessary for acquiring software

We have traced the cycle of information through input, processing, output, and storage. Each phase plays an important role in generating business, gaining a competitive edge, and improving operations and the bottom line. Information, then, is a *corporate asset.* The processing stages outlined how we can gather, process, and communicate this information among ourselves and to the outside world—processes vital to both daily activity and long-range planning.

For more than three decades, computers have been developing to the point where an organization of any size or any individual can use the tools of this technology. These tools allow people access to the right information in the right form and at the right time. Furthermore, they are accessible, easy to use, and inexpensive. Of all the advances in software, hardware, human resources, and communication, the most dramatic changes have been made in software. The software available today for either the personal computer or the large mainframe computer has many more applications and is sometimes more inexpensive than software available a decade ago.

Whether your organization is an industrial giant or a cottage industry, software is the key for creating, processing, and communicating information. Software ties together the other elements in information processing.

What Are Software Packages?

Software packages are the programs, or instructions written in computer language, that order the computer to perform specific tasks. The term was coined to contrast with hardware, the actual machinery of the computer system. Exotic computers are useless without software to make them work. A violin is just a beautiful object; but when Isaac Stern touched it with his genius, the result was brilliant music. A computer may be a beautiful piece of hardware, too; but it needs the genius of software to achieve a brilliant performance.

To say that a computer is useless without software (programs and data) is like saying that a book is useless without printing on its pages. The essence of a computer is the software author's information.

The Emergence of Software

Few people realized the potential of software during the early years of 1950–1965. People wrote programs to automate clerical tasks, to produce payrolls, and to keep track of airline seats. They perceived the function of software as a limited peripheral of computers.

The microcomputer industry began to form in the mid-1970s and since then has experienced rapid growth. Minicomputers and microcomputers increased. During the 1970s, however, the concepts and technology were heavily slanted toward the hardware. Software was designed to utilize the features of the hardware configuration and to establish the personality of the hardware. Because the software was *bundled,* or "built into," the hardware, early office automation buyers had to choose hardware to meet their specific application requirements. Their only alternative was to invest in costly customized programming. By 1980, millions of cars, microwave ovens, and copiers contained control computers. Interactive software spawned new industries. Word processing became practical. Computer games became popular, then faded out. However, they made a significant impact on computer graphics and "user-friendliness" or interaction between humans and computers.

Impact of Personal Computers on Software

Few products have generated as much attention as the personal computer. The small computer is designed to be used by a person rather than a company or institution.

Millions of personal computers were sold

by 1981, and the software for the personal computer had a ready market of new computer owners. These personal computer owners did not want to learn about computers, however: They wanted to use them. Nor did they want to worry about software: They wanted it to be invisible and to allow them to accomplish tasks that they understood without spending time learning complex programming or writing programs.

By 1980 *interactive software* for the personal computer satisfied a need for easy-to-use software. Interactive software created programs for generic (general) applications such as spreadsheets and word processing. It helped generate the sale of millions of small interactive computer systems to people who do not regard themselves as computer programmers.

For office automation, the idea of dedicated hardware to handle specific tasks, such as word processing, is a concept of the past. It was displaced by the emerging role of the more flexible personal computer and the availability of literally thousands of software applications packages in the marketplace.

In 1981, only one quarter of industry revenue was obtained from software. In 1985, worldwide sales of software reached $7 billion and in 1986 software sales reached $10 billion. The role of software has changed, and software product obsolescence has diminished. Hardware is still important, but decision makers must realize that software is increasingly able to evolve into a more powerful system without replacement of the hardware. The availability of low-cost, standardized software has contributed to the explosive growth of the information processing industry. However, by the end of this decade, software income is expected to outpace that of hardware.

Types of Microcomputer Software

There are three major types of software used on microcomputers: systems software, utility software, and application software.

Systems Software

Systems software includes operating systems, languages, and end-user tools. For our purposes, the only one we need to understand is the operating system. The *operating system* controls the basic operations of the computer, such as loading and running programs. It also stores and retrieves files on disks.

The operating system acts as a "traffic cop" for the computer system. It ties the terminal, printer, computer, and mass storage memory together. There are many disk operating systems (DOSs) on various pieces of hardware. You need only make sure which operating system (such as CP/M, MS-DOS, Apple-DOS, TRS-DOS, or Unix) will run your application programs. Software formatted on one kind of operating system will not run on equipment having another kind of operating system. Operating system software that is built into the hardware is called *firmware*. In a microcomputer system, the operating system instructions may be built in or contained on a diskette.

Utility Software

Utility software, a subsystem of an operating system, helps you use the operating system more effectively. In fact, more utility software programs come with the computer's operating system.

Utility software performs specialized, repeatedly used functions such as sorting, merging, or transferring data from one device to another. Because utility software packages are frequently geared to the technical needs of programmers, they are not easy for the average information worker to use.

Applications Software

Applications software is based on systems software. It is designed to perform specific functions or applications, such as writing payroll checks or updating mailing lists. Applications software gives the computer its "personality." There are general-purpose applications, such as accounting and word processing. Tailor-made packages may address the needs of a specific industry, such as manufacturing or retailing institutions. There are thousands of application programs available today. Summaries of some of the popular packages follow.

Spreadsheets. Some of the most popular software packages are spreadsheets. A traditional spreadsheet is a large columnar pad, such as an accountant's analysis pad, with vertical rows and horizontal columns. They are used to compare costs and revenues and to project future performance. When done manually, such tasks are very time consuming and tedious.

Spreadsheet software is an electronic worksheet for busy managers and professionals who can do all of the above tasks electronically. All calculations are performed automatically by the computer. When a figure is changed on a spreadsheet, the software instructs the computer to automatically and instantly recalculate the rest of the figures. Thus spreadsheets provide the distinct advantage of time saving and predictive power. Budgets, forecasts, income statements, cash flow statements and other financial models are common applications for spreadsheet software.

All spreadsheets are not created equal. Spreadsheets range from basic functions that allow you to enter data and do some calculations. Others are super spreadsheets that permit some data-base management and graphics. A unique spreadsheet feature has *macroinstruction* execution capability. In their most basic form, *macros* are simply stored collections of keystrokes that can be recalled with a two-keystroke command. These collections are stored in a spreadsheet cell or range of cells. In just a short time the computerized spreadsheet has become an established part of the business and financial computer culture. It is now a generic category, like word processing, data-base management, and graphics.

Word Processing. Word processing is the nucleus of any office automation application. It is something few offices can do without. Basically, word processing software provides the automated processing, revision, and manipulation of words by the computer. Moreover, it is fast and flexible. Most software programs are designed for easy-to-use command functions.

Over the past few years, the dedicated word processing market has been steadily eroded by microcomputers. Personal computers, running software word processing programs, can save money. First, the productivity of the operator increases. Second, personal computers are less expensive to purchase than dedicated word processors. Word processing software has sufficient flexibility and advanced features to meet the needs of all office workers. The workers include professional word processors who enter and edit documents forty hours per week, the secretaries, the managers, and the authors.

Word processing software programs offer a range of features. There are no-frills packages that offer all the features necessary to perform basic word processing. Advanced packages offer features for the heavy volume, experienced user. Generally, the less expensive word processing programs offer fewer options in formatting, editing, and lack extras such as list-merging functions. The selection of the word processing package that is best suited to your office applications depends on price, features, and design issues. These criteria were examined in Chapter 11, "Word Processing."

Accounting Software. Accounting software is made up of a series of accounting programs that automatically enter, calculate, and total financial documents and statements. Such programs include accounts payable, accounts receivable, general ledger, payroll, and inventory. Publishers of accounting software may have their systems certified by reputable accounting firms or independent certified public accountants. These individuals usually test the software packages and certify that the products perform in accordance with generally accepted accounting principles and auditing standards.

Modern programs are easy to use. The user simply enters the data and fills out the forms that appear on the screen. The cursor automatically positions itself on the line of the form that needs to be completed. Once that line is filled in, the cursor jumps to the next line, and continues moving to the next line until the entire entry is completed. The system is able to automatically store the information on the disk.

Accounting software programs vary in features and in specific accounting tasks. Some come with complete systems for all the major accounting functions, others come in individual modules. Accounting software computerizes the accounting tasks that used to take accountants hours to complete manually.

Communication Software. The real value of any microcomputer is its ability to send information from place to place, in a matter of minutes, not hours. Microcomputers can communicate with a worldwide, informal network of computers. They can tap into more information than one can find in most libraries. They can move complex documents across thousands of miles in minutes.

All a microcomputer needs is an ordinary telephone line, a modem, and the proper communication software. *Communication software* comprises the instructions that tell a computer how to perform its tasks. A modem and other communication systems were explained in Chapter 12. Communication software allows a microcomputer to communicate with a variety of other brands and models of microcomputers. It "talks" to mainframe computers and to subscription information services such as The Source and CompuServe.

Communication software may have some or all of the following features.

1. *Auto-dial:* If you have an auto-dial modem, the software will be able to dial the phone automatically.

2. *Auto-answer:* Software instructs the computer to answer incoming calls and allows callers to access the files on your computer system.

3. *Smart terminal:* Software instructs the computer to act as a "smart" terminal for any dial-up computer. It can automatically call up, log in, and enter commands to a host computer system. Moreover, no human assistance is necessary.

4. *File transfers:* Software instructs the computer to accurately exchange files with any other communication software or compatible program. This feature allows the user to exchange files with a wide variety of computer systems.

5. *Capture data:* Software instructs the computer to "capture" an entire communication section to a disk file. This feature allows the user to review or print the data at a more opportune time.

6. *Terminal emulation:* A terminal emulator is a type of communication software program that instructs the computer to hook up with a mainframe computer. It enables the microcomputer to "emulate," or look like, several popular communication terminals. This is important when someone is running programs on a host computer that requires a specific type of terminal.

In addition, communication software can be used for the following applications:

- Transfer of text or program files from a user's disk to almost any other computer regardless of the operating system and disk compatibility

- *Download* (call up and capture) data from a corporate mainframe, feed

them into a spreadsheet program on your own micro (Figure 14-1), and then *upload* the results to the mainframe
- Call up and capture data from one of the growing lists of data banks, such as The Source, CompuServe, Dun & Bradstreet, Dow Jones, or Western Union's Easylink
- Create your own electric mail service, to transfer text across the country or across town instantly

Electronic Outlines (Hypertext)

Software has taken on a new dimension in aiding the outlining process. *Outline software programs* are designed to help the user to place ideas in the proper order. Pure word processing programs do not facilitate creating or thinking. Outline software, however, is aimed at arranging segments or information into logical order. The electronic outline aids in the thinking process.

Text written on paper must be in a specific order. This chapter, for instance, has a particular order of paragraphs. The reader will normally follow it from beginning to end in sequential fashion. However, computer-based information is not bound by the restrictions of paper. Computers and electronic outlines enable the writer to look at various levels. Electronic outlining has also been referred to as nonsequential text, or *hypertext.*

The most obvious application of an outline program is the way a student outlines a paper before writing it. All of the advantages provided by outlines on paper apply to electronic outlines as well. As ideas develop they may be keyboarded randomly. The user may not know exactly where a segment of information fits into the total story. With these programs, ideas do not have to be organized into a linear outline. The system allows for the proper connection through a complex network. Through commands, the user organizes the various thoughts into a sequential pattern, if they are to be printed on paper. Individual ideas can be placed in proper perspective to other related concepts. Then, one can estimate the amount of detail to devote to each. The advantages of using an electronic outline include the ability to move a heading within the outline, and to move all subheadings and text along with it. Hypertext software is more advanced than the traditional data bank. The data banks permit the instant retrieval of documents, but do not allow movement from one document to the middle of another.

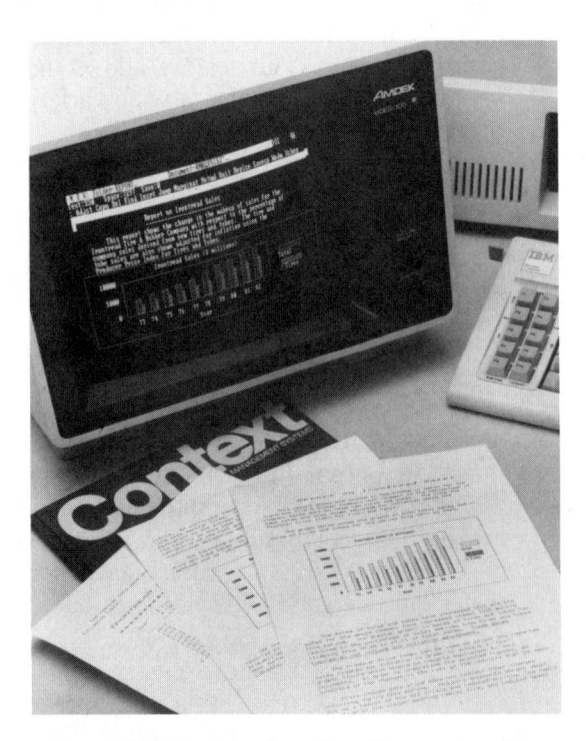

Figure 14-1. Downloading allows the user to write reports incorporating data accessed through telecommunications. Referenced data can be used in spreadsheets or graphs and included in the finished report without retyping.

Source: Courtesy of Context.

Data-Base Management

Data-base management software are programs that organize data bases into files and records and sort new files in logical order. The programs also selectively retrieve information. Applications range from sales leads, customer mailing lists, indexing archival storage, employee records, or students records. Database management packages can accommodate several billion records, make several files available, and sort at very high speeds. Database programs on microcomputers can be linked to mainframes to obtain access to information stored there. Special micro-mainframe software can instruct that information be transmitted from the mainframe to the micro, where it is stored on a diskette. The micro user can refer to this information at any time by inserting the data-base software into a PC. By retrieving data from the list, the data can then be merged into a report, letter, or other document. The applications for database management are limitless. Anywhere there is need for a list, data-base management software is able to manage it efficiently and economically.

Graphics

Graphics software helps users turn complex data and ideas into easily understood, presentation-quality charts. With a graphic software program, you can quickly produce a chart to communicate your message as clearly and persuasively as possible. Software is the driving force in business graphics and many vendors offer business graphics packages. With the continued emphasis on productivity in the automated office, graphics are viewed as one of the most cost-effective ways to measure and compare. As we remember from Chapter 10, specialized hardware, such as color monitors and printer plotters, must be used to support sophisticated applications.

CAD/CAM

As graphics capabilities become more available, the range of applications widens. These programs provide a creative force between business people and computer technology. Engineers use a special kind of graphics software called *computer-aided design* (CAD) (Figure 14-2) and *computer-aided manufacturing* (CAM) programs. These programs are used to design and manufacture new products. Combined with holographic techniques, CAD and CAM graphics can create three-dimensional models from two-dimensional designs in a computer. These designs and manufacturing programs comprise the largest segment of the computer graphics industry. They are used in drafting, architecture, engineering, and building manufacturing, tooling for automobiles, airplanes, and similar industrial products.

Time Management Software

The most important resource that managers have to manage is their own time. Time is an elusive factor that slips away during an executive's busy schedule. Too often, appointments overlap in a disorderly fashion and are sometimes forgotten completely. Notations on desk calendars or slips of paper are buried or misplaced. Personal computers and time management software can help executives make better use of their most valuable resource. Time management programs can help keep track of daily schedules.

When colleagues call to set up meetings, executives can display the "electronic calendar" on the computer screen. If time is available for the meetings, they key in the appropriate box. In this way, the calendar is always up to date. The system can also be used to set up a meeting among several on-line managers who use the same time management program. Users instruct the computer to communicate with others on-line to find a time when everyone is free for a meeting. The computer then checks each schedule and records the time chosen for the meeting in each manager's personal calendar.

Project management software is closely related to time management software. Project

Figure 14-2. CAD (computer-aided design) software enhances the design, drafting, and analysis capabilities of graphics programs.

Source: Courtesy of Harris Corporation.

management software packages are designed to handle crucial items: time and resources. At first, project management software (for the personal computer) was able to handle only one project. Now, it can be used for several projects and it can allocate resources among them. Project management software can be useful to anyone involved in a number of tasks, deadlines, and resources.

Financial Data Software

Stock market and financial data have been available to computer users from information retrieval services, such as the Dow Jones Retrieval Service (Table 14-1). These services, discussed in Chapter 12, allow users to connect with a data base over telephone lines and respond to stock queries one at a time. The data can be saved but often must be reentered into electronic worksheets by hand. Users are charged on-line services.

New financial data software packages go beyond merely reporting stock and financial data; they alert users to key price movements of a favorite stock. The software can revalue an entire portfolio of several hundred stocks as their prices change, and automatically analyze the implications of new stock price levels on a user's financial spreadsheet.

The market for such a software program consists of hundreds of thousands of avid stock market investors, small brokerage firms, and financial advisers. It is ideal for computer owners who already have integrated packages, but do not have access to the giant computer systems of the major Wall Street companies.

One such product, Signal by Lotus, uses no

TABLE 14-1. Financial data bases

	CompuServe	Dow Jones News/Retrieval	National Computer Network	The Source	Warner Computer Systems
Price quotations:					
Bonds	X	X	X	X	
Commodities	X	X	X	X	X
Options	X	X	X	X	X
Stocks:					
Current	15-minute delay, instant quotes for $35 a month	15-minute delay; instant quotes for $18.50 a month	Instant quotes only	20-minute delay; instant quotes for $20 a month	8-hour delay; no instant quotes
Historical	1973 to present	1978 to present	1979 to present	Past 12 months	1975 to present
Reports available:					
Company	X	X	X	X	X
SEC	X	X	X		X
News	X	X		X	
Financial analysts	Corporate analysis, earnings estimates, economic projections, commodity analysis	Corporate analysis, earnings estimates, economic projections	Earnings estimates	Corporate analysis	Earnings estimates

telephone lines to transmit data (Figure 14-3). It uses the technology of instantaneous broadcasting through FM radio stations.

The data are gathered from the exchanges, beamed to a satellite, and then distributed by FM radio stations on an unused portion of their radio signals. Using Signal software, for example, the computer constantly sifts through radio signals for stocks it is programmed to retrieve. The figures on 20,000 listings are updated every 8 minutes, unless a trade—which is recorded instantly—takes place. The data can be transferred to a spreadsheet with just a few key strokes, and the compatible Lotus 1-2-3 integrated program instantly converts it to graph form.

New, advanced software technology allows instantaneous broadcasting of stock market prices, options, and commodities quotations. The computer user can access the latest financial data without paying time charges.

In addition to data bases, other types of financial software include a *portfolio management program*. These programs keep records of your transactions, as well as display your portfolio's structure and update its market value. Some financial software programs do a lot more, such as calculate your return on each security.

Beside portfolio management programs, investment software tends to fall into two broad categories. The smaller group consists of *investment advice programs*. These programs combine fundamental and technical analysis to project stock prices and to tell you when to buy and sell.

The other broad category of investment software includes programs for fundamental and technical analysis of stocks. If a user bases investment decisions on charts, it is helpful to have a computer with graphics capabilities. A variety of financial data bases are available.

Knowledge-Based Systems

Business-related applications software such as spreadsheets have changed the way corporate America does business. Improved versions of spreadsheets enable the user to enter numbers into a spreadsheet and ask "what if?" These

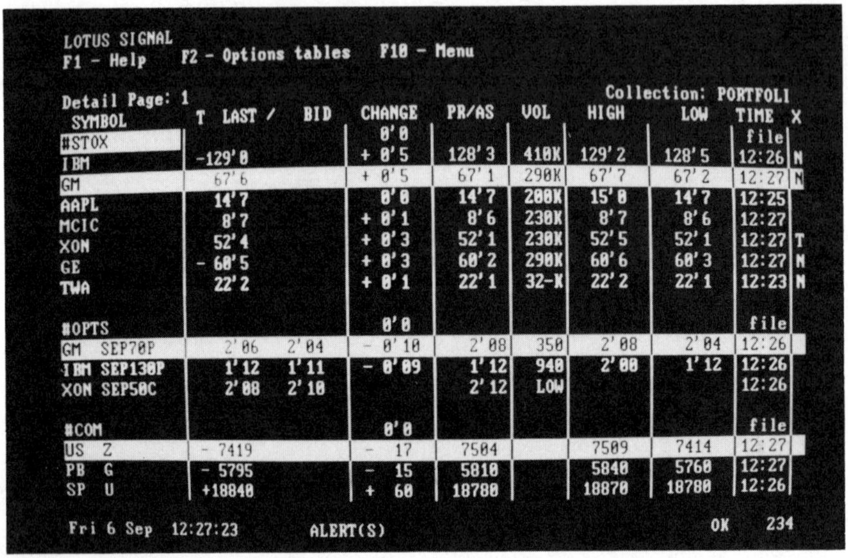

Figure 14-3. Signal monitors the stock market and updates portfolios.

Source: Courtesy of Lotus Development Corporation.

software solutions, however, were based solely on numbers. While the bottom line has always been considered financial, few high-level decisions are made on quantitative factors alone. Strategic planners began to look for what-if answers based on natural language or plain English responses. As a result, software was developed to handle some of these applications. This was known as *knowledge-based software*. Knowledge-based refers to a system's ability to perform tasks at the level of the human expert. Researchers in the specialized field of *artificial intelligence* (AI) have been working on ways to discover rules that could be used to program a computer to think and reason much the way a human being would. Although the refinement of this goal has not yet been achieved, it now appears that simulated reasoning—as opposed to real reasoning—will soon be packaged as a software program.

An *expert system* is one application of the embryonic technology of artificial intelligence. These programs are beginning to make their way out of the lab and into the microcomputer software market. An *expert system* software program simulates the kind of "data base" that someone with a particular expertise stores in his or her head. It uses many of the principles that might be employed in making a decision or recommendation to a client, patient, or customer. Whether this new software is called decision support, artificial intelligence, or a knowledge-based system has not been universally decided. There is a bright potential for advice and information delivered via knowledge-based systems. Applications can be as varied as legal, medical, automotive repair, or office systems. For all its promise, however, the area of expert systems software will never attain the level and power of the human mind. The intuitive, creative element of human thinking is what makes it unlikely that computers will replace people. Knowledge-based software will become a useful tool to provide a consistency for human decision making. The user of this software must be careful not to become overly dependent on this or any other software system.

Vertical Markets

The previous section discussed only a small portion of applications software. The list of software is constantly growing and evolving as new professions and specialized cottage industries emerge. There are an estimated 20,000 to 34,000 software packages available to meet the business needs of personal computer users. These packages cover the essential requirements of such broad categories as accounting, data base management, and spreadsheets. Now publishers of software packages are looking to expand their offerings by developing a vertical or custom-tailored market. The groups of people who comprise these markets have specialized needs. For example, a vertical market can be as huge as lawyers and doctors or as small as rock group managers or steam boat captains. Many specialized careers, professions, and small businesses have specific needs that are not wholly satisfied by more generalized programs.

Thousands of programs are available. Finding the correct software requires researching software catalogs such as:

> *The Software Catalog:*
> Elsevier Publishing Co.
> P.O. Box 1663
> Grand Central Station
> New York, NY 10163

> *Datasources:*
> ZIFF-DAVIS Publishing
> One Park Avenue
> New York, NY 10016

Other resources are on-line data banks, computer trade shows, and current magazines.

Custom Software

If standard applications packages and vertical market sources do not meet their needs, users are now turning to custom software. Basically, *custom software* refers to creating or designing your own software package. Early attempts to create custom software required hiring programmers to write specific programs. Some users learned basic programming and wrote their own custom software packages. Now, you do not have to be a programmer to customize your software. A special software program known as a *keyboard redefiner,* or *macroprocessor,* can make this customization process simple.

Keyboard enhancers, or *definers,* allow the user to modify the "content" of any individual key, as well as certain combinations of keys, on a computer keyboard. With a keyboard enhancer, a user can make one key stand for another, or produce a string of characters that encompass an entire document or a string of commands. The latter task best exemplifies macro technology discussed in a previous section.

An example of a simple macro is a writer using a word processing program. When ending a paragraph, a macro can be programmed to produce a period, two carriage returns, and a tab for the next paragraph. All of these actions can be accomplished with only one keystroke, instead of five.

Macro programs and keyboard enhancers have the following advantages in developing custom software programs:

- *Single key:* A single key can record lengthy keystroke sequences and play them back at the touch of a single key.

- *Screen protection:* Some programs make the screen go blank after a predetermined time of screen/keyboard activity. This feature protects your monitor's precious phosphor.

- *Security:* Through keyboard enhancers, confidential files and documents can be protected. File encryption assigns passwords that protect documents from unauthorized individuals.

Integrated Software

Spreadsheets, word processing, data bases, and graphics have become the most popular applications programs. However, users grew tired of handling numerous diskettes, squeezing large manuals onto their bookshelves, and learning hundreds of commands just to perform four or five separate applications.

As a result, *integrated software* came into popular existence with the arrival of Lotus 1-2-3. Now every major vendor offers a line of integrated software. Integrated software is a comprehensive software system with basic computer functions such as word processing, data bases, communications, spreadsheets, and graphics on one disk.

With integrated software the same data can be used for a variety of applications. The individual applications work together without the user removing and inserting separate disks. The information that is entered in one section can automatically be transferred and incorporated into the other sections when appropriate. With integrated software, users can press a few keys and perform multiple applications instantly and interchangeably.

Because of the integration of spreadsheets, graphics, and data bases, it is possible to quickly and easily generate what-if applications. They include projection of budgets, control of inventory, scheduling, and analysis of stocks and bonds. You can include the information in a letter or report and send it via computer anywhere in the world. Integrated software allows the user to view and make changes in letters, graphs, and spreadsheets simultaneously, on the same screen.

Many of the computer functions included in an integrated package are powerful and comprehensive. They can stand on their own in a separate disk. The strength of integration lies in what users can do when combining two or more of its capabilities.

Integrated software packages may be the complete software system for office applications. The packages provide the user with the power needed to solve today's problems and offer unlimited potential for the future.

Software Selection

Software firms specializing in specific applications and industries are developing integrated systems for all sizes of computers that will permit companies to link their operations. They are investing enormous resources to develop and maintain dynamic applications. Also there are software packages on the market with innovations that increase the ease with which the nonspecialist can use a computer.

Software packages are available for practically everything a business or professional user can dream of doing. However, they are not always easy to find. The best software might be the worst for your organization. It must be effective to meet your organization's precise business needs. It must be efficient and offer enough productivity to match the cost. How do you find information about software packages? What software should you buy? What do you want the software to do for you? Where do you buy software? These are some of the questions that must be answered to initiate an intelligent selection. The software selected for your firm's personal computers depends on several guidelines.

Gather information necessary to compare microcomputer software packages to meet the requirements on an organization's checklist. This can be done by watching presentations and demonstrations, contacting other users of a system, or reviewing a package's documentation.

Once you have a general idea of the kind of software you want, go to several computer stores and see what's available in that category. Software distribution is still in an evolutionary stage. The sales staff in retail computer stores, originally aimed at the home market, may not be experienced enough to translate your needs into the name of an appropriate software package. Not all software is available in such outlets. Many useful packages are sold through consultants, software or hardware manufacturers, or through the mail. Once you have identified your software needs and narrowed the field down to the packages that can best fill those needs, you are ready for the next step. Find the software that will run your hardware and make sure it meets the following criteria:

1. *Compatibility:* Is the software compatible with your hardware and operating system?

2. *Functions and features:* Does the package have the functions and features necessary to meet your application needs, such as strong single function application, ability to customize, integrated package, security, and communication?

3. *Ease of use:* Is it easy to use? Does the software come with "help screens"—trouble-shooting instructions built into the software so that you do not have to refer to the instruction book when you are in trouble?

4. *Documentation:* Is documentation available? Is the user manual that covers operating instructions and service useful? Can you understand it? Does it have indexes, topical headings at the top of each page, illustrations, tables, copies of explanations, and reproductions of screen menus? Documentation is one of the most important and most overlooked areas of software evaluation. It is important because it serves as the primary source of information for using and understanding a software application package. Fortunately, most good software packages come with documentation. Unfortunately, most instruction books are not as good as the software. Look through

the book to see how complete it is and whether you can follow it without an interpreter.

5. *Service and support:* Does the software have a reasonable warranty? What should you do when the software program does not work? How much help can you get in learning to use the program and working your way out of problem areas? Are there sources at the software company or elsewhere who can offer support? Can you reach them easily?

6. *Costs:* The initial purchase price of the software may be only the beginning of the costs incurred in acquiring a software package. Other costs are associated with software selection:
    a. *Conversion cost:* Replacing existing applications with new and better systems may involve the transfer of data by having users reenter the information from old programs to new programs.
    b. *Training cost:* What is the cost of training? Is it worth a reasonable fee for your staff to call a company with questions as they arise?
    c. *Support costs:* Many vendors will support their systems for an annual fee, usually some percentage of the purchase price.

7. *Update:* What are the prospects for expansion or enhancement? Old software is constantly being improved upon and reissued in new versions. You should never buy software without a guarantee that you will be notified about updates and will be able to purchase them at a reasonable cost.

Selecting the correct software involves the same painstaking and grueling strategies as selecting other components of information processing. Although the total cost of microcomputer software is significantly less than that of hardware systems or mainframe software, the evaluation and selection process should be no less thorough. In the long run there will be rewards in choosing good software. Training is simple if the package is easy to use. The dependency on support will be minimal. Features and functions should match your applications. And, finally, acquisition of the right software means selecting with vision for growth and change.

Summary

Software is the bridge between the machine and the user—the tool that directs the power of the computer. Moreover, software is defining today's information issues.

Instead of the emphasis, of past years, on building better and more powerful machines, the emphasis now is on harnessing the full power of the existing hardware through improved software design.

The three main types of software are systems software, applications packages, and application utilities. Systems software consists mostly of operating systems such as CP/M, MS-DOS, Apple-DOS, and UNIX. These programs provide overall direction for the applications programs such as word processors and spreadsheets. They tell those applications programs where files are, and so on.

Integrated programs combine basic computer activities such as word processing, spreadsheet analysis, data management, and sometimes graphics or telecommunications.

Expert systems—computer programs that use human traits such as logic to solve problems—will serve as the vanguard for the advance of artificial intelligence into the office. The most popular microcomputer software programs are word processing followed by spreadsheet packages and integrated software.

There are software packages available for practically every business and professional application. The selection process should follow an intelligent, well-planned approach. Define your needs, select appropriate applications software to meet those needs, and follow well-defined criteria.

Review Questions

1. Define software. Describe the analogy of software to hardware, as an artist to brushes and canvas.
2. What effect did the personal computer have on the growth of software?
3. What is the difference between systems software and utility software?
4. What are macroinstructions? Which application software is suitable to macroinstructions?
5. What criteria do publishers of accounting software use to certify their programs?
6. List some of the applications that can be used with communications software.
7. Contrast the functions of downloading and uploading.
8. What is the primary function of data-base management software?
9. How are graphics related to CAD and CAM software applications?
10. What are expert systems? Should a user depend solely on the answers from expert systems? Explain.
11. Define keyboard enhancers and keyboard definers. How are they used in custom software?
12. Describe integrated software. Which computer functions are usually included in an integrated program?
13. What precautions should users take before they acquire software in computer retail stores? Through the mail?
14. What does documentation, as applied to software mean? Why is it important in the selection of software?

Projects

1. You have acquired a communication software package and a modem to upgrade your personal computer for communications. As a large multinational consulting firm you have branches in all of the world capitals including Paris, Rome, Berlin, London, and Tokyo. You find that you are missing important messages because of the difference in time zones among geographic regions. In order to receive messages, your employees would need to wait around after hours, in the middle of the night. You learn that there is a special software program that can alleviate this problem, one that allows you to send and receive information between your microcomputer and others, without anyone present at either end. Briefly describe such a program. What are its features? Identify and evaluate some of the popular commercial programs.

2. You are part of a research team to investigate the current trend of software use. You are assigned the following questions as part of the survey:
    a. What are the five most frequently used generic applications for microcomputer software?
    b. What are the five most frequently used microcomputer makes and models used by the participants in this survey?
    c. What are the three most frequently used operation systems used by the participants in this survey?

   You are to conduct this survey by mail questionnaire within your community. Describe the total number of respondents, the types of business, and the method of collecting the data. Present your findings in a brief summary report. Create bar charts to supplement your findings.

3. As a new doctor who has recently joined a medical group, you are interested in computerizing the medical group's operation. Before purchasing a computer, you justify to the other doctors how an intelligent medical software program will improve the health group's operation. The

software must
a. handle an unlimited number of patient names and visits
b. store, retrieve, and sort information
c. handle multiphysician codes within a practice
d. handle cash transactions at the time service is delivered
e. produce patients' monthly balances
f. automatically place your message on past due accounts
g. maintain a data base for clinical research

Research current software packages to determine which system best meets your needs. Write a summary report on a complete medical office software program. Include the features and functions of a system that comes closest to increasing the efficiency and productivity of your clinical practice.

4. You are the president of a newly formed software consulting firm. As part of your service, you employ programmers who customize special applications. You have been asked to design a specialized program that will improve the management of temples and synagogues. Investigate the unique applications involved in running the practical side of temples. What are the main computer functions in temple management? Describe how you would design a complete program that incorporates the correct applications in one integrated package?

CHAPTER 15

Storage and Retrieval

After reading this chapter, you will understand
1. The records management process
2. The distinction between a centralized filing center and a decentralized system
3. The categories of filing equipment and supplies
4. The concept of micrographics
5. The meaning of reduction ratio and its effect on selection of a micrographics system
6. The terms COM and CAR
7. The various storage media, such as floppy disks to optical disks

In the preceding chapters we have explored the benefits of advanced technology and the miracles of Space Age gadgetry. Despite all these claims, organizations are still fighting to manage and control the burgeoning growth of paper and electronic information. All of the devices and advanced systems are directed toward one end—providing essential business information when and where it is needed. Like any other valuable corporate asset, therefore, information should be subject to sound business practices and controls (Figure 15-1).

Records management is a process that includes the creation, storage, retrieval, retention, protection, and disposal of all vital records. A *record* is any form of recorded information—paper, microfilm, or computer disk. Thus, records management involves controlling records from their creation to their destruction. Each procedure provides an important phase in a records management system. During the active phase, quick and frequent access to records is needed. When access is no longer critical, records are stored at a records center, either in-house or at a commercial facility. When they are no longer needed, records are destroyed.

A Paper-Based Society

The paperless office may exist in the future, but we are still very much a paper-based society. Paper has been, and in many cases remains, the primary information-storage me-

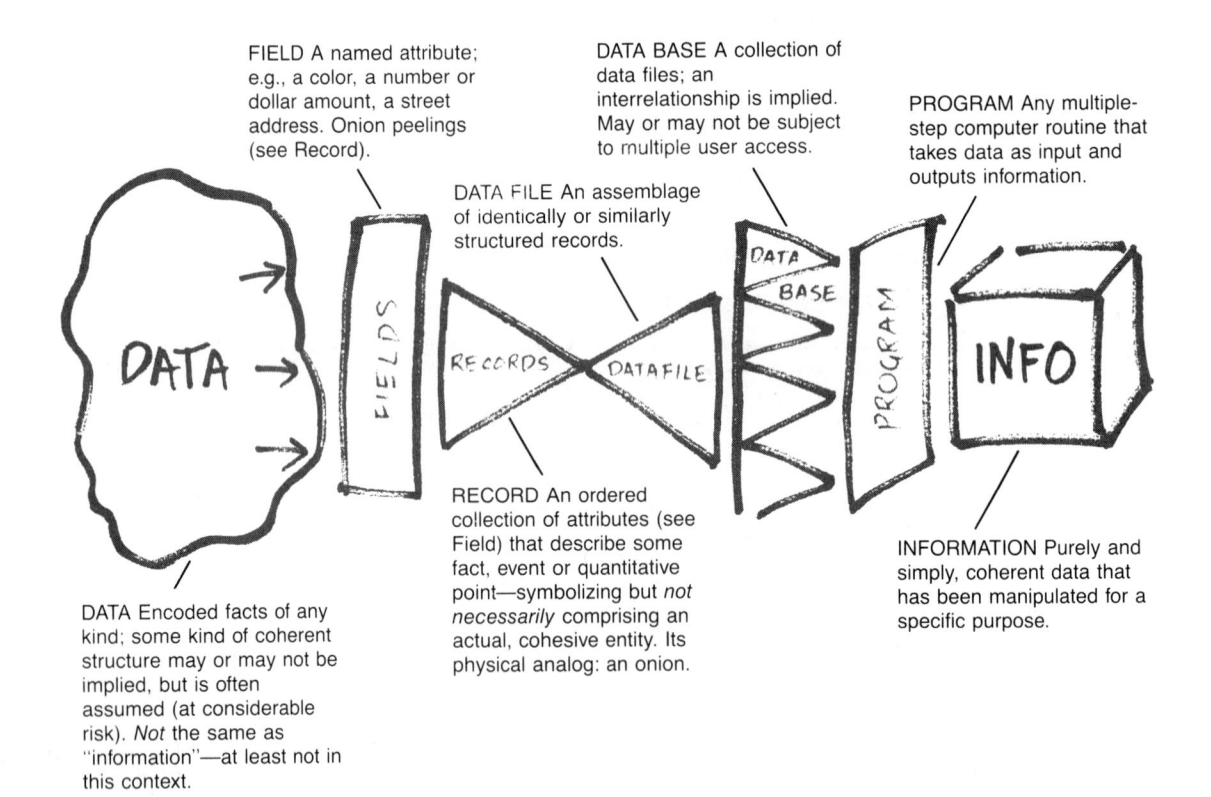

FIELD A named attribute; e.g., a color, a number or dollar amount, a street address. Onion peelings (see Record).

DATA BASE A collection of data files; an interrelationship is implied. May or may not be subject to multiple user access.

PROGRAM Any multiple-step computer routine that takes data as input and outputs information.

DATA FILE An assemblage of identically or similarly structured records.

RECORD An ordered collection of attributes (see Field) that describe some fact, event or quantitative point—symbolizing but *not necessarily* comprising an actual, cohesive entity. Its physical analog: an onion.

DATA Encoded facts of any kind; some kind of coherent structure may or may not be implied, but is often assumed (at considerable risk). *Not* the same as "information"—at least not in this context.

INFORMATION Purely and simply, coherent data that has been manipulated for a specific purpose.

Figure 15-1. A real world glossary of data-base terminology.

Source: Reproduced with permission of Roger Gorman, Reiner Design, New York.

dium for most organizations. International Data Corporation estimates that nearly 1 billion pages of business-related information is processed each day in the United States by 50 million white-collar workers.

About 18 million clerical workers are engaged in sending and retrieving documents (or images of them) from a variety of filing systems. The fact that there are 18 million clerical workers should not be surprising since every single day, American business creates 370 million pages of new business documents, generates 1.9 billion pages of computer output, and makes 1.9 billion copies.

The growth rate will average nearly 5 percent a year through the mid-1990s. That means that the total number of new documents produced annually will grow from roughly 1.6 billion pages in 1984 to 1 trillion pages in 1995, or by 60 percent.

If that is not enough, consider that some portion of these new documents is being added to the trillions of pages already stored in files across the country.

Coping with the Avalanche of Papers

How can organizations deal with the fallout from the information explosion of paper? As businesses produce more paper than ever before, the price of storing it climbs. It is important that each organization establish a well-managed system of hard-copy storage.

Filing Systems

A file is a way to store documents or other materials in an organized and standardized fashion so that they can be retrieved by following some logical procedure. There are two types of filing systems. The first is a *centralized filing system* whereby all active files are located in one or more central areas, usually called a *record center*. The center may hold the records for one department, several departments, or the entire firm. Keeping all or part of a company's active records in a central file area saves considerable time for information workers. In addition, space and filing equipment is no longer wasted storing duplicate records. Generally, the file center is managed by a records manager and file clerks. Because a few well-trained professionals maintain the records, fewer documents are misfiled or lost. Producing records for litigation is simplified when records are consolidated and well organized. Records are sent to a records center situated in an area readily accessible to most employees when workers are finished with them.

Filing Equipment and Supplies

In-house storage requires an intelligent selection of equipment and supplies. There are numerous vendors, who supply a wide variety of equipment to fit filing applications of every size and description. Managers responsible for acquiring this equipment should be familiar with the following types of filing equpment.

Vertical Files. A vertical file is a filing cabinet with one to six drawers. The vertical file is considered the standard in practically every business office. Its drawers pull out from the front of the cabinet. These files are called vertical because papers are filed upright (lengthwise) in drawers instead of lying flat (horizontal) in the bottom of the drawers.

Horizontal Files. Horizontal files are used for storing documents in a flat (horizontal) position. Oversized documents such as maps, engineering drawings, blueprints, and schematics are stored in horizontal filing cabinets. The length and width of the drawers are as large as the documents to be stored. The drawers are shallow since only a few documents are stored in each.

Lateral Files. Lateral filing cabinets offer faster and easier access than heavier vertical pullout drawers. Every folder is visible, and you can pull out folders to mark repositioning space.

Visible Files. Visible files make information accessible at a glance. Visible card files are available in the form of trays. They lie flat horizontally in a cabinet, on revolving racks, or in loose-leaf binders.

Rotary Files. Rotary files are units that store records around a common circular hub. Documents may be removed or added by rotating the file to the desired location for access. Rotary files have both desktop and floor models. Desktop models are generally used to store address cards. With these systems, holes are punched in the bottoms of the cards so they can be easily removed.

Open Shelf Files. Open shelf files are suitable for an active central-files room. If space is limited, compact files that move on tracks can accommodate more records. This shelving, which is arranged in rows with aisles on each side for access, can hold file folders and/or boxes containing folders or other materials.

Mobile Files. Mobile files use sliding shelves placed on tracks in order that the shelves can be moved together when not in use (Figure 15-2).

This concept can be compared to the familiar household clothes closet, where garments are squeezed together in order to gain more closet space. Instead of leaving 6–12 inches between every hanger, the hangers can be moved laterally to open the space required to select a garment. In the same manner, carriages in a bay can be shifted to expose only the required records, creating an aisle for retrieval or deposit. These can be operated with handles or cranks or by electricity and are mounted on tracks for easy access.

The design in mobile systems has advanced rapidly in the past few years. The transition from small, rolling push-type shelves to sophisticated manual-assist shelves to large, electric, motorized systems offer many new options. Engineering makes it possible to extend mobile double-sided carriages to 68 feet in length, and 13 feet in height, creating a system capable of holding 90,000 pounds of records. These systems can be customized to fit

Figure 15-2. Mobile filing system.

Source: Courtesy of Acme Visible Records.

any office situation, and their range of application is virtually limitless.

Transfer Files. Transfer files are made of cardboard or steel boxes (Figure 15-3) and can open on either the top or the front. These files are used to transfer documents from one location to another. They are frequently used to store inactive records that have been transferred from the active files.

Automated Records Management Systems

Paper-based systems are expensive and continue to rise in cost as paper and space charges escalate. As a result, organizations are prompted to explore automated systems for managing records. Paper-based filing has been making way for other more high-tech forms of records management, including micrographics, electronic storage, and optical-disk storage.

The transformation from a paper-based system to an automated system is a difficult one for a company that has collected files for many years. The conversion task is a massive undertaking as there may be hundreds of thousands of active and semiactive documents in the system. Before we discuss the actual conversion process and offer guidelines and strategies, it is necessary to differentiate between some of these automated electronic storage systems.

Micrographics

Micrographics is the process of photographing paper records to miniaturized records. This system entails photographing documents in a greatly reduced size on microforms (Figure 15-4). A *microform* is the film or medium used to store the image or document. There are several different types of microforms in use today. They include roll film, cartridges, jackets, aperture cards, microfiche, and ultrafiche.

Roll Film. The original microform, roll film is normally used to store information sequentially. A 100-foot roll can contain up to 4000 images. Microfilm rolls may be either 16 mm or 35 mm. The 16-mm rolls are used for standard business documents, such as letters and reports. The 35-mm format is generally used for filming graphic materials and large documents.

Cartridge Microfilm. Cartridge microfilm is a more efficient use of the roll. A cartridge can contain up to 4000 letter-sized documents. Unlike roll microfilm, which usually requires hand loading on retrieval units, cartridges are self-threading. Cartridges are convenient to handle, and they protect the film and make it possible to find a specific image among the 4000 in just seconds.

Jackets. Microfilm jackets consist of two pieces of thin, clear plastic that are joined together at various points to form channels. Rolls of microfilm are cut into strips and inserted into these channels, either manually or

Figure 15-3. Transfer files.

Source: Courtesy of Esselte Pendaflex Corporation.

226 Part 4 / *Software and Storage: Components and Technology*

Roll Microfilm

Cartridge Microfilm

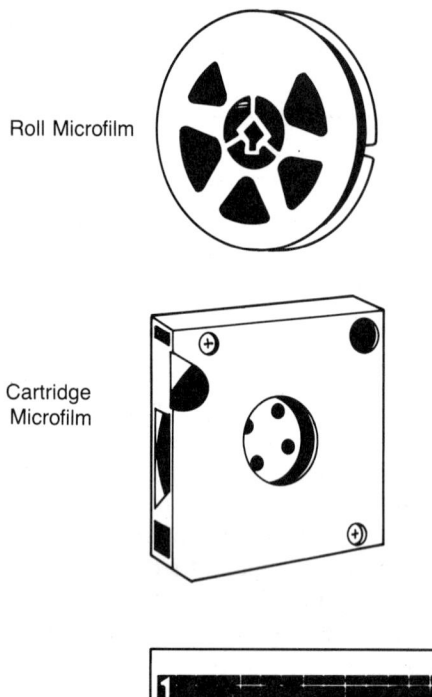

Microfilm Jackets

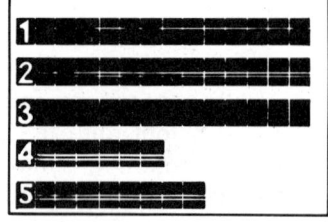

Aperture Card

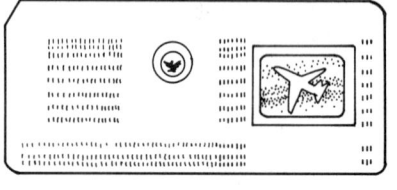

Microfiche

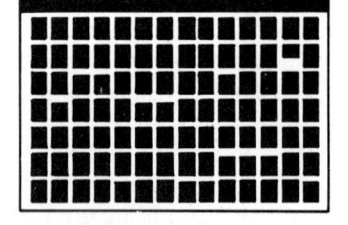

Figure 15-4. Microforms.

by machine. A microfilm jacket is like a file folder; instead of holding paper documents, it holds strips of film. These strips can be removed and replaced with new strips when the information on the existing film becomes outdated. A standard 4 × 6-inch jacket contains up to 75 images. A jacket can be visibly titled for quick, easy reference. Jackets are easily duplicated for mailing, security, or reference.

Aperture Cards. An aperture card is a card, usually 3¼ × 7⅜ inches. It has a rectangular hole (aperture) for the insertion of a microfilm image. Most aperture cards contain one 35-mm film chip, but some can hold up to 8 images.

Microfiche. Microfiche are sheets of 105-mm film mounted on a 4 × 6-inch card (Figure 15-5). One microfiche contains up to 420 images in a grid pattern. Microfiche can contain title or index information that may be read without magnification. One fiche may

Figure 15-5. Microfiche tray sets.

Source: Courtesy of Dennison National Company.

contain a 100-page report, an insurance policy, or a book.

Ultrafiche. Ultrafiche goes beyond microfiche in miniaturization. Ultrafiche is basically a microfiche with a much higher reduction (90 × or higher).

The number of images that can be housed on one fiche varies according to the *reduction ratio*. The reduction ratio expresses the number of times the size of a record is reduced photographically.

The most common size of microfiche (4 × 6-inches or 105 mm × 148 mm) can hold 98 letter-size pages at a 24× reduction. At a 42× reduction, 325 letter-size pages can be housed on the fiche, and 420 pages can be accommodated at a 48× reduction.

With ultrafiche a reduction ratio of 150× can be achieved. With this rate, over 3000 letter-size pages can be stored on one ultrafiche.

Micrographic Processing

The transition from paper-intensive filing systems to a mix of hard copy, micrographics, and electronic storage is being facilitated by products that store a variety of media. When a firm makes the decision to convert its paper documents to microfilm, there is generally a backlog of hundreds of thousands of documents. Careful planning is needed during this conversion process—often a tedious, multiyear effort. Three operations may continue simultaneously: (1) the preparation of records into the new system, (2) the maintenance of manual files that have not yet been converted, and (3) the handling of new documents using the automated systems.

Some firms decide to leave the existing paper files intact and microfilm only new documents as they are added to the files. Other companies decide to tackle the enormous job of microfilming existing files. Throughout the conversion process, workers should have a minimum amount of inconvenience. A labor force must be assembled and trained to microfilm the existing files.

An organization can send its files out to a microfilm service bureau to be filmed, it can train its own staff, or it can hire temporary labor to handle the job.

The Conversion Process

The first step in micrographic processing is the preparation of documents for filming. There are two ways in which microforms can be created: source documents and computer output microfilm. *Source documents* are the original paper documents that are then filmed with a microfilm camera. *Computer output microfilm* (COM) is a process that converts computer-generated, machine-readable data into human-readable information on microform. It is created by transferring data that are stored in host computer memory (on-line) or on magnetic tape (off-line) directly to microfilm or microfiche (Figure 15-6). COM and other micrographic processes will be discussed in greater detail in later sections of this chapter.

Microfilm cameras for recording documents come in many sizes and types. The choice depends on the microform being used and the volume of information to be photographed.

Once filmed, microfilm must be processed. A number of cameras have self-contained processors. Some firms send film outside to be processed, much as you would send out a roll of photographic film. If only one copy of the microform is needed, the processing is the final step in the filming process. Additional copies can be duplicated from the master set easily.

Retrieval

Retrieval equipment can be as simple to operate as a push-button telephone. Microfilmed documents can be selected from among thou-

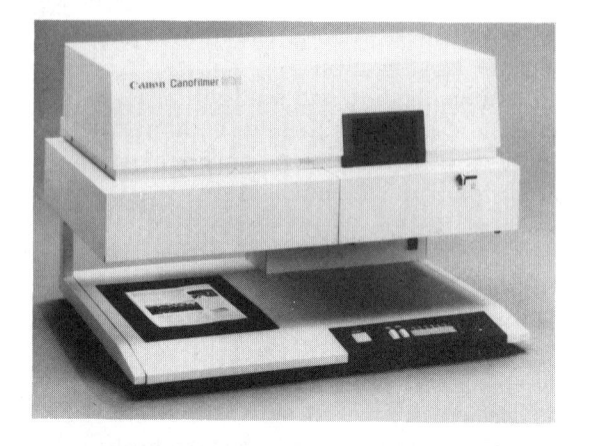

Figure 15-6. Microform camera/recorders.

Sources: Courtesy of Canon U.S.A., Inc., and Bell & Howell Company.

sands of images in seconds with the touch of a finger. Microform readers are used to view the microimages. These readers enlarge the images on the microfilm to a readable size and display it on a screen. To obtain a paper copy of the image, a reader/printer must be used (Figure 15-7). Reader/printers are similar to readers, but they have the additional capability of being able to produce a paper copy of a microimage.

One of the important recent developments in micrographics is the use of bond print or plain paper technology for reader/printers. The use of plain bond, rather than zinc oxide or dry silver paper, results in greater convenience and superior quality. Plain paper reader/printer printing costs, however, are greater than plain paper copier costs. This is because plain paper microfilm causes higher toner consumption and therefore is more costly. Users should carefully compare and select the print process best suited to their needs.

Recent technology also makes it possible to use mainframes or minicomputers with highly automated micrographic systems.

Computers and Micrographics

Prior to the integration of computers and microfilm, micrographics were considered archival technology. Computerized data had to be distributed via high-cost paper printouts or on-line data terminals. Through the use of computers and software, micrographics now provide random and speedy retrieval of information. These key factors make micrographics a highly efficient tool for records manage-

Figure 15-7. This microform reader/printer uses bond paper for better copies at a lower cost.

Source: Courtesy of Bell & Howell Company.

ment. The combination of data processing and micrographics has spawned two hybrid technologies: computer-output microfilm (COM) and computer-assisted retrieval (CAR).

Computer-Output Microfilm

Computer-output microfilm (COM) is a process whereby the digitized information stored in a computer's memory is transferred to microfilm. COM can help reduce labor costs and provide for more timely distribution of data. This computerized photographic process eliminates the need to print hard copy then photograph it on microfilm. Data, text, or graphics are stored on a magnetic tape and sent in coded digital form to a decoding device. Then, they are photographed and converted into images on a tape-to-film machine at a very high speed. COM images can be printed at more than ten times the speed of impact printing. It is possible to produce 10,000 pages per hour from a laser printer.

Computer-Assisted Retrieval

Computer-assisted retrieval (CAR) systems use the computer to index and retrieve COM files of randomly microfilmed information and documents with speed and flexibility. CAR allows local and remote simultaneous retrieval and updating of image and computer data information from two separate data bases. These systems provide the user with the instant indexing capability of computers and the inexpensive document storage capability of microfilm. Some CAR systems also include software that provides basic office automation applications, such as word processing and electronic messaging.

The Kodak Image Management System (KIMS) (Figure 15-8) and the Kodak Ektaprint Electronic Publishing System (KEEPS) are examples of how computers can merge with micrographics. These concepts improve image and information capture, storage, processing, and reproduction capabilities. They are designed to make the processing and management of information more cost-effective and more productive.

Specifically, KEEPS is a document imaging system for in-plant and in-office publishing. KIMS integrates multiple-stage media such as films, magnetic disks, and optical disks. The system can store, retrieve, and transmit document images over a local area network (LAN) (Chapter 12).

Benefits of Micrographics

The use of micrographics instead of paper to store and retrieve important documents has several benefits:

1. Cost Savings. Use of micrographic technology in an office reduces information storage and distribution costs, while significantly improving access to vital data. As an example, 4000 pages of information can be stored in a 4 × 4 × 1-inch space on a microfilm cartridge.

2. Easy Indexing. Since microfilm images are in permanent sequence, seldom are documents misfiled or missing. *Indexing* is simple since the user creates the retrieval system, which can correspond to existing paper files. Computer indexing has helped merge the roles of records managers and data processing personnel.

3. Efficient Management of a Record's Life Cycle. Micrographics has helped manage the life cycle of information more efficiently. Information has a life cycle. The first 90 to 120 days are usually the *active* days. During this time, on-line access is preferable. Once the record's life cycle has entered the *inactive* phase, however, off-line access is better to conserve file space. CAR systems are ideal

Sample KIMS Configuration

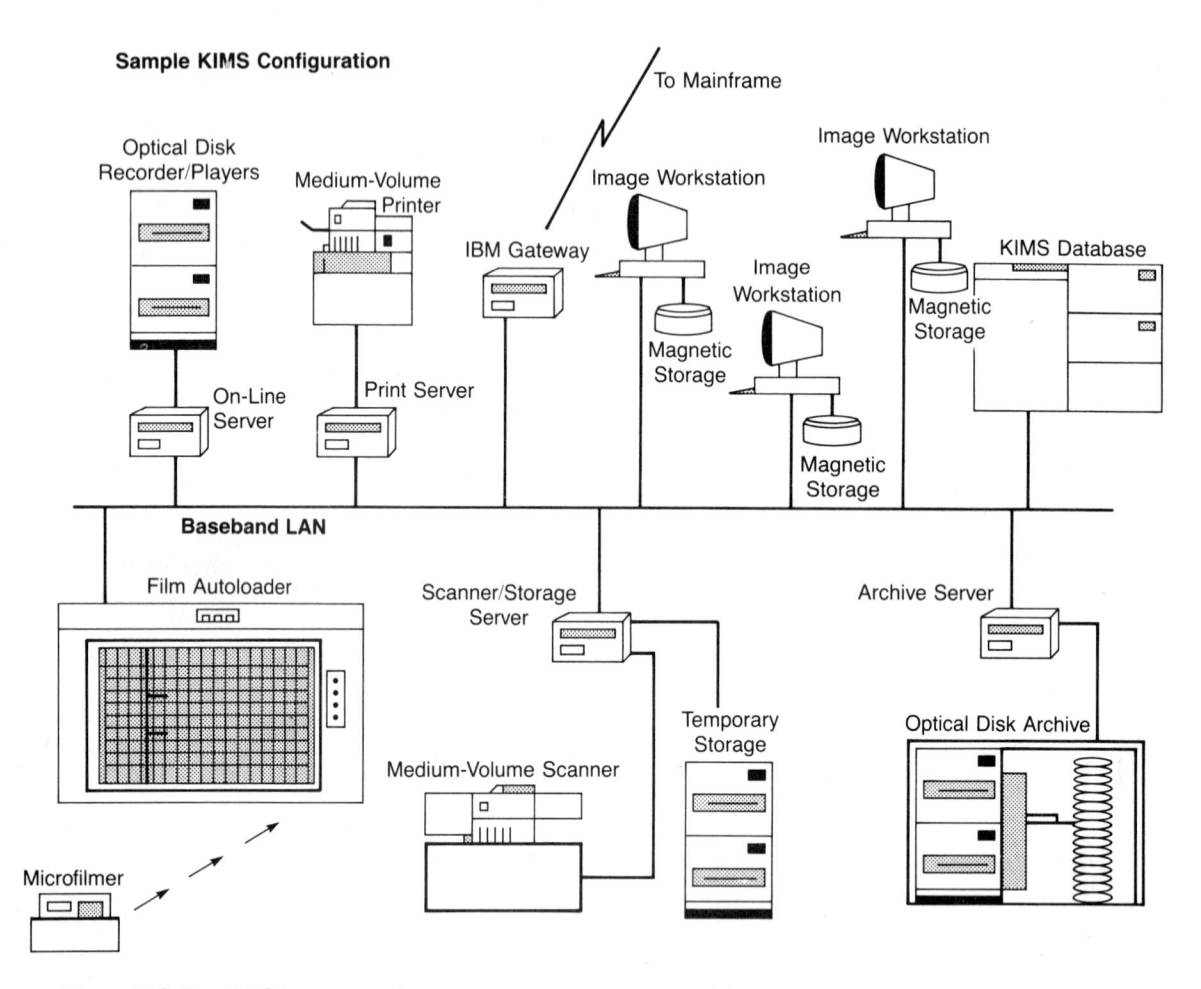

Figure 15-8. The KIMS is a system that merges computers and micrographics to store, process, and reproduce documents and images.

Source: Courtesy of Eastman Kodak Company.

for handling a record's life cycle. Placing records on- and off-line electronically is much easier and faster than it is in a paper-based system.

4. Security. Microfilm ensures the security of records. A duplicate microfilm for working records can be produced inexpensively. Then, an original can be stored in a secure place, away from such hazards as fire and water. Confidential records can also be more easily maintained with a micrographic system. Because the amount of space taken up by documents is dramatically reduced when converted from paper to microfilm, security and control of the files are more effectively monitored. In addition, microfilm records will remain in good condition for years, even when they are referred to frequently.

Micrographic Applications

Micrographic applications are found throughout the spectrum of all business and professional activities. The signs of an inefficient paper filing system can probably be seen in any office. One doctor, for example, has floor-to-ceiling file cabinets. Sometimes he waits five to ten minutes while his secretary hunts for a patient's file. In many cases, documents are lost. Workers sit around and wait for files. It all adds up to an incredible amount of wasted time and money.

The largest users of micrographic systems are those with large data bases that require frequent access. The banking, financial service, insurance, and utility industries were pioneers in converting paper records to film for compact storage and easy access. As a result of their success, thousands of applications emerged. Any large corporation can use micrographics to help increase worker productivity and speed retrieval time.

With the advent of portable and lightweight desktop microfiche readers, small businesses and professional people are now automating their records. Doctors, attorneys, bookstore owners, and real estate brokers, as well as libraries, hospitals, and medical centers are using micrographics. The portable reader enables a salesperson's carrying case or a service representative's van to become a mobile office. Parts catalogs, price lists, technical bulletins, pictorial displays, engineering drawings, and service procedures can be moved from city to city easily.

New Technologies in Storage and Retrieval

Micrographics, like other segments of the information processing field, have been undergoing tremendous changes. The marketplace has evolved to the point where new solutions will soon be needed to accommodate its impending growth.

Floppy and hard disk storage systems were discussed in Chapter 8 as separate parts of computer components. It is presented here again, to provide a perspective for understanding and evaluating newer storage technologies. *Floppy disks* are still the most common mass storage medium (Figure 15-9). They are inexpensive and offer unlimited space, but the user must not mind taking floppies in and out of his computer at frequent intervals.

One floppy disk will hold the equivalent of about 180 paper pages of information. An 8½ × 11-inch page of text, double-spaced, contains about 2000 characters. Today's floppy disks can contain as much as 3.2 megabytes or over 1600 typewritten pages of information.

But there has been a shift in user patterns and style that makes floppy disk technology seem old and cumbersome. Consumers are demanding hard disks. As a result, manufacturers are responding by providing more storage capacity at lower cost. For example, hard disks originally had a 5-megabyte capacity. Manufacturers discovered that it cost no more to make a 10-megabyte disk drive than it did to make a 5-megabyte one. Ten megabytes hold more than twenty-seven IBM PC floppy disks' worth of information—almost 5000 double-spaced pages. Consumers gained twice the memory for essentially the same price.

Hardcards

The latest device to enhance hard disk technology is a miniaturized 10-megabyte hard disk drive that fits on a standard plug-in board (Figure 15-10). This drive can be inserted into any expansion slot on a variety of personal computers. The new hard drive system, called the *Hardcard,* was developed by Plus Development Corporation of Milpitas, California. It consists of a 3½-inch hard disk drive unit that can be inserted into the computer's row of internal expansion slots. A software program tells the computer how to use the new storage device. The user loads the program into the machine's memory and is ready to ex-

Figure 15-9. Floppy disks, or diskettes, are the most popular storage medium for PCs.

Source: Courtesy of Eastman Kodak Company.

pand the computer's storage capability. A built-in file management program enables users to sort through the 10 megabytes of data—equal to about 6000 typed, double-spaced pages—quickly and easily. This latest achievement in miniaturized hard disk can easily double the storage capacity of a desktop personal computer and may soon be built into lap-size computers.

Computerization of storage and retrieval

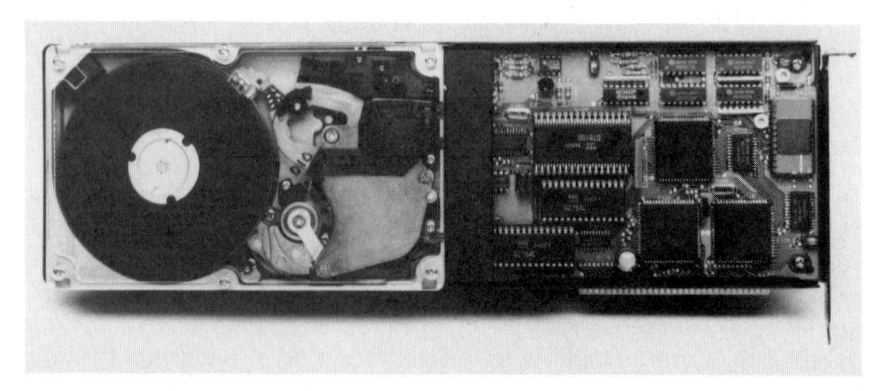

Figure 15-10. A 10-megabyte hard disk drive on a plug-in card (shown with exposed head disk assembly).

Source: Courtesy of Plus Development Corporation.

systems has provided exciting new technology. The following are newer systems that have improved the storage and retrieval tasks through integration of office functions.

Optical Disk Storage and Retrieval Systems

Ongoing advances in technology are refining optical disk-based systems. Now users can squeeze a roomful of information into a cup (Figure 15-11). The optical disk system's greatest feature is its storage capacity. Most magnetic disk drives used today are capable of storing kilobytes or megabytes (thousands or millions of bytes) of data. But the new optical memories can store gigabytes—billions of bytes. On just one side of the larger disks now being manufactured, it is possible to store 100 times more data than can be stored on the highest-capacity magnetic disk and 2,500 times more than on the lowest. Each retrievable can be viewed on a monitor in a half-second. Optical disk systems provide on-line mass storage and retrieval of up to 20 million page images.

Optical disks resemble standard phonograph records except they are slightly thicker and sturdier and their surface has a metallic sheen (Figure 15-12). The term *optical* refers to the fact that these disks store data, for later retrieval, in digital form by laser technology. Information is burned onto the disk's surface, creating the necessary bits and bytes that are familiar to a computer's processing unit.

Optical memory disk recorders allow users to record in their own offices, letting the user control the accessibility. It is like having a private data bank. Confidentiality is protected.

A doctor, for example, could keep all patients' records, the contents of many medical textbooks, and a pharmaceutical data base on one disk. An attorney could store a backlog of cases for twenty-five years on one optical disk and still have room to add new cases for the rest of his career.

Today, the average businessperson working

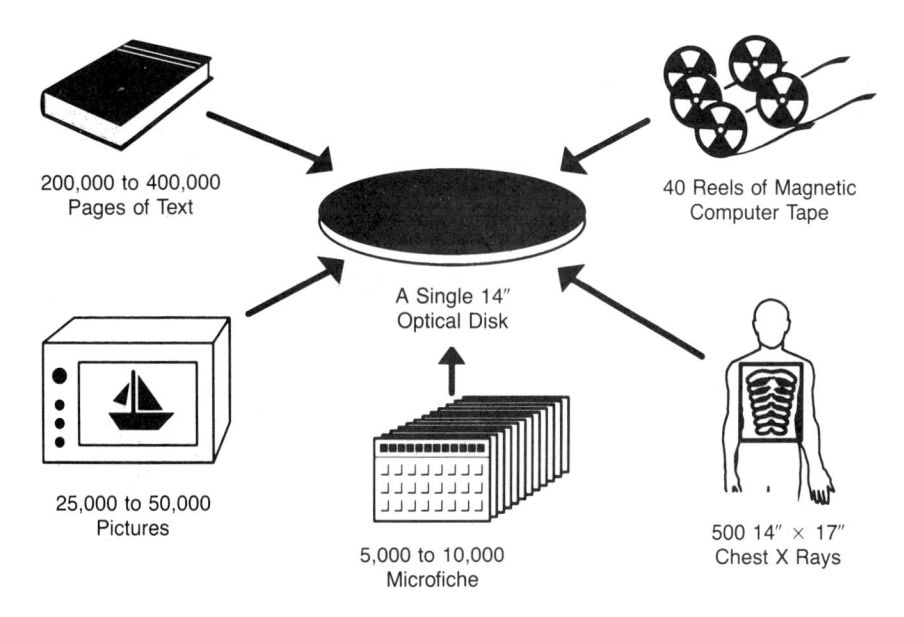

Figure 15-11. The storage capacity of an optical disk.

Source: Courtesy of E. I. du Pont de Nemours & Company.

Figure 15-12. A 12-inch optical disk.

Source: Courtesy of Maxell Corporation.

for thirty years consumes over 1 million sheets of letter-sized paper weighing 8500 pounds. Scattered on the ground, that paper would cover an area five times as large as Yankee Stadium. With laser optical technology, a person can store the same information in an area smaller than a hat box.

Peripherals have their own printers to make hard copies and a facsimile machine so users can transmit stored information to receiving units anywhere in the world. Previously, optical disks could enter data but the data could not be changed, which limited the optical disks to sorting archives. Now, optical disk systems have been developed whereby data can be moved, changed, and erased just as they can on magnetic disks.

The difference between the optical disk and microfilm is that the optical disk is not in human-readable form as the microfilm is. It is read off a disk by a laser and then converted to human-readable form by a computer terminal.

Compact disks (CDs) are part of the optical disk technology that provides for the storage of huge amounts of data on one small disk. A 5-inch compact disk can store the information on 1500 floppy disks or 250,000 typed pages.

Superchips: A New Generation of Microprocessors

The microprocessor is a silicon chip that acts as the brain of a computer (Figure 15-13). It revolutionized the computer and accelerated the information age. As technology progressed, these tiny silicon chips enabled computers to calculate and store vast amounts of information. The earliest microprocessors could only process 4 bits of information at a time. That was barely enough to run an electronic toy or a handheld calculator. However, they had tremendous appeal because they were inexpensive and could accomplish simple tasks for one-tenth of one-hundredth the cost of big machines.

The first personal computers were made possible by 8-bit microprocessors. Later 16-bit chips were developed. New generations of 32-bit microprocessors are emerging. The new microprocessors will give low-cost desktop

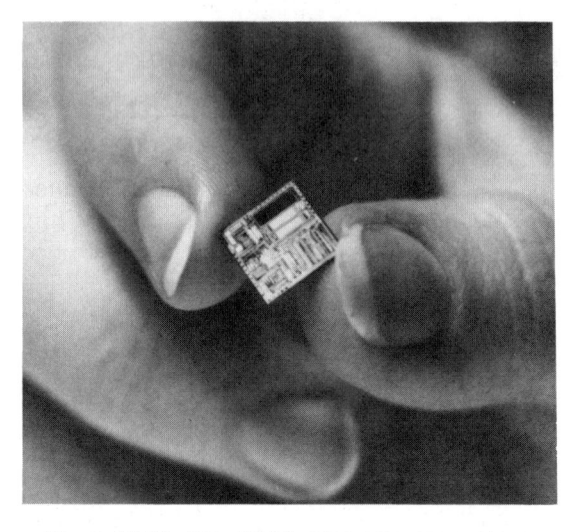

Figure 15-13. This 32-bit microprocessor chip can make more than a million calculations per second.

Source: Courtesy of AT&T/Bell Laboratories.

workstations and personal computers the speed and storage capacity they need to do the big jobs formerly reserved for expensive minicomputers and mainframes. A 32-bit microprocessor is able to process thirty-two pieces of information simultaneously and execute up to 8 million tasks in a second. Not only can these *superchips* manipulate information at great speeds; they can also store massive amounts of information in the computer's memory. Motorola's 68020 microprocessor, for example, can execute between 2 and 3 million instructions per second. Superchips will not only bring processing power to computers but will also increase their storage and memory. In addition to superchips, two other related technologies have expanded their product ranges to include high-density memories: magnetic bubble memory and gallium arsenide. *Magnetic bubble memory* consists of spots on a thin film semiconductor. The memory stores data in magnetic domains, or bubbles, in a thin film on a garnet chip. This storage technique uses a *nonvolatile* technology that enables the computer to retain data stored in its memory even if the power is shut off. Magnetic bubble memory's nonvolatile nature is particularly important when problems with energy and power supplies can mean frequent blackouts and power surges, which can destroy stored data.

The compound *gallium arsenide* is another advance in microelectronics that is an alternate to silicon-based microprocessors. Chips based on gallium arsenide (GaAs) offer operating speeds many times higher than even the fastest silicon-based devices. Moreover, this compound can operate in wider temperature ranges. It consumes less power than high-speed silicon devices and provides much higher resistance to radiation.

Galium arsenide is already the preferred technology. It is used in a number of high-speed communication systems and some industrial and scientific instruments. Declining prices and ease of manufacturing will soon open a variety of new markets to this compound.

Data and Information Backup

Whichever storage system is used, words, numbers, and images are susceptible to obliteration. For all their virtues, magnetic, hard, and optical disks are subject to damage from the environment, power failure, and human handling. For these reasons, backup is a must (Figure 15-14). *Backup* is simply making a copy of data that have been stored. There are several ways in which this can be done.

1. When the mass storage available to a personal computer is small, a floppy disk is an acceptable way to backup data. However, as many as twenty floppy disks may be needed to back up a 10-megabyte hard disk. This process could take 45 minutes and you would have to sit feeding the system the whole time.

2. Information stored on a hard disk can be backed up to a magnetic tape on a mirror-image or a file-by-file basis. *Mirror-image*

Figure 15-14. This tape backup kit includes the controller, software and tape cartridge, everything the user needs to install it in an IBM PC/XT/AT or compatible computer.

Source: Courtesy of Priam Corporation.

backups record all the data on a disk and are ideal for archiving large amounts of data such as accounts receivable and payroll records. *File-by-file* backup transfers only selected data from a hard disk to tape. This process can be used to selectively copy smaller data files such as memos or letters.

Tape backup systems use a process known as streaming. *Streaming*-tape subsystems usually use a ¼-inch tape cartridge. The user simply inserts a ¼-inch tape cartridge into the machine and types in a command at the keyboard. The tape cartridge copies the entire hard disk in 8–10 minutes.

3. Another backup method uses removable hard disks. A few vendors are developing small, removable cartridge drives. They can be removed to back up or completely replace fixed Winchester (rigid) disks. These removable hard disks match fixed rigid-disk rates in storage capacity and are far faster than floppy disk or tape options. However, the fixed/removable units are not as popular as the removable-only systems for personal computer applications.

No matter which backup medium is being used, backup is a chore that many computer users ignore. As a result, many vendors are trying to make backup systems simpler. The growing use of large application programs and multiuser systems have increased the amount of mass storage needed. This highlights the importance of safeguarding the information stored on hard disks. To guard against accidents, information workers must develop the habit of backing up data as part of their daily routine.

Media Conversion

An employee in Department A has been told to send a report stored on magnetic disk to Department B so its computer can update its records. A problem arises when Department B tries to use a diskette on its computer, which is different from Department A's. Department A has a different PC model than B. How can this problem be resolved? Send Department B the one hundred-page printed report and let it keyboard the information into its computer? That seems like a waste of time. The answer lies in magnetic media conversion.

Magnetic media conversion is the process of translating data from one medium to another. The actual conversion process involves a combination of hardware and software. The original data diskette is inserted, and its data are taken and converted into a generic (general) format. From that generic format, the conversion program sends out the data in a specific format to the new medium: a diskette, a magnetic tape, or a hard disk. Therefore, information from a personal computer, for example, can be shared with a minicomputer and a mainframe as well.

Strategies for Implementing a Records Management Program

Records management pertains to the organizing, storing, distributing, retrieving, and disposing of information. A good records management program is responsible for controlling each phase in the life cycle of a record (Figure 15-15):

- The creation phase
- The storage phase
- The retrieval phase
- The retention phase
- The disposition phase

Records management, like other information processing components, must be planned and implemented to meet the needs of the organization it serves. Selecting the best records management program will be made easier if the following guidelines are addressed.

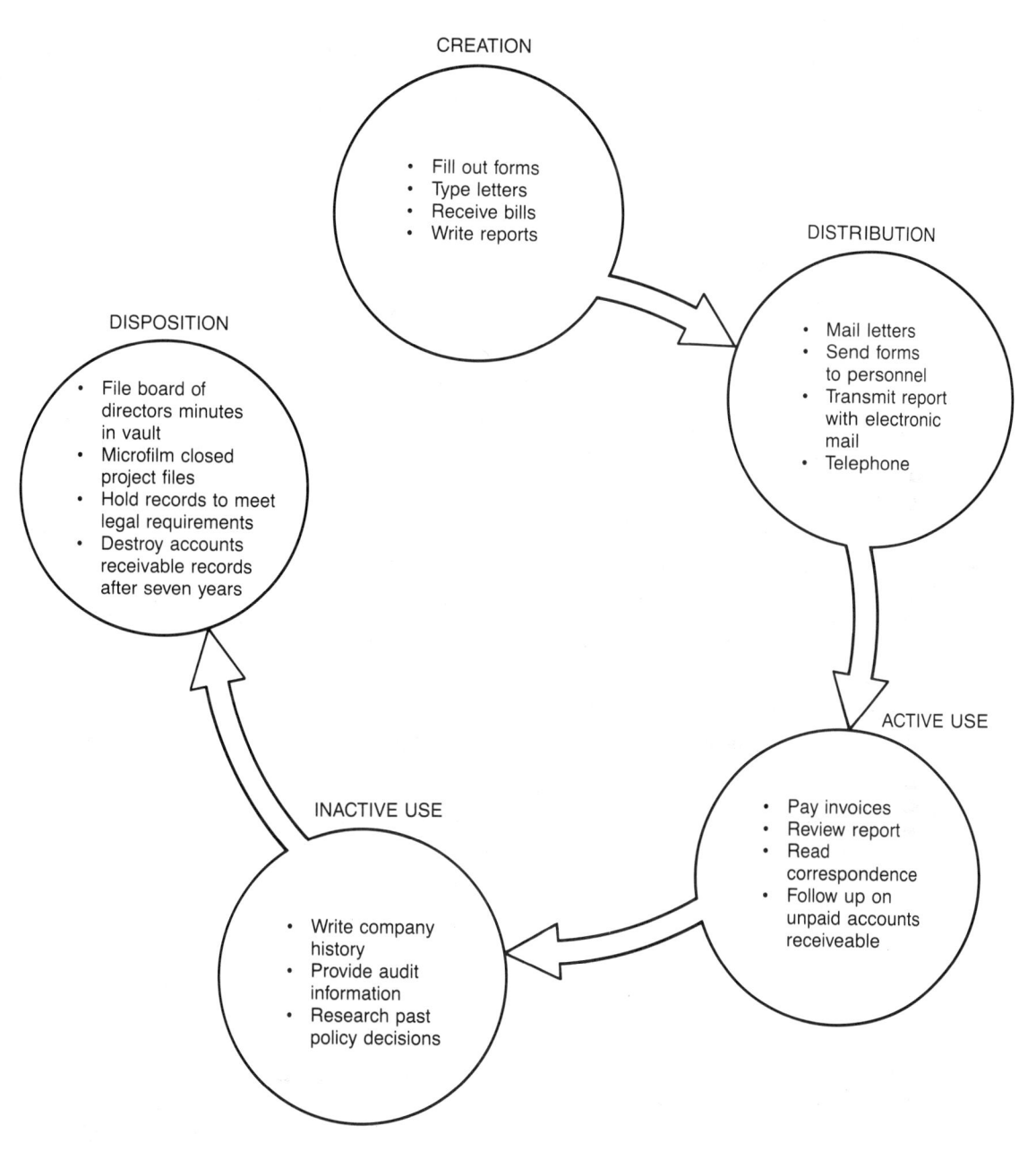

CREATION
- Fill out forms
- Type letters
- Receive bills
- Write reports

DISTRIBUTION
- Mail letters
- Send forms to personnel
- Transmit report with electronic mail
- Telephone

ACTIVE USE
- Pay invoices
- Review report
- Read correspondence
- Follow up on unpaid accounts receiveable

INACTIVE USE
- Write company history
- Provide audit information
- Research past policy decisions

DISPOSITION
- File board of directors minutes in vault
- Microfilm closed project files
- Hold records to meet legal requirements
- Destroy accounts receivable records after seven years

Figure 15-15. The life cycle of typical business records.

Source: Reprinted with permission of *Office Systems '85*.

1. Inventory Present Records. Analyze the type of information that must be filed. Identify vital records. Once the types of records are identified, list the titles that fit specific categories.

2. Volume. Conduct a records inventory, or survey, of present activity and volume of each. Just how many documents are there? How do you find out? Do you take an actual count?

3. Establish Filing Procedures. Define the steps in the creation and filing of documents. A centralized records center, for example, can be created with controls for keeping records.

4. Retrieval. Set up steps that provide immediate access to the information requested. For example, if users complain that information is not easily or quickly available, a micrographics program or a records center could be installed to correct retrieval problems. Color-coded records can alleviate retrieval difficulties.

5. Retention. Establish a retention schedule. Develop a classification system for retaining records. Is the present retention schedule satisfactory? How should authorization for purging documents be handled during the conversion process?

6. Select an Appropriate System. Select the system that best serves the needs of your organization. It should be easy to use and maintain. Hardware technology must be compatible with the installation and expansion of the management programs.

7. Personnel and Training. Decide who will install and maintain the system. Conduct education and training procedures. Prepare a records management procedures manual.

8. Security. Records should be protected to prevent unauthorized copying, theft, and destruction and to avoid the hazards of water, dirt, and fire. Records should never be discarded in a container for general refuse. Shredding or recycling of material is recommended.

9. Storage. Whether the records-storage area is in a large facility or a company basement, systems should be adopted that provide a clean repository for material.

10. Disposition. Decide how long to keep records, vital or otherwise. Abide by legal requirements. Establish a policy of finally destroying unwanted material. Overcome the tendency to save things. Often the retained document, instead of helping, takes up valuable storage space.

This chapter has focused on the benefits of advanced technology for modern records management. Despite innovations, businesses are still struggling to manage and control the growth of paper and electronic information. It is therefore more important than ever to design and implement an intelligent records management program. Whatever system is selected, from paper files to optical laser disks, these general guidelines should be considered when designing and operating a records management program.

Summary

As computers become faster and more powerful, on-line usage grows at a rapid rate, doubling almost every two years. Today's business environment is dependent on improved storage and retrieval systems that provide fast and efficient access to information. The technology of mass-storage media evolved from paper tape and punched cards to the almost exclusive use of magnetic media on microcomputers. Equipment is available to utilize magnetic cards, diskettes, cassettes, Winchester disks, optical disks, and the newest technology—compact disks.

Since its introduction to the commercial market in the late 1940s, micrographics has undergone a number of changes, the most recent of which marry microfilm to computer systems. The result is the ability to digitize microimages as an alternate storage medium. *Computer output microfilm* (COM) devices transfer computer-generated data to microfilm. *Computer-assisted retrieval* (CAR) systems use the computer to index and retrieve COM files of randomly microfilmed information and documents with speed and flexibility.

The microimage methods that combine films with electronic technologies serve to create sophisticated storage and retrieval capabilities.

Review Questions

1. Define records processing. What is a record? What is meant by the active process?
2. In the midst of an information age, where electronic computers and telecommunications thrive, why is the use of paper still on the increase?
3. Compare a centralized filing center with a decentralized system.
4. Can you list some advantages in a lateral filing system?
5. What are mobile files? Why are they attractive to buyers selecting a file system?
6. Define micrographics.
7. Why is cartridge microfilm an improvement over roll film?
8. What is meant by a reduction ratio? What form of microform has the highest reduction ratio? The lowest?
9. What can be done with present paper records once an organization decides to convert to a micrographics system?
10. Briefly compare COM with CAR.
11. Describe how a portable reader can help a traveling sales representative or a computer repairperson.
12. Compare the storage capacity of floppy disks with that of hard disks. What makes floppy disks cumbersome and inconvenient to use as a storage medium?
13. What is an optical disk? What is the range of storage capacity of optical disks?
14. What are some of the reasons that make backing up stored data important?
15. Describe magnetic media conversion.

Projects

1. You are a secretary in a medical group facility. Three doctors, six nurses, an X-ray technician, and two secretary-receptionists serve 2100 patients. All patient records are stored in vertical paper file cabinets. Over the past several months, the doctors have been complaining that they have to wait too long for access to patient records. Documents within patient file folders have been missing and are often not up to date. You suggest that the medical group convert to a micrographics system for better records management. The doctors do not know anything about micrographics and ask you to submit a report. Research and write a report. In your report, briefly define micrographics and list some advantages over the present paper-based system. Suggest how the conversion process can be implemented without disturbing the current filing system.
2. Research current literature in the micrographics reader and reader/printer markets. Design a product comparison chart. Give details about the many readers and reader/printers available on the market today.
3. Phyllis Adams of Fieldston Manufacturing Company has a problem. She has been told to use the personal computer to create a file of

sales records in her department. She must show unit amounts and dollar amounts sold each week across the nation, by state.

That is not the main problem. Rather, after a report is printed from it, the file is to be sent to the accounting department so that department's computer can update its records. In itself, dispatching the diskette is not a problem. The trouble arises when accounting tries to use the diskette on its computer. Phyllis' department has an IBM PC and the accounting department has an Apple Macintosh. The diskette from the IBM is not the right size, much less the right data format, for the Apple.

Should Phyllis send the accounting department a hard copy of the seventy-page report and let it keyboard the information into its computer? This would involve two worker-days a week to rekeyboard the information. Can you think of a better way to transfer the information? Describe a technology that solves this problem in a more efficient and cost-effective way.

CHAPTER 16

Supplies

After reading this chapter, you will understand
1. The difference between brand name supply manufacturers and private label manufacturers
2. The choices in selecting a source for supplies
3. The methods for protecting and storing supplies
4. The variety of supplies and their applications
5. The ways to control the costs associated with copying

Planning the integrated office involves more than acquisition of equipment and selection and training of operators. The initial capital outlay for office information systems is substantial. Often overlooked, however, are the costs associated with the day-to-day operating expenses. They include keeping the equipment "up and running," finding the proper supplies, selecting a reliable source, and exercising sound handling procedures.

Once you have purchased your basic computer system and peripherals, the expense of the supplies, over the life of the equipment, is usually far greater than the cost of the equipment itself. Even a small personal computer can cost nearly half its original price in ribbons, diskettes, and printwheels in the course of a year.

Supplies are extremely important to the quality and quantity of output. American offices spew out millions of pages of printouts, photocopies, and letters each day. Offices consume hundreds of thousands of ribbons, printwheels, and related supplies as computer systems hum, whirl, beep, and blink in their nonstop pace of processing information. Supplies and accessories consist of the following:

1. Media, such as diskettes, on which the information is stored.
2. Ribbons
3. Printwheels
4. Cleaning products
5. Disk storage equipment
6. Copyholders
7. Paper and forms
8. Photocopy consumables (paper, toners, and replaceable parts)

Managing and Controlling Supplies

The deluge of paper used and supplies consumed can become overwhelming to an organization that does not have some plan or strategy for control. Feeding the modern integrated office places tremendous pressure on management to control the purchases of the supplies and to establish a lid on the inflationary spiral of offices costs.

Depending upon the application and on the volume of work generated, a large percentage of the office budget goes for ribbons, diskettes, forms, magnetic media, and the many other products that churn out information on paper.

The Marketplace for Supplies

Computer supplies come from a mixed assortment of brand name manufacturers and "private label" distributors, all vying for a share of this lucrative market. The office supply field is vigorous and teeming with competition. "Big league" brand name suppliers are plentiful. Manufacturers may also subcontract private label products from a "secondary manufacturer," who in turn may have a private label attached to the product.

Evolution of Supplies

A whole new line of supplies has entered the marketplace. As new hardware technology is introduced, manufacturers of supplies and accessories are constantly trying to keep up with, and even anticipate, equipment trends. There is a large growing office automation aftermarket.

Correction fluid, for standard typewriters, ink for mimeograph machines, paper tape for calculators, magnetic cards, and 8-inch disks for early word processors are examples of diminishing markets. These supplies flourished when their companion products were the staples of the office work place. However, the life cycle of high technology supplies is much shorter than the life cycle of conventional office supplies. Many of the new products are technically obsolete by the time they are on the market because of the speed of technological advancement. Today's vendor of supplies must keep pace with the technology, constantly revising or changing products to satisfy consumer needs.

Selecting a Vendor

Word and data processing supplies are very important elements in today's office. Selecting the proper supply item is essential to obtain optimum machine performance. Product quality and reliability are vital in acquiring supplies. There are several ways to buy supplies, including the usual first choice: from your hardware vendor, or the *original equipment manufacturer* (OEM). To many consumers, this appears to be the safest source for supplies. The theory is similar to that of a new car purchaser who buys parts directly from an authorized dealer. However, this is not the only source for supplies, and often it is not always the most economical place to buy.

OEM supplies generally cost more than supplies purchased from other sources. Moreover, the hardware vendor's first priority is the sale of big ticket items. The prices of information processing supplies continue to decline as more non-OEM vendors enter the market. Since new suppliers offer good-quality products at lower prices, there is a steady trend by users to purchase their supplies away from the original equipment manufacturer. In addition, as users become more familiar with automation supplies, they tend to view them as commodities. As with any commodity, users shop for price and delivery if the performance and products are similar among brands.

The consumer, therefore, has several choices in selecting a source for supplies.

1. Original Equipment Manufacturer (OEM). The OEM option is best suited to medium- and large-size companies because of quantity purchasing power. The user is ensured proper media, particularly if special formats are required. As a rule, a warranty agreement that implies that the media are covered with the equipment is given. The disadvantages of buying from an OEM are lack of support for media sales due to the small percentage of corporate sales, high selling price, and long wait for product delivery.

2. Media Manufacturer. Media manufacturers usually specialize only in office automation supplies. They do not sell computer systems or hardware components. These vendors offer consumers warranties, toll-free numbers, and brand recognition. Their products are usually sold through independent office product dealers.

3. Independent Office Product Dealers. Independent office product vendors sell products manufactured by someone else. This distribution channel offers the consumer convenience, immediate delivery, and a wide selection of products. Once a consumer establishes a positive relationship, such dealers can assure familiarity and convenience, which can be a great asset when emergency needs arise.

4. Mail Order Vendors. Mail order vendors conduct their business by catalog orders through the mail. They may manufacture some of their products or sell brand names or OEM products. The buyer must be careful in making selections of products pictured in catalogs to be sure that the products selected are of good quality and compatible with their hardware. The advantages of direct mail buying are easy ordering procedures, low prices, and a large range of supplies. Toll-free telephone ordering and same-day mailing services help many catalog vendors respond quickly to consumer needs.

5. Computer Store. Local computer stores are a convenient source for purchasing small quantities of supplies. This source offers the buyer convenience of shopping within a few blocks of the office. Care should be used, however, when relying on the advice and counsel of computer store sales personnel. Computer store salespersons' knowledge of media and supplies may be limited to those products compatible with the systems they sell.

6. *Media Consultant.* A media consultant who specializes in supplies technology may be a good source if your supply needs are extensive and complex. These specialists may be former members of manufacturers' sales representatives who are now in their own business. They may have extensive knowledge of a local market and an expertise in matching supplies to hardware technology. When dealing with an individual supplies specialist, the consumer should be sure to ask for references and check them out.

Also, the consumer should test a sample product or two before buying in large quantities.

General Guidelines for Purchasing Office Supplies

Quality, consistency, and prices are the most important factors to consider in selecting supplies. Quality should be the first priority because office systems can have precise tolerances, and faulty, expendable products can shut down your system, costing time and money. Whichever supply source (vendor) is selected, the National Office Products Association (NOPA) offers the following guidelines to help consumers buy supplies intelligently.

Experience. Does the supplier have enough experience to be able to give sound product advice, as well as advice on office procedures? Can the vendor keep you informed of new ideas, products, and techniques for improving office productivity?

Sophistication. Is the supplier keeping up with new technologies and systems? Does he or she offer training and guidance in the use of automation supplies?

Trained Personnel. Are the supplier's employees experienced enough to provide sound product advice and proper selection for your needs?

Service. Does the supplier take time to get to know you and your needs? Does she or he volunteer ideas to help you cut costs and improve productivity?

Stability. Is the supplier financially sound? Can you depend on an ongoing source of supply?

Inventory Breadth. Does the supplier stock the products you need in sufficient depth to fill your ordinary requirements quickly? Is there easy access to a wholesaler's or manufacturer's warehouse that can ship products quickly?

Flexibility. Can the supplier meet unusual demands when required to do so?

Value. Does the supplier offer good value for the prices charged?

Location. Can the supplier deliver quickly and respond to service calls in a timely manner?

Care and Protection of Diskettes

Selecting a reliable vendor and purchasing quality supplies, however, cannot guarantee that problems will not arise. Users must take measures to back up their investments by developing in-house procedures for properly receiving, storing, and handling supplies. Diskettes are generally known to be sensitive and subject to easy damage if mishandled.

When a shipment of diskettes is received, it

should be inspected upon delivery. Make sure the item as well as the shipping ticket correspond exactly to your order. The following are suggestions to protect diskettes:

- Keep diskettes away from magnetic objects.
- Do not write on labels after they are affixed to the diskettes.
- Do not put paper clips, rubber bands, or adhesive tape on or around diskettes.
- Do not eat, drink, or smoke when handling diskettes.
- Never touch exposed media surfaces.
- Protect diskettes from direct sunlight, as well as from too much heat, cold, or humidity.
- Do not stack heavy objects on the diskettes.
- Back up especially important material on another disk.
- Always store and file diskettes carefully.

Storing Diskettes

Diskettes must be safely stored and filed so that they are protected, organized, and accessible. The proper handling and storage of magnetic media are important factors in the automated office and should not be ignored. The following facts should be taken into consideration before deciding which filing system is best for your organization's needs:

- The size of the library
- The amount of activity
- The level of protection required

Cabinets and safes are generally used for several types of diskettes housed in one location and those used for archival purposes. However, desktop filing systems are mostly used to store media used more often. A variety of devices are designed to make the storage of magnetic media simple, efficient, and safe. They range from plastic boxes, to vinyl page holders, to elaborate cabinets.

Filing systems should be flexible and adaptable to meet present as well as future needs (Figure 16-1). Storage systems must be available for the needs of one or more operators. In some cases, the systems must be moved from one location to another. In addition, the storage device should accommodate an indexing system that can be understood by everyone in

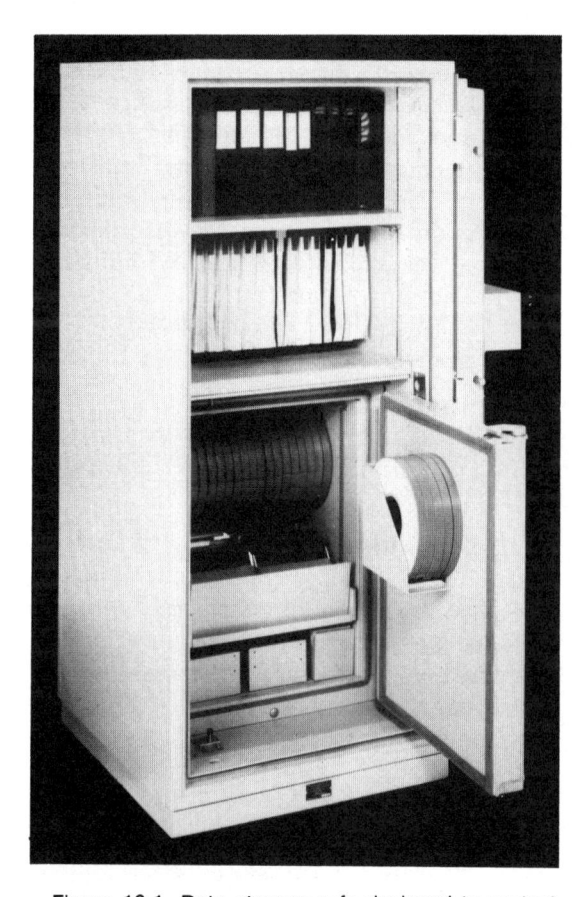

Figure 16-1. Data storage safe designed to protect vital computer-generated business records from fire, theft, or explosion.

Source: Courtesy of Acme Visible Records.

the work group. Storage devices should include locks that prevent unauthorized people from gaining access to confidential information.

Magnetic media should be stored in cabinets that offer maximum protection against fires. Fire can destroy a diskette in seconds. At just 125° F, diskettes distort and vital data are lost. Conventional insulated file cabinets may provide paper protection, but they cannot provide adequate protection for diskettes. Specially designed safes for superior protection against fire and heat levels above 125° F should be used.

Diskettes can be damaged or distorted in other ways too. If the relative humidity exceeds 85 percent, distortion wipes out the information contained on the diskettes. Dust and sunlight can also be damaging to a computer system. Moreover, accidental magnetic exposure erases information stored on diskettes. An organization's business life depends upon records and information. Without taking precautions to protect information on stored magnetic media, records can be destroyed forever.

Protecting and Storing Other Supplies

Ribbons and printwheels are the vital supply items that determine the quality of an organization's hard copy. Once information is stored on the media, often the next step is to produce a printed copy of that information. How well or poorly this hard copy is printed is in part determined by the quality of the ribbon and printwheel used.

Ribbons

There are basically two categories of ribbon: fabric and film.

Fabric ribbons are made of cloth that is looped and loaded into the cartridge to place a continuous cycle of ribbon in front of the printwheel (Figure 16-2). As it is used, the ribbon gradually loses its density, signaling the user that it is time to change the cartridge. When the cartridge is changed, there is no direct contact with the ribbon itself. Fabric ribbons are the least expensive type of ribbon and tend to last the longest.

Film ribbons are made of film or carbon. They also come wound reel-to-reel in a cartridge and are easy to load. Film ribbons may be of the *multistrike* or *single-strike* variety. Multistrike ribbons produce a high-quality print image. High-capacity multistrike cartridges make several passes through the ribbon offering good quality and savings. The singlestrike ribbon and the multistrike ribbon come wound reel-to-reel in a cartridge. They offer the highest quality of impression and are suited for applications that require professional-looking output.

Although film ribbons produce a clean, sharp impression, they tend to wear out quickly. Fabric ribbons, on the other hand, are

Figure 16-2. A fabric ribbon cartridge.

Source: Courtesy of Xerox Corporation. Diablo is a registered trademark of Xerox Corporation.

generally less expensive than film and tend to last longer; however, impression may be insufficient for certain applications.

Printwheels

The printwheel is another factor that determines the quality of hard-copy output (Figure 16-3). Printwheels are made of either plastic or metal. Plastic printwheels are of three types.

1. *Mono-plastic wheels* are made of a simple plastic and are the least expensive.
2. *Reinforced plastic wheels* are slightly higher priced than mono-plastic wheels. They are constructed of plastic mixed with another substance to make them last longer.
3. *Dual-plastic wheels* are the most expensive and most sturdy of the plastic variety. They are made of one kind of plastic and then molded with a harder plastic on the outside.

Metal printwheels are the most expensive on the market. These devices, by virtue of their metal coatings, are extremely tough; they offer a very high-quality print image, especially when used with a multistrike ribbon. Both plastic and metal printwheels offer 10-pitch and 12-pitch wheels.

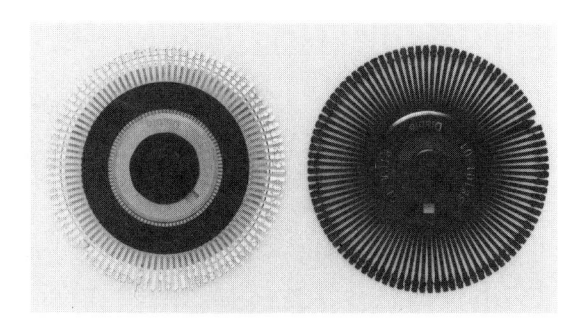

Figure 16-3. Printwheels may be either metal (left) or plastic (right).

Source: Courtesy of Xerox Corporation. Diablo is a registered trademark of Xerox Corporation.

Shelf Life of Supplies

The dates stamped on a package of cheese and a jar of mayonnaise purchased in a supermarket give the consumer an indication of the recommended shelf life of the product. Information processing supplies usually include a similar date stamp. This stamp identifies the shelf life of the product and should be taken into consideration when buying diskettes, ribbons, and printwheels. Ribbons and some floppy disks are no longer good after sitting in your cabinet for six months unused.

Users may still buy supplies in large quantities at discount prices, but they should try to arrange with the vendor to have a portion of the shipment delivered each month. This system will ensure that the user will have fresh supplies on hand. The dealer may not give the full discount, but the buyer may receive a price break that falls somewhere between the high price for buying a dozen disks and the low price for buying a gross outright.

Cleaning Supplies

If the quality of the printed paper output is not up to expectations, it could be that the system needs a simple cleaning. Supplies for cleaning printwheels, such as special liquid cleaners and brushes, are available.

Paper Products

Current surveys show that the information age has not reduced the amount of paper generated but increased it. The computerized office revolves around paper, and computer systems have spawned the computer printout as the ultimate "consumable" supply. Selecting computer printout forms should be done just as carefully as selecting computer hardware and software.

Business Forms

Types of business forms include the following:

1. *Single-copy forms* are paper products con-

sisting of single sheets (not bound) and are the most commonly used office forms. They consist of letterheads, interoffice memos, purchase forms, and financial statements.

2. *Carbonless and carbon-coated forms* permit impressions from copy to copy without the use of carbon. The image is made when special coatings on the back of one sheet and the face of the following sheet are brought together under pressure.

3. *Units sets* are business forms that are bound together in one packet (Figure 16-4). They may consist of letterhead, carbon paper, and copy paper. Carbons and copies may come in different sizes, weights, and colors. The snap-out feature of unit set forms permits easy removal of carbons. Unit sets may also be constructed by using carbonless or carbon-coated papers.

4. *Continuous forms* are assembled in a continuous strip and are designed to use with a *forms tractor* (Figure 16-5). A forms tractor or paper feeder area peripheral device attached to a printer guides the paper through the printer automatically. Hundreds of documents can be printed without operator intervention. A *sheet feeder* is another peripheral device that automatically feeds paper into the printer. The sheet feeder is designed to feed cut sheets of paper automatically into printers.

Continuous paper forms may come in the following varieties:

- Continuous letterheads, memos, business forms
- Continuous envelopes
- Continuous letterheads mounted with carbon copies and carbons.
- Continuous unit set forms using carbonless paper.

Innovations in continuous form papers eliminate the ragged edge "computer form" look. Specially designed continuous form letterheads and envelopes have smooth edges, with no visible perforation ties (jagged edges)

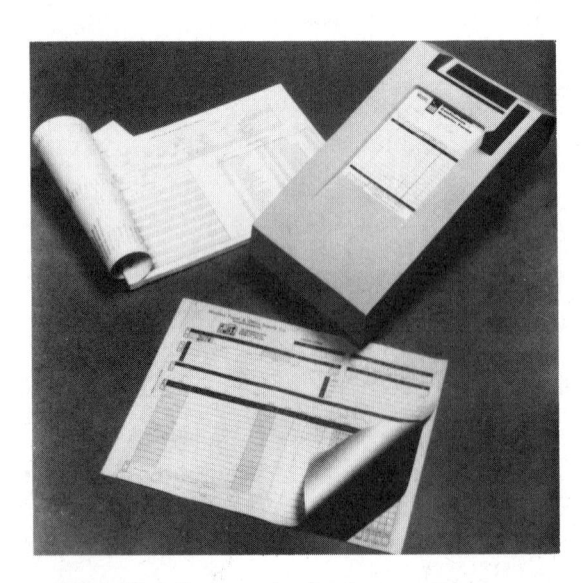

Figure 16-4. Snap-apart unit sets.

Source: Courtesy of Moore, Business Forms & Systems Division, Glenview, Illinois.

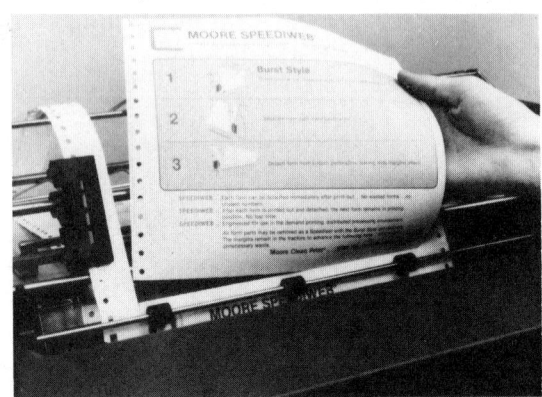

Figure 16-5. Continuous forms fed through a forms tractor.

Source: Courtesy of Moore, Business Forms & Systems Division, Glenview, Illinois.

when margins are removed. Printout forms should have various shadings and colors that do not tire the eyes.

Paper supplies should be selected, handled, and stored with care. It may be desirable to gain quantity discounts by ordering a large quantity and storing the unused portion until needed. However, if stored too long under adverse moisture conditions, the glue used to mount the paper may deteriorate, which can cause the printer to jam. Therefore, in addition to the quality of the paper product itself, careful storage is a consideration that should not be overlooked.

Copier Supplies

Copying costs in most organizations are now totally out of proportion to actual need. Since the introduction of the first plain paper copier in 1960, usage of copiers continues to increase annually. Every day American businesses create 370 million pages of new business documents, generate 1.9 billion pages of computer output, and make 1.9 billion copies. With a growth rate averaging nearly 5 percent per year through the mid-1990s, the total number of new documents produced annually will reach one trillion pages in 1995.

In an environment where paper generation is adding to the cost of business, there is a crucial need to control procedures for copying and to use intelligent buying strategies for copier supplies. There are specific operating expenses associated with every copy made on any copier. Copier costs can be of two kinds: *invisible* and *visible* costs. Invisible costs are costs directly attributable to wear and use of the machine and its working parts. Major replacement parts of a photocopy machine are the photoconductor (drum), fuser rollers, and lamps. These major replacement parts can be covered by a service contract and do not fall under the category of "consumable supplies."

The everyday visible costs of copiers are the photoconductors, toners, and paper.

1. The *photoconductor* consists of the drum and the surface of the copier. These units are the heart of the copier. Most modern copiers use optical system photoconductors.

2. The *toner* is the black or colored powder that renders the latent image visible. In addition, the toner is composed of finely ground particles of carbon and colorless thermoplastic resin.

3. *Paper* is the major supply cost for copier consumables. Thus selecting brands and grades of paper are important decisions. Anyone involved in the selection process should understand and consider such factors as physical properties of weight, *caliper* (thickness), stiffness, smoothness, electrical resistivity/conductivity, brightness, opacity, and shade.

The office supplies field has become crowded with vigorous competition. In a highly competitive environment, the (office) consumer will be the ultimate beneficiary. As quality and performance are perceived to be comparable among brands, the consumer will shop for price and delivery.

Organizations must begin to plan and implement strategies in order to protect their investments in supplies. Supplies are gaining an important place in an organization's hierarchy of budget items. Intelligent purchasing and handling of supplies can help ensure that the cost remains within the organization's allotted budget.

Summary

Information processing supplies include the media such as diskettes on which the information is stored, ribbons, printwheels, cleaning products, storage equipment, and paper. The proper quality and reliability are vital when it comes to supplies. Quality and consistency are two most important factors to consider in selecting supplies. Users need high-quality supplies because the cost of running a system ineffectively is always higher than the cost of supplies.

Magnetic media must be safely stored and filed so that they are protected, organized, and accessible.

Review Questions

1. What is the difference between brand-name supply manufacturers and private-label manufacturers?
2. List the choices in selecting a source for supplies. What is the main disadvantage of purchasing supplies from an original equipment manufacturer?
3. Briefly describe provisions to be taken to store diskettes to protect them from fire.
4. What is the difference between a film ribbon and a fabric ribbon?
5. What actions can users take to make sure that quantity shipments of supplies are fresh?
6. What is continuous form paper? How is it used in a computer system?
7. Describe the invisible costs associated with copying.
8. What is the major supply cost item for copying? What factors should be considered in purchasing this item?

Projects

1. The law offices of Drayer, Spier, and Harworth have recently installed twelve personal computers and twelve word processors. The cluster of twelve computers and twelve word processors shares three printers. The production of multiclient letters has been slowed by operators' manually inserting letterheads and envelopes into the printers one at a time. As office manager, you suggest the use of paper handling devices such as sheet feeders and/or continuous form tractors to automate the output process. Your recommendation is to attach a sheet feeder to each printer. Draft a proposal to justify your position. Describe how productivity is lost through inefficient paper handling. Cite the advantages of a sheet feeder for applications such as multiclient letters and similar documents. Include in your proposal brochures from a sheet vendor and a paper supplier.

2. Bob Lloyd is the information manager of a large insurance company. One of his main problems is storing all the diskettes within his work group. The volume runs into the hundreds. Mr. Lloyd found the diskette boxes provided by the vendor quite inefficient. He decided to write to other manufacturers of filing supplies and equipment to obtain information about their products.

    After you refer to several office systems journals, make a list of three manufacturers and compose a letter to each. The letter should describe the needs of the department and the type of storage files required.

3. Gather information on the three brands of diskettes that would be compatible with the computer used at your college. Create a chart or matrix comparing three different brands of diskettes. Include such headings as manufacturer, size, model, special features, and price per box of ten.

PART FIVE

Safety, Security, and Environment in the Office

CHAPTER 17

Preventive Maintenance, Service Plans, and Disaster Recovery

After reading this chapter, you will understand

1. The importance of cleaning and caring for computer equipment
2. The effect of electrical power surges on computers and stored data
3. The way that uninterruptible power systems can provide protection for computer systems
4. The value of computer protection plans
5. Disaster recovery plans
6. The need and procedure for backup storage
7. Hot sites and other off-premise storage facilities

Preventive Maintenance

Today's computer systems are extremely reliable. The internal electronic circuits are designed to stand up to the rigors of daily pounding for extended periods. Nevertheless, even the best computer system will require repairs. Taking care of your computer will require a little extra time and effort, but the preventive maintenance can pay off in uninterrupted service and savings on repair bills.

Preventive maintenance for computer systems means cleaning, lubricating, and otherwise maintaining the various components of the machine. Fortunately, only a few areas of a computer need regular cleaning.

Cleaning the Disk Drive

The disk drive is one of the most vital components of your computer system. It is the point where the floppy disks are inserted into the chassis, and where the vital communication between the floppy disk and the computer takes place. When a drive fails, data are lost. The most common problem occurs when dirt, dust, and oil from the environment are transferred onto the read/write heads of the disk drive unit. It does not take much for this to happen, because the tolerances in the disk drive are low.

Special head-cleaning diskette kits let you clean the read/write heads on disk drives. In just a few seconds, without any disassembly, mess, or bother, the heads can be completely cleansed of dirt, dust, and magnetic oxides. Disk cleaning kits vary. A popular version is the wet/dry drive cleaner. *Wet/dry cleaners* typically use a white fiber disk that is sprayed with a cleansing solution, slipped into a disk jacket, put into the drive, and spun for a few revolutions. The spinning action and liquid cleanser dissolve dust particles, grease, and oxides collected on the read/write head.

How often users should clean disk drives depends upon the amount of usage. If the disk drives are used every day, then once a month is a good cleaning interval. If they are used occasionally, then every six months would probably suffice.

Using cleaning kits correctly and buying the proper disks for your drive will reduce the risk of drive problems significantly.

The Screen

The same contaminants that can distort data in a "dirty" disk drive can also collect on a computer's display screen. Poorly maintained computer screens produce blurred, hard-to-read characters; cause errors; and contribute to eye strain. The screen should be gently cleaned with a soft, lint-free cloth that is dampened with a household glass cleaner. Never spray cleaning solution directly on the screen itself. Excess fluid could drip into other parts of the computer and cause damage. Antistatic solutions should be used on the screen to reduce static electricity. The best long-term solution is simply to keep the monitor off when not in use and, hence, not build up static electricity.

However, when using a computer, it is a good idea to leave it on throughout the day. When walking away from the screen for more than a minute or two, it is a good practice to lower the brightness level all the way. Leaving the same image on the screen for long periods of time can cause the image to burn into the monitor's inner surface. The damage done during prolonged exposure of the same data on the screen is called *etching,* and it is irreparable.

The Keyboard

Eating snacks or consuming drinks around a keyboard should be avoided. Food particles or drinks can fall into a computer's keyboard and can cause internal components to malfunction. The surface area on and around the keys can be cleaned with anything but a very harsh cleanser. Dust and dirt usually form between and underneath the keys. To eliminate grime, simply wipe the keyboard once a week with a lint-free, nonwoven, polyester cloth dampened with a gentle cleaner. A moistened cotton swab or soft brush can clean between the keys, and a can of compressed air (with a narrowed nozzle) can effectively blow out anything lying low underneath the keycaps.

Printers

Because the printer produces the final document, its performance is critical in creating a favorable impression on those who read it. Ultimately, all office work is judged by the final document—the printed word. It should be clean, crisp, and consistently uniform in shading. Anything less detracts from your organization's image. The printer is an integral part of your computer system, and printers, like computers, must be clean to perform properly. A small vacuum cleaner or a can of compressed air will get the dirt out. It is also advisable to apply a single drop of lubricating oil to a dot-matrix printer's *carrier bar* (the metal rod upon which the print head slides back and forth).

For character printers, the key to sharp, letter-quality printing is proper care of the daisy wheel or type element. Special font brushes can remove dirt. For more thorough cleaning, there are special type element cleaning kits and products. Others, like ink-jet or laser printers, are more delicate and should be cleaned by an authorized service technician.

The printer's platen should also be cleaned periodically with either a dampened, lint-free cloth or with a specially designed platen cleaner. A once-a-week cleaning for printers should be practiced to ensure uninterrupted performance.

Environmental Hazards to Computer Systems

Environmental conditions, such as static, dust, smoke, humidity, and other airborne contaminants, can be harmful to computer systems. Users must generally be aware of the effects each of these environmental forces has on a system. They should take preventive measures to reduce the possibility of a failed system and minimize the effect of such a failure should it occur.

Static and Dust

Office workers may contribute to static build-up just by walking across the floor. Walking may generate 5000 volts or more of static electricity. During the winter months, static may build up to 10,000 volts. By walking across the carpeting and touching a terminal or desk, the computer may "crash." This results in a blank screen and loss of memory. Computer "crashes" or "static bombs" can raise havoc with deadlines, work schedules, and productivity. Few events could be more devastating than to have a complicated multipage document wiped out and then face the task of re-keyboarding the entire document.

Humidity

Static is caused by a lack of humidity in the air. To combat the static problem, antistatic carpeting or mats can be installed in the work area. This will limit static to well below the levels of sensitivity of computer systems and most electrostatically sensitive components.

In additon to solving the static problem, antistatic carpeting can help to eliminate dust conditions by gathering and holding particles until they can be removed. Without carpeting, the loose dust floats freely or is swirled into floppy disk drives with disastrous results.

The quickest approach to combat static is to spray the area around your computer with antistatic spray. A more permanent solution would be to install a humidifier. A humidifier helps control the moisture in the area, reducing or eliminating the problem of static electricity.

Heat

Excessive heat is another environmental problem that can affect your system. Too much heat can cause chips to burn out. If the temperature rises above 100°F, for example, the circuits within the computer chassis malfunction. Most computers contain internal fans. These fans circulate the air inside the computer to keep it within a safe operating temperature. Microcomputers are not as sensitive to heat as the old mainframe computers, which were housed in almost sterile temperature-controlled rooms. It is highly recommended, though, that all computer work areas be completely air-conditioned. Excessive cold can be harmful to computer systems and supplies as well. Temperatures below 32°F. can be harmful to floppy disks. Disks expand and contract with abrupt changes of temperature. A warped disk will prevent the drive head from staying in contact with the read/write head.

Smoke

Smoke is one of the most harmful contaminants to electronic equipment and especially floppy disks. As indicated in the Chapter 5, many offices are enforcing stricter smoking codes and even complete bans. Nonsmokers are demanding a healthier environment. An antismoking policy not only helps achieve a healthy "human environment" but also serves as an effective preventive maintenance measure for computer systems. To reduce the amount of smoke-related dirt in the air, a table-top air cleaner should be used. The contaminants in the air around the computer will be drawn into the air cleaner and pass through a filter that catches the particles.

Power Failures and Power Surges

No matter how clean the environment or the internal components of your computer system may be, computers are subject to the menace of power line surges and power failures. The levels of power-related problems include voltage drops, voltage surges, and complete blackouts. Overvoltages are far more damaging than voltage drops and are of two varieties:

1. *Spikes* or transients typically last from a nanosecond (a billionth of a second) to a microsecond (a millionth of a second).
2. *Surges* are longer overvoltages, lasting into milliseconds (a thousandth of a second).

Microcomputers require a clean and steady power flow. These systems operate over a very limited input voltage range. This means that an uneven electrical power surge, of only a few milliseconds in duration, can knock a system out of service and require extensive maintenance. Managers must recognize that power is as important to a computer system as hardware, software, and supplies. Damage resulting from power problems can be costly. There are a variety of ways to keep your system running.

Uninterruptible power systems (UPSs) are devices that protect your computer system from the irregularities of commercial power. UPSs isolate the critical load and supply clean, reliable, uninterrupted power during any on-line power problems.

Uninterruptible power systems consist of the following components:

1. A *rectifier* (or charger) that converts alternating current (AC) from a wall socket to direct current (DC).
2. A *battery* powerful enough to run the computer and all its peripherals by itself.
3. An *invertor* that converts DC energy to AC again and feeds, or loads, the computer. These components allow the AC lines to go through a double conversion before getting to the load; therefore the computer is never powered directly by the incoming line.

An uninterruptible power supply (UPS) is the best protection against surges and power failures because it isolates the computer from the power line.

Another preventive measure is alloting your system an outlet of its own. Appliances such as air conditioners and televisions can have an ill effect on computers sharing the same line. A dedicated power line lessens the chance of voltage irregularities.

Turning the computer system on and off several times during the day is not a good practice. It tends to wear down the switch and to put thermal stress on solder connections and circuits. When a computer is turned on and off, you subject the circuit to the same sort of abuse as when the power company causes brownouts, surges, or failures. Some organizations leave computers on twenty-four hours a day, seven days a week. They believe that a stable electrical environment is better than the continual stress of heat and power surges.

The best prevention against the danger of irregular voltage is to back up copies of floppy disks regularly. This practice will minimize the amount of data irreplaceably lost through power irregularities.

Service Plans and Maintenance Contracts

A computer *maintenance agreement* is an insurance policy for a computer. It protects the customer from unpredictable repair costs and may guarantee a specific response time from the service vendor if an equipment failure does occur.

It has been well documented that one of the primary considerations in selecting a computer system is reliability. Today's electronic devices generally perform well, but even the best systems will eventually need repair. Before this occurs, users should select a service or maintenance plan that will be suited to their needs. When computer systems fail, users generally want them fixed as quickly as possible.

Who Will Service the Computer Systems?

The growth of the computer market has increased the need for fast, reliable computer service. The computer repair business is divided among computer manufacturers, retailers, repair companies, and do-it-yourselfers.

1. *Manufacturers* provide maintenance and service plans for their customers. Manufacturers usually repair their own equipment. As the market shifts away from mainframes and minicomputers, product manufacturers will lose a share of service revenues to third-party service providers and customer self-service plans.

2. *Retailers* and franchised computer chains service the products they sell. Some stores sell maintenance agreements in which the dealer will fix a corporate customer's machines in the dealer's localities around the world.

3. *Single-vendor service* is a service concept whereby one vendor maintains all of a customer's equipment regardless of the manufacturer. The benefits of the single-vendor service concept are the elimination of finger pointing among vendors and better coordination of diagnoses and repairs.

    The single-vendor concept has gained acceptance with vendors as well as users. The single vendor is responsible for fault diagnosis and isolation, management of a subcontractor who makes the repair, and service contract administration.

4. *Third-party maintenance companies,* also referred to as *mixed-vendor repair services,* maintain a variety of computer systems and peripherals. They offer several key service options, such as on-site carry-in, pickup/delivery, and shipping. On-site servicing means that a field engineer will visit the user. Most third-party maintenance companies require a minimum twelve-month service agreement for on-

site repairs, but time and materials should be covered under the monthly fee.

Third-party outfits include corporate giants such as Xerox and TRW, Inc. Other third-party firms are chains such as Computer Doctor, Inc., and local mom-and-pop neighborhood repair shops.

Most computer repair companies try to sell contracts that provide annual service for a fixed fee. The repair companies try to persuade customers that service contracts will save them money. Having a computer fixed is an expensive proposition. Labor charges generally range from $46–110 an hour. However, service contracts are expensive, too; they typically cost 8–10 percent of the actual price of the computer system. For example, an annual service contract for a $5000 computer system would typically cost $500 for on-site repairs. Less expensive plans in which the repair company picks up and fixes equipment in the shop are available. Also, there is a customer carry-in service.

Whether the business of selling computers rises or fails, the business of repairing them continues to boom. The spread of computerized operations throughout the American business place is requiring a growing army of repair workers on call day and night to keep the computers humming. It is a growth industry that will continue to be in demand.

This trend will translate into more jobs for technicians. These repair workers, the unsung heroes of the high-tech age, are part of a $22-billion-a-year industry that is growing at an annual rate of 14 percent. Much of the business comes from servicing information processing equipment, such as business mainframe computers, minicomputers, and point-of-sale scanning equipment used in supermarkets and other stores. According to the Bureau of Labor Statistics, the number of computer service technicians employed in the United States will reach over 125,000 by 1995, almost double the number a decade ago.

Built-In Backup Systems

As technology advances, service and maintenance plans may shift from outside repair services to built-in diagnostic systems, and better standards for environmental tolerance and reliability. Each computer system may contain a built-in battery, plus a highly sophisticated power supply that prevents loss of your system in case of a power failure.

Computers are already beginning to diagnose the problems of other computers. *Remote diagnosis,* for example, enables a service computer, miles away from another mainframe, to use a telephone hookup to "peek" into the mainframe's circuits and alert its operators to an impending breakdown. This service is offered by most manufacturers of the largest computers and has won praise from repair engineers as well as computer owners.

Other protection plans may offer a built-in backup system. This means that each unit contains its own built-in duplicate to keep it going if a failure should occur; also it can protect your information from loss or damage due to mechanical or electrical problems.

A *fault-tolerance* computer system is another method to prevent a computer from "going down." Such systems incorporate both a primary set of components or *processor modules* and at least one complete backup set. When any of the components from the main computer system fails, a backup component takes over. Organizations especially vulnerable to crashes can install a second or even a third set of backups so that the computer keeps on computing, even when several components fail at once.

Backup systems, remote diagnosis, and fault-tolerance computers are part of a new trend in computer maintenance. These "electronic doctors" are helping to shift computer maintenance from a labor-intensive business, which is very expensive, to one that is capital-intensive. Computers will be used more and more to diagnose and repair other computers.

Checklist for Computer Maintenance

Sensible care of an organization's computer system is the best defense against breakdowns. Computer care is neither difficult, expensive, nor time-consuming. Throughout the previous sections we have detailed preventive maintenance measures. Some of these involve outside sources and perhaps are elaborate and expensive. The following basic maintenance checklist is neither of these. A few minutes of simple maintenance each week will keep an organization's computer system and peripherals running longer and smoother. The following is a summary of steps to be taken to care for computer systems:

1. Do not place computer or computer equipment by open windows where sunlight, dirt, changes in temperature, and humidity can affect performance.
2. Cover computer terminals and equipment at night.
3. Discourage operators from smoking or eating at their workstations.
4. Use a dedicated (separate) power line for your equipment.
5. Use antistatic rugs or mats under your computer workstation.
6. Clean disk drives at regular intervals using recommended materials.
7. Maintain proper air conditioning and humidity levels.
8. Purchase computer supplies that are compatible with your system. Choose quality before price, especially for printer ribbons, printwheels, and diskettes.
9. If you are not sure how to repair your computer system or peripheral, do not attempt to fix it yourself.
10. Avoid oiling or spraying your computer system. Oil will clog the machine and collect dust.
11. Unplug all equipment before cleaning. Vacuum the printer at regular intervals.

Computer Catastrophes

We have discussed the value of information as a corporate asset in a previous chapter. Information, in the form of a company's computer records, can be more valuable than cash, and the reliability of its computer equipment more vital than the performance of some key executives. Corporate officials usually have many assistants capable of taking over in an emergency; computers and their software programs do not.

Computer disasters and catastrophes arrive in many forms. They include fire, flood, earthquake, explosion, hurricane, and even sabotage.

Disaster Recovery Plan

Any company whose survival depends upon the use of the computer should have a *disaster recovery plan* (Figure 17-1). The objective of a disaster recovery plan is to assure that an organization's operations, employees, and assets effectively survive the impact and consequences of a catastrophic event. Disaster recovery planning is a fundamental responsibility of senior management. Unfortunately, very few organizations that rely on computers have established such plans. Some executives avoid implementing disaster recovery plans because they offer no immediate return on investment. However, far-sighted managers feel that disaster preparation should be treated as a necessary business expense.

A disaster recovery plan is more than just a replacement of computer components. It should be an organized action strategy that attempts to minimize the impact of the disaster. The objectives of a disaster recovery plan should

1. Establish an alternative contingency processing environment after a disaster.

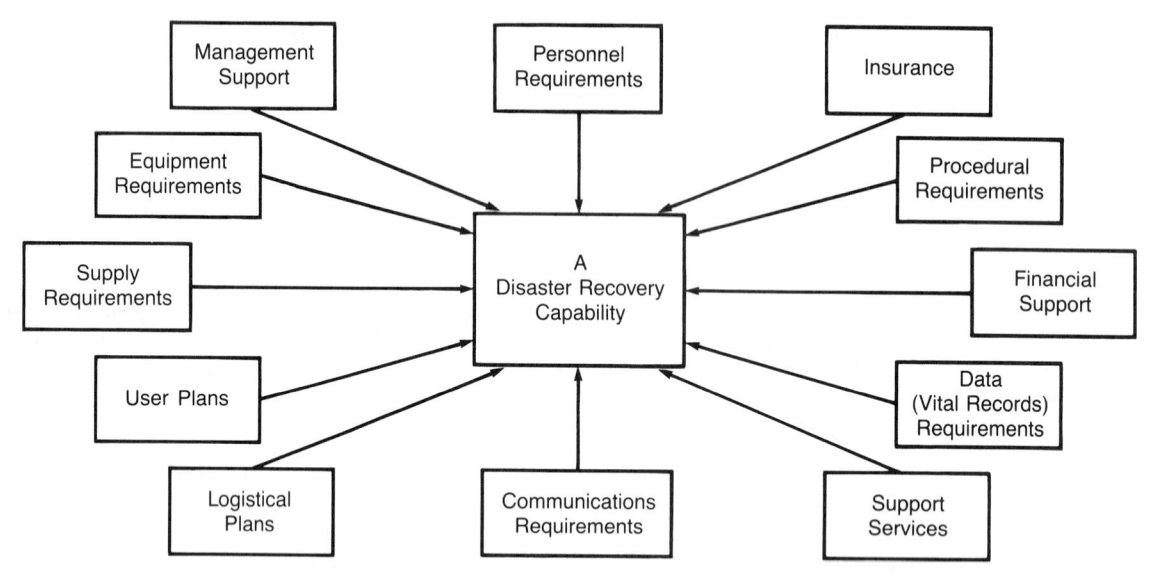

Figure 17-1. Many components are involved.
Source: Reprinted from INFOSYSTEMS, November 1984. Copyright Hitchcock Publishing Company.

2. Create an environment to operate and maintain the organization's automated systems.
3. Refurbish the primary processing site.
4. Return the total computer operations to the primary site.

The plan should include time periods when operations can be resumed. In the short term, critical application processing must resume within a defined interval. In the intermediate term, total application processing must be restored and returned to standard operating procedures. In the long term, all application processing must return to a new or refurbished primary information processing facility.

Developing a Plan

A thorough disaster recovery plan must clearly and consistently define the objectives, resources, responsibilities, and tasks to be used during and after an event that renders an organization's computer operation inoperative. Developing a disaster recovery plan involves the following steps:

1. Define which applications are critical to the organization.
2. Prepare an outline of the plan.
3. Estimate the cost of the processing loss by application.
4. Define the critical resources needed to restore processing operations.
5. Establish a time frame for restoration of total processing.
6. Document the final plan in a manual.
7. Present and review the plan. When the plan is finalized, store it in a secure place.

Strategies

There are several approaches to carry out the key objectives of a disaster recovery plan. Organizations must choose the most appropriate strategies.

Backing Up Data. Backing up magnetic media containing vital information should be standard operating practice for oganizations that use computers. However, most backing up is usually *on-site* backup. Unfortuantely, an on-site backup has one obvious drawback: it is possible to lose both the backup and the original in a disaster.

Off-Site Storage. On-site safeguards such as data backup, uninterrupted power supply, and environmental protection are vital preconditions to a disaster recovery program. But, one of the best safeguards for corporate information is a program incorporating *off-site* storage.

Chapter 15 discussed off-premises storage as a low-cost alternative to storing unused, but necessary, files. Off-site storage has the additional advantage of safeguarding data and information. Having access to computer media stored away from the primary computer means that crucial data files can be reconstructed in the event of master file loss or destruction.

An off-site storage facility (Figure 17-2) should be far enough away from your company to avoid simultaneous exposure to disaster, yet close enough to allow fast recovery.

Figure 17-2. Tape vaults in an off-site storage facility.

Source: Courtesy of Off-Site Storage, Inc.

A good facility should include most of the following:

- A sealed floor to eliminate ground water leakage in the storage area
- An alarm system wired directly to the police station for backup security against intrusion
- A heat, cooling, and humidity control system
- Fire protection, including a vault sprinkler system and an inside vault ion detector
- Frequently changed combination locks on all vault doors
- Proper storage racks/holders for both tapes and disks

Clerical Support. Clerical support can temporarily maintain vital records and may replace computer operations in an emergency. However, only those operations that are not computerized will benefit from such human support.

Outside Adjustment. The use of a third-party service to have mail delivered by overnight express will help to ensure timely processing of data if users are working in a remote site.

Alternate Site. Moving processing functions to an alternate site is a strategy that more and more companies are using in their disaster recovery plans. Specialists in this area provide a service for companies struck by a major disaster—an opportunity to continue business as usual until the destroyed property is usable again. The following are examples of such places.

A *hot site,* or dedicated contingency center, is a computing facility equipped with operational replacements such as CPUs, peripherals, media storage devices, telecommunication equipment, environmental controls, and security. They are unoccupied offices, ready for companies to move into and operate in after they have suffered a disaster. Hot sites are provided by companies who arrange this service for a fee. Prices for hot sites vary according to their equipment and contract specifications. Although expensive, a hot site enables an operation to be up and running in hours, not days, after a disaster.

A *cold site* is similar to an empty, unfurnished apartment into which you move after you have been forced out of yours. It is an empty shell, which provides fully prepared, but not computer-equipped, rooms for emergency use. It does contain air conditioning, electricity, raised flooring, communication lines, and adequate security. Subscribers to this service pay a fee for the rental of these sites. When disaster strikes, arrangements must be made with computer vendors for rapid shipment of complete systems to the cold site. Cold sites are available on a first-come-first-serve basis, meaning that concurrent disasters could restrict usability.

Many companies on Long Island were shut down by power outages as a result of Hurricane Gloria in September, 1985. Because of the total number of subscribers to cold sites and their geographic density during this widespread disaster, there were not enough cold sites to serve all the companies that were shut down.

A *mutual aid* or reciprocal *agreement* can be made between companies to provide a part of their facilities for use during a disaster. This alternative site is usually a friendly, nearby, noncompeting company. It usually does not involve paying a fee or contracting to a third-party off-site company. This arrangement is analogous to staying at a friend's house after being burned out of yours. It is a prudent, cost-justifiable interim solution. However, the following questions should be answered before moving in:

- Do alternate sites have compatible or equivalent equipment?
- Are software systems compatible?
- Is sufficient processing time available?

Transferring Data Off-Site via Telecommunications. Telecommunications provide another protective measure for ensuring continuous operation when a catastrophe is imminent. The weather bureau's reports of forthcoming hurricanes and tornadoes may provide ample lead time to safeguard data. Hurricane Gloria was tracked as it began its journey off the coast of western Africa. Radio reports gave residents along the East Coast of the United States warning to take the necessary precautions. Boats along the harbor were secured, windows were sealed, and in some cases, residents were evacuated. Some businesses transferred their important records and documents to remote branch offices via telecommunications. Customer accounts, sales reports, and orders were sent by high-speed modem from one branch to a remote branch away from the path of the storm. By using electronic telecommunications in this matter, vital data—people and machines—were transferred to a remote site.

The selection of alternate-site routes depends upon the complexity and volume of a user's computer operations and the proximity of the site. The mutual aid concept would be a good approach if your computer system consisted of personal computers and dedicated word processors. The cold site environment, however, is better suited for large computers within a controlled environment. A cold site is far too costly for small-scale companies using only PCs and word processors.

Organizations must tailor their disaster recovery plans to suit their own operations and applications. They must carefully analyze all the key elements described and use one or a combination of approaches. In addition, there is no assurance that any plan will work unless it is *tested*. Various approaches should be tested on a continuing basis as the plans are developed, so that any significant change in conditions can be compensated for.

Human beings working in an information environment cannot avoid the ravages of fires, hurricanes, or other acts of God. However, organizations can recover and survive these catastrophes if advanced planning and preparations are carried out effectively.

Summary

Preventive maintenance for computer systems means cleaning, lubricating, and otherwise maintaining the various components of the machine. Fortunately, only a few areas of a computer need regular cleaning.

Environmental conditions such as static, dust, smoke, humidity, and other airborne contaminants can be harmful to computer systems. Computer users should take preventive measures to reduce the possibility of a failed system and minimize the effect of these environmental forces.

Computer users face a variety of power problems. Protection against these power irregularities is vital. Managers should take steps to ensure that their computers are not damaged should the electricity go awry.

Any organization whose livelihood depends upon the use of the computer should have a disaster recovery plan. Disasters such as a fire, flood, earthquake, explosion, windstorm, or sabotage can wipe out an operation. A disaster recovery plan should assure that an organization's operations, employees, and assets effectively survive the impact and consequences of a disastrous event.

Some specific approaches in carrying out a disaster recovery plan are (1) backing up data, (2) selecting off-site storage facilities, and (3) using clerical support.

Review Questions

1. Why is it important to clean disk drive heads?
2. Briefly describe how display screens should be cleaned. What is etching?
3. How do workers create static? Why is this hazardous to computer systems?
4. Differentiate between spikes and surges.
5. What is an uninterruptible power system (UPS)? What does it include?
6. Why is the practice of continually turning your computer system on and off harmful?
7. What is a computer maintenance agreement? Does every computer owner need one? Explain.
8. What services do third-party maintenance companies offer users?
9. Describe remote diagnosis. How does it compare to fault-tolerance computers?
10. What are the major catastrophes that will render computers inoperable?
11. What is a disaster recovery plan? List the objectives of such a plan.
12. Most organizations back up data stored on magnetic media on their own premises. What is the problem with this? What steps should be taken after backup is complete?
13. Compare a hot site with a cold site.

Projects

1. You are considering the purchase of twenty personal computers for your organization. A major consideration in your selection is the maintenance agreement offered by the vendor, the manufacturer, or the computer store. Investigate three types of maintenance agreements that will cover your computer.
    a. Original equipment manufacturer
    b. Computer retail store
    c. Third-party maintenance company
    Summarize the terms and conditions of each. Include the cost, the coverage, and the length of the contract.

2. You are the manager of The Bayview Agency, a franchised travel agency. You are about to enter into a mutual aid *alternate site* agreement with The Ocean Spray Travel Agency. Both travel agencies are in the same community, use the same computer systems, and employ six to

ten agents. The hours of operation are 9 A.M. to 5 P.M. The president of your agency asks that you draft this agreement in writing. Draft a proposal for a mutual-aid alternate site agreement as part of a disaster recovery plan. Include the names of the companies, their addresses, types of equipment, numbers of employees, hours of operations, financial terms, if any, and all other provisions that you feel would be necessary.

CHAPTER 18

Information Security

After reading this chapter, you will understand

1. The concept of information as a valuable resource
2. Computer crime and some of the reasons why people commit it
3. Who are hackers
4. The effect of law on computer crime
5. The broad areas of computer security
6. Computer security devices and technologies that protect information
7. Software piracy and methods to thwart its spread
8. The growing spread of worldwide terrorism and measures that can guard against it

The tools and technologies of the information age have brought about new levels of power, independence, and productivity to office workers. Personal computers, workstations, and information networks are allowing companies to compete on a higher level and at a faster pace than ever before. However, with so much information being created and dispersed, the information age has also brought enormous security problems. The freedom of access to information through networks, shared systems, and mainframe data bases has created new problems. It has opened the door to infiltration of computer business records and a whole series of computer abuses and crimes.

As we progress further into the information age, *security* of information becomes an increasingly difficult problem. More PC keyboards will sit beneath the fingertips of American workers. At a finger's touch, workers can link up to extensive banks of information, examine mainframe-based information, or copy it (download it) on their own disks. With this new-found freedom to information access, an ever-increasing volume of valuable and priceless company information is vulnerable to tampering, theft, and loss.

The problems of information security and confidentiality are not new to business. But the increased use of computers has added some new dimensions to information security.

Computer Crime

Computer crime ranges from computer abuse, computer-assisted crime, computer-related crime, to simple computer crime. All are incidents associated with computer technology in which a victim suffers or may suffer loss, and a perpetrator by intention profits or may profit.

The computer, in a computer crime, is usually the center of the criminal acitivity, it is the *instrument* or the means of commiting a crime.

The types of computer crimes may be described in the following ways:

Theft. This pertains to the theft of the physical computer hardware and components. There is an active market for stolen personal computers and their internal components.

Accessing, Changing, and Disclosing Information. Information crime involves the acquisition of valuable information via computer such as confidential product formulas, mailing lists, or other information.

Financial Crime. Financial crime involves the taking of funds through a computer, by transferring monies from one account to another or by other devious fund transfer means.

Software Piracy. Piracy of software involves copying copyrighted or company software for private use. If employees distribute the packages they copy to others this action can be considered a violation of the copyright law.

Vandalism. Vandalism is intentional damage to a computer, computer system, or software by either physical destruction or alteration. This may take the form of bulk erasing, copying files and programs for later use outside the company, or destruction of any of the physical components of a computer system.

Outright vandalism and theft of physical properties are easier to categorize and prosecute than some of the other types of computer crimes such as the taking of information or services.

Who Commits Computer-Related Crimes?

The computer may be the instrument used, but it is the *person* not the computer, that commits the crime. Bank fraud, stolen military documents, and accessing national data banks are stories that make headlines. Actually, these stories are only the tip of the iceberg of high-tech and white-collar crime (Figure 18-1).

Modern white-collar criminals do not carry weapons or look sinister. They work in offices and use a computer as their weapon. They feel that copying software, accessing confidential material, and tapping into subscription-paid data bank services are not crimes or dishonest activities. Actually, making an illegal copy of a program is just like stealing money from the people who created that program.

Hackers: A New Form of Electronic Trespassing

Hackers are individuals who enjoy the challenge of trying to enter electronic computer networks without valid authorization. They use the tools of data communication: a computer, telephone, modem, and software. The hacker simply dials a local telephone number of the data network and begins to access the passwords by creative trial and error until a successful connection is made. Hackers may enter large subscriber networks such as schools, hospitals, or military organizations. They do it for fun, profit, or just outright maliciousness.

However, contrary to popular belief, more computer crime is perpetrated by insiders, or employees, than outsiders, or hackers. According to a survey of Data Processing Management Association members, most of the computer abuses occurred in their workplace, and only a fraction of these offenses were commited by outsiders. Many of the offenders cited in the survey were information workers, such as programmers, data entry clerks, and systems analysts. They were motivated by "ignorance of professional conduct," "misguided playfulness," "personal gain," and "malicious revenge."

Computer Crime and the Law

Information as a Valuable and Tangible Asset

A criminal who robs a bank takes money in the form of dollars, coins, or financial papers. There may not be the same criminal offense when a criminal takes money from a bank by using a computer. What was transferred was not something tangible, like dollar bills or coins, but rather electrical impulses representing money. Money has an assignable value. Physical assets such as automobiles, jewelry, houses, and furs also have a monetary value.

Unfortunately, most people regard information differently. It is nearly impossible to assign a tangible value to different kinds of information, so we tend to regard it as if it had little or no value. Yet, we know information is critically important to virtually every organization, whether that information is company trade secrets, military reports, business records, strategic plans, inventories, or customer lists.

It is for this reason that computer abuse, such as the stealing of information, is more difficult to analyze in traditional terms. The logical category for prosecuting computer abuse might appear to be under the *law of theft*. However, traditional criminal statutes, such as larceny, have been inadequate and ineffective in prosecuting computer crimes. The law of theft implies that *property* capable of being stolen must be tangible and that such property must physically change possession. It is, therefore, extremely difficult to prove computer theft under the law of theft unless the abuser physically removes a tangible item.

It is essential, however, that there should be a recognition of the harm caused by computer crime so that there is a uniformity of opinion regarding what constitutes computer crime. Computer law varies from state to state. The unauthorized transfer of funds may not be considered stealing in some states. Courts and

What makes the computer trespasser our newest public enemy?

In the old days criminals looked like criminals. And they toted submachine guns to prove it.

Today, there's a new breed of criminal. His weapon is the personal computer. His target: the corporate computer.

Unfortunately, this criminal is unwittingly assisted by hackers who make computer trespassing seem innocent. It is not. Computer crime is costing businesses millions of dollars a year. And the problem is growing.

GTE Telenet is concerned about protecting your computer. The trouble is that fewer than half the states have laws governing computer crime. And there's no directly applicable Federal legislation.

Laws won't stop unauthorized access to your computer. But stiff penalties will help.

What can you do? If you agree with the need for such laws, write or call your local, state and Federal legislators.

We want to help you. For a report on what states have these laws now, and the status of proposed legislation, call toll-free 1-800-TELENET.

GTE Telenet

© 1983 GTE Telenet Communications Corporation

prosecutors of computer crime must always answer the question of which law, if any, applies to the activity they want to prosecute.

The Need for Clear and Uniform Computer Crime Law

Confusion over what constitutes a computer crime increases the likelihood that more computer abuses will be committed. It also leaves the victims of computer abuses unprotected. Each time a prosecutor declines to prosecute a computer crime because the law is too vague or does not cover that particular crime, the people damaged usually clamor for specific legislation to prohibit the type of activity involved.

Legislation Efforts

Legislation designed to control the spread of computer crime has been introduced in a variety of ways. Some laws are still under consideration at the state and federal levels. One of the most innovative approaches to the computer abuse problem is the Florida Computer Crimes Act of 1981. It creates two new classifications of computer offenses: an offense against intellectual property and an offense against the authorized computer user.

The Florida law recognizes computer abuse as a separate criminal activity within a range of degrees and penalties. By penalizing many types of computer abuses, such as unauthorized access, the statute shifts the focus away from the monetary value or tangible value, which is inherent in conventional theft laws.

In 1985, Congress expanded the scope of computer legislation that will affect the security of office systems. This action was based on Chapter XXI of Title II of the Comprehensive Crime Control Act of 1974, which makes it a felony to trespass into a computer system owned or operated by the federal government, to modify, use, or destroy data. It is a misdemeanor to trespass into, and obtain information from, a system operated by a financial institution or credit bureau.

Over forty states have passed special computer crime laws intended for specific prosecution of computer abuse and misuse. There are similarities in some of these laws, but the particular acts necessary to make the activities crimes, by definition, are not the same.

Each year, U.S. corporations lose an estimated $3 billion to computer criminals. Passing laws against computer crimes is one way to halt the spread of computer intrusion, but it is not the only way.

A coordinated effort, supported by the public, is required to halt the spread of computer crime. Industry, government, computer users, and consumers must join together to urge lawmakers to provide stronger legal protection to deter computer crime.

Methods to Ensure the Security of Information

Organizations can take a variety of practical steps to protect vital information and to minimize the risk of computer crime. There are two major means of securing information. The first to consider is *physical* security. This pertains to the place in which the data files, processing equipment, and software reside. The second means pertains to securing data through *controlling access*.

Figure 18-1. GTE Telenet's advertising campaign drew responses from hundreds of people who supported the company's efforts for more effective legislation against computer crime.

Source: Reproduced with permission of GTE Telenet Communications Corporation.

Figure 18-2. A locking workstation.
Source: Courtesy of Acme Visible Records.

Physical Security

A security policy must be established that will protect the computer installation in its entirety, as well as within the computer (Figure 18-2). Locks, closed circuit television monitors, security guards, key cards, hand prints, and electronic eyes are some of the means used to physically protect information. Computer hardware components should use special locking devices. Disks should be stored in locked files, and backup copies should be made as a routine procedure to protect against loss or tampering. A simple means for providing security for hardware and software is the use of furniture and fittings that allow terminals to be bolted to desktops or locked into secure cabinetry.

Controlling Access to Information

A good system for controlling access to the computer and its computer information consists of using a number of different products, services, and resources. The following are some of the methods that are used to control access to computer information.

Passwords. Passwords can be used to control the access to terminals, files, records, or even fields within a record. In a password system, users must enter the appropriate password to gain access to the data for which they have been cleared. Multiple levels of passwords can provide entry to different layers of information in a corporate data base. Password use has become very common in computer systems. The following practices should be observed when implementing a password system:

1. A user should avoid choosing a password with a personal reference that may be known by others. Names of spouses, children, pets, nicknames, street addresses, birth dates, and the log-in ID are obvious references that might be associated with the user in the password file.
2. A password must be long enough so that a key search would take an exhausting amount of time.
3. Passwords should not appear in clear text (uncoded) in the computer or during transmission.
4. Passwords should not be visible on the screen when keyed in by an operator.
5. Users should change the password periodically.

Common sense is important in establishing guidelines for password protection. If a password log-on procedure is too complicated, it may deter users from using the system. The best approach is to use passwords to create a hierarchy of entry and progressively more complex entry codes as the information becomes more sensitive.

Call-backs. Another method being used to authenticate the user is the *call-back.* Special call-back modems use the principle of pairing a legitimate password with a legitimate phone number. Both must match for access. The user calls in with the password and the call-back modem system gives an instruction to hang up. Then the call-back modem calls back the user. If the user gives the incorrect password, entry is denied.

Passwords and call-back systems are being incorporated into modems.

Encryption. Another well-known method of ensuring computer security is the use of encryption. *Encryption* is a process that "scrambles" data when they are stored or transmitted. Data so treated become unintelligible without a data "key." When the encrypted data are sent to another terminal, the user can convert the ciphered data back into computer data, or *intelligent text.* The use of encryption can be a complex process. It should be used only for data that are highly confidential and require utmost security. Less critical data can be secured in other ways.

Labeling and Disk Backup. Backing up disks has already been discussed as a common-sense measure to safeguard disks in the event of loss through disaster or other types of destruction. It is also important as it pertains to basic computer security. Data backup is also an important safeguard should an unauthorized user access and change a file or document.

Security Software. Special security software programs designed by in-house programmers or purchased commercially are additional measures of protection. Such software programs are designed to prevent unauthorized data input and manipulation electronically. Security software programs can also audit computer use. This procedure is known as an *audit trail.* When the program is activated, it helps keep track of who is accessing what data, when, and how often. This technique, normally incorporated into mainframe resident software, notifies supervisors of any unusual activity that merits investigation.

Personal Signature Recognition. The user must write his or her name with a light pen after logging on. If the signature matches with an authentic sample in the computer memory, the user has gained access.

Voice Print Recognition. This procedure is similar to the personal signature recognition, except that the user identification is by voice. A special voice recorder is contained in the computer system. After logging on, the user is requested to repeat a sentence. If the pattern of speech (speed, volume, pitch, high tones, low tones) agrees with the speech sample in memory, the user gains access.

Post Security. Post security is a clearing operation that ensures that only proper data and information have been stored. Once the user terminates the connection, the system screens the messages and gets rid of obscene remarks.

Not every system or device is appropriate for all organizations. Those responsible for implementing security systems must weigh the costs of suffering a loss. Then, they can determine the value of each method and develop a complete security plan that is correct for their situation. In order to be successful, computer security has to be an on-going management concern. It is, therefore, necessary to review and update all aspects of a security program periodically. No system is 100 percent secure, but each can contribute to keeping business information in the right hands. For any security system to work, there also has to be an atmosphere of trust and support throughout the organization. Positive attitudes, ethics, and a spirit of cooperation must prevail for the continued security and protection of a firm's corporate data.

Software Piracy: Illegal Copying of Software Programs

Software piracy involves the unauthorized copying of software programs that have been published and copyrighted. Software publishers are the principal victims of this illegal activity. It is estimated that as many as ten illegal versions of every piece of commercial software are copied (duplicated). This translates into millions of dollars of lost revenue for software publishers.

Unauthorized duplication of copyrighted material began with the introduction of copying machines. These devices allowed people to make quick and inexpensive reproductions of printed works. Similarly, high-quality consumer recording equipment has undeniably damaged the record and tape industry. The videocassette recorder (VCR) allows people to make unauthorized copies of motion pictures. In the same way, the proliferation of microcomputers has provided an easy and quick way for users to make illegal copies of software.

However, the unauthorized copying of micro software is quite different, in terms of dollar value, from other types of duplication. The cost differential between buying a legitimate copy (about $80) and making an illegal copy of a popular video movie (about $5 for a blank tape and $3 for video rental) is not as great as the difference between buying a legitimate software program and making an illegal duplicate. For example, *Jazz,* the integrated software package from Lotus Development Corporation, is priced at $595. A blank disk may cost less than $5. The savings realized by illegal copying of software can be substantial.

Pirating software hurts the microcomputer software industry as well as honest consumers. It drains profits from the companies that sell and service the programs and forces the price up, ultimately affecting the consumers.

Antipiracy Policies

The software industry has mounted a campaign to fight illegal copying of their products. They are taking a tough legal stance to prevent their customers from making unauthorized copies of their computer programs. Software publishers feel that the situation will not improve until legal actions are taken against firms that knowingly violate copyright laws. In the past, many user organizations did not actively support antipiracy policies. Managers felt that they already had enough to do without spending time ensuring that software vendors' copyrights and license agreements were not violated. Illegal disks often proliferate quietly through an organization as one person makes a copy for a coworker or friend, who makes an additional copy for someone else. Although these activities seem harmless to many users, they not only are creating serious problems for the mirco software industry but can create serious legal problems for organizations that allow copying. Court actions and fines for violators have prompted organizations to recognize and respect software copyright laws. As a result, an increasing number of user corporations are beginning to instill and promote antipiracy policies.

Technological Advances Against Copying

In addition to legal remedies, technical advances in copy protection have been developed to protect commercial software. Though copy protection can be approached effectively from both a hardware and software basis, until now, software solutions have proven most popular. Copy protection is a piece of computer code written into a software disk to prevent unauthorized copying. The two popular methods are both "lock" methods that involve the application of noncopyable signatures on the software diskettes. One system uses an electromagnetic mark, and the other uses a physical mark that identifies the "original" and prevents it from being copied.

The Association of Data Processing Service Organizations (ADPSO) is developing a hardware solution to piracy. This protection device is known as a *fuse box* or a *key ring*. The system contains a portion of an applications program embedded on a microchip. Unless the proper key is inserted in the box, an applications program will not run. Figure 18-3 illustrates the way the system works.

Although hardware-based copy protection seems to offer the most comprehensive protection, the method has not yet proven practical on a large scale, despite considerable interest. This system requires in-depth cooperation between hardware and software producers. It promises to offer some user frustrations, too, since office software will not run at home and vice versa.

Using disks with copy protection is cumbersome, especially with large networks. These devices inhibit processing, cause printer malfunctions, and cause hard disks to crash. Software vendors are now issuing special disks that allow users to remove the copy protection, provided they purchase disks in quantity and demonstrate to the software vendor that they have an enforceable software piracy policy.

Terrorism and Sabotage

The variety of computer crimes, such as illegal copying and theft of physical property and software, are sometimes characterized as victimless crimes. These crimes do not result in bodily harm or injury to people. There is, however, a new security menace permeating the world. It takes the form of devastation and violence brought about through international terrorism, sabotage, hijacking, and kidnapping. The perverted motives of people with causes move them to destroy and wreak havoc upon people, property, and a society that opposes their philosphies, ideals, and goals.

Today's headlines tell stories about terrorists destroying foreign embassies, hijacking planes and ships, kidnapping foreign diplomats, and sabotaging government facilities.

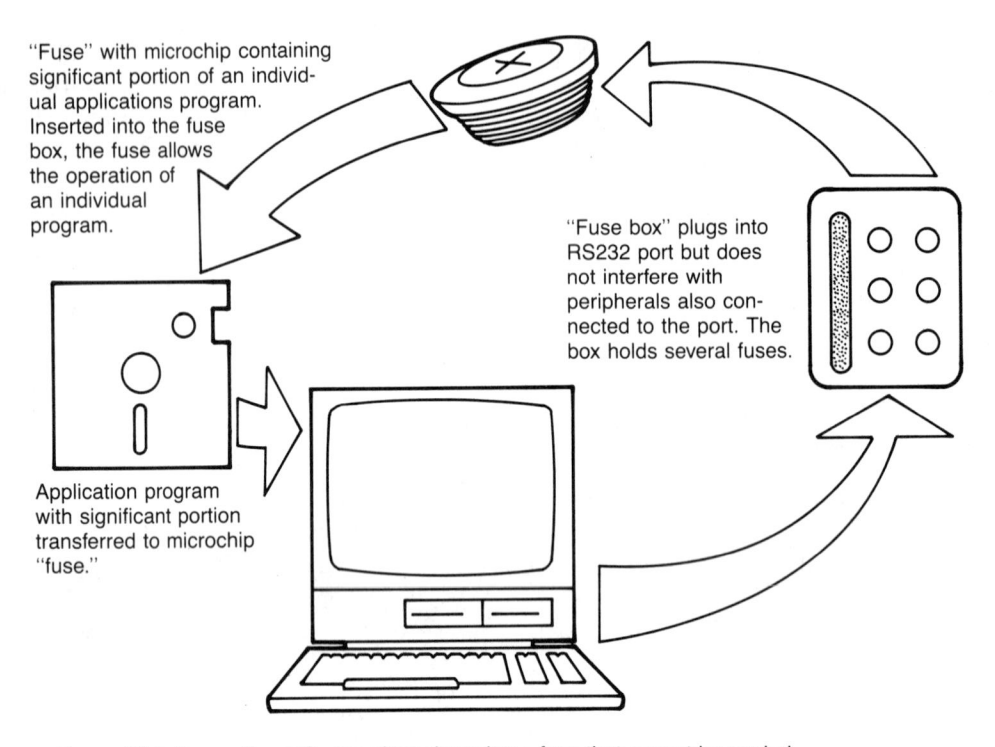

Figure 18-3. Preventing software piracy by using a fuse that cannot be copied.

Source: Copyright 1985 by Cahners Publishing Company, Division of Reed Holdings Inc. Reprinted with permission from *Business Computer Systems*, February 1985.

The acts of spies who pass along government secrets to our enemies are also insidious crimes that breach the security of our nation. Security in today's environment is changing. And the means and levels of violence these criminals are willing to achieve have grown to become an international problem.

Data centers have been the targets of some terrorist attacks. The United States has grown dependent upon complex data processing systems for its institutional well-being. Although the greatest terrorist threat is in Europe, the prospects for terrorism are increasing within the United States. Radical organizations have become more sophisticated. Their targets are those that would have the greatest impact on American policy. The potential to destroy large military and corporate computer and data processing centers looms as a very feasible possibility.

Protection Against Terrorism

Tightening up security around possible targets can never be a sufficient answer. There are just too many targets—if not a cruise ship, then a tour bus, or a computer center, or a boat, or a plane. However, information managers who need physical security cannot wait for the perfect system.

Tightened security has become the focus of concern in all aspects of commercial, residential, and recreational life. Cruise ship lines, airports, embassies, homes, offices, factories, schools, and military facilities are implementing new security measures and high-tech devices for the protection of people, property, and information.

Effective answers to terrorism must be achieved through the coordination of federal agencies and the cooperation of world governments through penetration, preemption, and punishment. In the private sector, most corporations have not established specific preventive measures against terrorism. The question of the way industry should respond to the terrorist threat domestically has not yet prompted security personnel to act. However, many corporations have established physical security measures to protect computer centers and sensitive information. Physical security methods employed at well-run computer centers include the following:[1]

1. Housing computers in one room and peripherals in a different room
2. Low building profile
3. Secure building perimeter
4. Emergency preparedness
5. Designated areas where smoking and eating are prohibited
6. Minimizing traffic and access to work areas
7. Physical access barriers
8. Remote terminal security
9. Universal use of badges
10. Alternative power supply

As the acts of terrorism, break-ins, espionage, and computer crimes escalate, strong physical security is needed within all organizations. The critical asset of an organization is usually its information. This information resides within the computer systems. Organizations must protect their "information assets" with the same safeguards that they protect their tangible assets such as cash, equipment, and physical property. Banks and corporations would never leave their vault doors open. Leaving the data center doors open would be just as bad.

Finding Solutions: A Total Effort

Computer crime and related acts of computer abuse such as fraud and deception are growing at an alarming rate. Creating an effective deterrent will require a combined effort by all sectors of society. It must be a committed approach to a long-range solution. Legal remedies and high-tech devices cannot solve the problems alone.

Attitudes and Ethics

We must begin to educate people and instill a sense of ethics and moral responsibility about the importance of information and data security. We cannot continue to deal with offenders with inconsistency and leniency. We cannot condone computer abuse of any kind—even the innocent pilfering by young hackers. Respect for property, whether tangible or intangible, must be taught at an early age.

New technology brings with it a new set of values, ethics, and etiquette. It also changes the way we behave toward each other. We must not only learn the technical aspect of new systems and devices, but develop behavior that conforms to socially accepted practice. Etiquette reflects what is proper and improper to do. In the absence of a specific law against an activity, positive ethics and etiquette of computer use should serve to guide the computer user. It is clear that we need to develop a whole new set of guidelines of ethics and etiquette pertaining to computer security.

Education: Expanding Public Awareness

There must be a massive public information program to alert people to the issues of secu-

---
[1]Courtesy SRI International, Inc.

rity, the need for vigilant protective measures, and the prosecution of offenders. Corporations, manufacturers, and educators must address the issue of computer security education.

Corporations need to draft and publish policies on data security to cut down on casual "sharing" practices within the workplace. They should train their employees to be aware of the need for data security.

Computer manufacturers should include literature explaining the proper use of computers. Vendors of hardware and software should alert the public to the importance of data security.

Educators must instill a proper attitude toward ethical behavior in general and respect for the integrity of data in particular. Educators can set the tone in a subtle manner about etiquette and personal privacy. The concepts of computer security should be taught as a unit in every computer or business course. Students in hands-on application classes should be taught to use computer security systems and software.

The proper use of computers combined with respect for information must be conveyed, not only to students, but to all of society.

Summary

Computer crime takes the form of (1) physical theft; (2) accessing, changing, and disclosing information; (3) financial crime; (4) software piracy; and (5) vandalism. The computer criminal does not fit the usual sinister profile. He or she may work in an office and feel that there is nothing wrong in copying or accessing confidential information. The law is vague in its perception of whether or not information should be considered a valuable asset. It is for this reason that computer crime and the theft of information are difficult to prosecute. Uniform and standard laws should be enacted to recognize the seriousness of computer crimes.

There are two methods of securing information—physical security and controlled access security. Passwords are a popular method to control the access to computer systems. In a password system, the user must enter the appropriate password in order to gain access to the data. Software piracy involves the unauthorized copying of software programs that have been published and copyrighted. Pirating of computer programs hurts the microcomputer software industry as well as honest consumers.

Terrorism is a new form of security menace. Attacks against people, facilities, and computer centers are the methods of terrorists. Effective measures to counter terrorism must be achieved through the combined efforts of world governments.

There should be a sense of urgency to educate people about the need for moral responsibility in dealing with information and data security. The proper use of computers and respect for information must be stressed to all members of the information work force.

Review Questions

1. Distinguish between theft of computer hardware and unauthorized access to information.
2. What are some of the reasons why people commit computer crimes?
3. Define hackers. What are some of their computer abuses?
4. Why is it easier to prosecute theft of physical property than theft of information? What does *the law of theft* imply?
5. Describe the current legal climate for computer security legislation as a whole. What can be done to improve the situation?
6. What are the two broad areas of computer security?
7. List some of the precautions when using password security.
8. Describe call-backs. What special piece of equipment is needed?

9. Briefly describe audit trails, personal signature recognition, encryption.
10. What is meant by software piracy? In what way does it affect software publishers?
11. Describe one device to thwart software piracy.
12. Why would terrorists attack a computer facility?
13. Why have American corporations been slow to establish effective preventive measures against terrorism?

Projects

1. Visit a corporation within your community that has a computer security system. Gather information pertaining to its system and present your findings in a report. Include the name of the organization, the product or service, and a general description of its physical and internal security system to protect information. Does the company have a policy against software copying? If so, describe it.

2. Compile information about computer security software programs. Design a table or matrix that compares six leading software programs. Include categories such as vendor, title of program, special features, and type of hardware needed.

3. Analyze the physical building security of your school. Describe some of the methods employed. Include suggestions that would strengthen your school's basic protection plan.

4. Visit the library and find a recent court case about the legal actions taken against firms that knowingly violated copyright laws by illegally duplicating software. Submit a written report describing the case and the final disposition.

CHAPTER 19

Ergonomics

After reading this chapter, you will understand
1. The meaning of the term *ergonomics*
2. The way to plan for the health, comfort, and safety of the office worker
3. The variety and applications of furniture and components in the automated office
4. The effect of lighting, color, and noise on the worker
5. The regulation of air quality, temperature, and humidity at comfortable levels

The office is the center of activity for information workers, whether they comprise a small work group or a large information center. Whatever form it takes, the office is a place where ideas and knowledge are exchanged. It is also a point of convergence—a place where individuals, ideas, and technologies come together.

It is the organization that creates the physical and philosophical environment that helps people work better by supporting who they are and what they must accomplish. Each individual's unique needs, tasks, tastes, and job requirements must be integrated with the technology to record, transmit, and process information. The health, safety, and comfort of workers must also be considered in designing an office environment.

The tools of thought and information must be intelligently supported, adapted, and managed. The science that addresses the relationship between people and machines is called *ergonomics*.

Defining Ergonomics

Stated simply, *ergonomics* is the study of human beings at work. In the United States, the discipline is called *human engineering*. The term *ergonomics* originated in Europe (after the Greek words *ergon* and *nomos*, which mean, respectively, *work* and *laws of*). The goal of ergonomics is to make the things people use and the places in which they use them as safe, healthy, and comfortable as possible.

As office environments become more involved in complex technology, the need to consider the human part of the human-machine equation becomes critical. This is partly because automation, such as video display terminals (VDTs), can bring new physical (and mental) problems for workers, such as backaches, headaches, eye strain, and earaches.

Ergonomics: A New Mandate from Management

Ergonomics has become a high priority in strategic planning for a growing number of organizations. There are two major reasons why ergonomics should be included in any strategic plan for acquiring new office equipment and machines. One is that money spent improving the physical design of offices garners an excellent return as a result of better employee performance. The other is that, because of new health and safety laws, there may be no other choice.

Perhaps the most important reason management should consider an ergonomically designed environment is the well-being of the office workers. Poorly designed desks, chairs, and electronic devices have been associated with absenteeism, stress, eye strain, and physiologic discomfort.

Designing the Workstation for More Comfort and Greater Productivity

With the growing use of electronic technology the potential for increasing office productivity has never been higher and the need for precision and total alertness has never been greater. Since the late 1960s, designers and engineers of office equipment have been studying the sight, sound, and motion requirements of people who use electronic equipment in offices.

Ergonomists (people engaged in the design of furniture, machines, and environment) have influenced the production of desktop computers and their VDTs. Their primary concerns are to design workplaces that enable workers to function most productively. An ergonomist considers the interaction of lighting, acoustics, furnishings, privacy, equipment, and placement. All of these aspects make up the total environment for the worker and are discussed in this chapter.

Video Display Terminals

The single piece of equipment that most symbolizes office automation is the video display terminal (VDT). No matter how sleekly designed, computer workstations are ineffective if they are not also comfortable for the user. Although people come in all shapes and sizes and have different requirements, early designs of computer systems were highly standardized. The one-size-fits-all mentality provided inexpensive and cost-effective mass production of machines. However, it did not address the need for variations because of human differences. People had to bend to accommodate a so-called standard technology, when it should have been the other way around.

Ergonomists recognized the need to design *adjustable systems* to accommodate human variations. Manufacturers have considered their advice, and a number of advances have been made.

The Keyboard

The most obvious change in display terminals over the last few years is their physical appearance. First-generation display terminals combined the keyboard and screen into one unit. Today, nearly every display terminal made includes a detached keyboard. Thus, the operator can move the keyboard to virtually any spot on the desk or onto the lap, if desired (Figure 19-1). Many models incorporate angle, rotation, and height adjustments. In addition, keyboards are designed in a "low-profile" thinner version of the standard keyboard, which includes the following special features:

1. Sculptured Keytops. Keytops are individually sculptured to allow the fingertips to strike each key squarely, so the operator is less likely to make errors. Each key's pressure-response action and "click" provide positive input.

2. Palm Rest. A large palm rest extends below the keyboard.

3. Color-coded Keys. Special function keys are color-coded to eliminate confusion during the early learning periods.

4. Minimum Function Keys. Minimum function keys provide for simplicity. Using a single clear key to erase the entire screen is preferable to coding several keys for a single function.

5. Repeat Keys. Frequently used keys, such as hyphens and dashes, should repeat when held down to save key strokes.

Competitive VDT manufacturers are starting to take advantage of the way ergonomics can differentiate between various makes. The mouse, which makes moving a cursor on the screen as easy as using a pencil, bypasses the keyboard for many maneuvers. Even more interesting are the touch screen controls; the user simply places a finger on the VDT screen

Figure 19-1. An ergonomic terminal.

Source: Courtesy of Televideo Systems, Inc.

to issue the computer orders. To help office workers accomplish even more, an increasing number of "handless" cursor control products is becoming available. The foot-controlled mouse and the NOD cursor control allow the user to position the cursor with simple head movements.

The Screen: Eye Strain and Other Health Hazards

Since the mid-1970s there has been considerable concern regarding potential health problems associated with the effects of prolonged VDT use on workers. Prompted by workers' complaints of eye strain, dizziness, nausea, and skin irritations, agencies such as the National Institute for Occupational Safety and Health (NIOSH) have conducted studies on the potential health hazards presented by VDTs.

Although critics still charge they are hazardous, up to now the results have shown that VDTs do not emit any radiation known to be harmful. Various studies, however, claim that a majority of VDT users report physical problems in using monocolored phosphorus screens during normal working days.

The Harvard Medical School estimates that more than half of VDT operators suffer from eye irritation, fatigue, or difficulty focussing. These issues have led national unions and workers' associations to campaign for legislation to improve working conditions for VDT operators.

As a result of these claims, the vendors have responded by designing new terminal enclosures that they hope will bring more acceptance for their products. Factors such as color, type of VDT, display design, and size of characters can have different effects on the worker.

The *size of the characters on the screen* varies among brands and among countries. European standards call for letters and numbers at least 2.6 millimeters (about a tenth of an inch) high; the United States has no standard.

The *quality and design* of screens vary from coarse, flickering CRTs to ultra-high-resolution screens whose clear, sharp-edged characters are formed by no fewer than a million dots, called *pixels.*

A good display terminal also has separate *contrast and brightness* controls that are accessible to the operator.

The *contrast control* determines the brightness of the background in relation to the characters. The *brightness control* determines how much light is emitted from the characters themselves.

Display *color* is largely a matter of personal choice and the environment in which the VDT is used. The largest distinction is between light characters on a dark background (positive contrast) and dark characters on a light background (negative contrast). Green-on-black is no longer the only color combination available. The Apple Macintosh and most Digital Equipment Corporation machines, for example, have black lettering on near-white backgrounds. This combination is less searing to the eyes than green-on-black.

According to Sandy Parks, president of Color Charisma, *color* may be having a psychological effect on the worker as he or she views it on the screen. For example, if the color blue slows heart and respiration rates, Parks reasons that it might be beneficial for operators who spend the most time in front of a VDT to use a blue image because time seems to pass more quickly. Similarly, yellow stimulates creative thinking, and green, the best choice for the majority of users, combines the effects of blue and yellow. Green promotes logical thinking and dialog and has a calming effect.

A problem troubling eyesight even further is *glare.* The VDT screen is suspected of causing glare for two reasons: (1) The monitor screen acts like a mirror, reflecting indirect lighting from a variety of often hard-to-find sources. (2) Since the screen is tilted upward, it is likely to reflect the light from bright windows or long rows of overhead lamps. These factors intensify the glare on the screen and hamper an operator's performance. New VDT screens combat glare with matte surfaces instead of shiny ones. An etched or matted screen reduces sharpness but usually will not improve contrast. A better alternative would be to use filters that fit over display screens (Figure 19-2). These filters enhance contrast and reduce glare.

Eye strain, according to the National Institute for Occupational Safety and Health, is a problem for 80 percent of all VDT operators. When operators do not strain their eyes, they are more alert. This helps to reduce errors and increases productivity. Correct lighting will be discussed in a special section of this chapter.

Overall Size

Ergonomic research has also centered on the spatial requirements of the complete workstation. The cabinet design should be small enough to leave the user plenty of work space. *Footprint* size pertains to the amount of space a device takes up on a desk surface. This is an important factor in companies where desk space is limited. Computer workstations should be lightweight to permit easy transfer to another office or desk by any member of the staff. Terminal vendors are designing new systems that take up a minimal amount of space. One terminal has a 9-inch screen that is portable; the keyboard folds up into the terminal face and locks. Some systems occupy less desktop space than an open looseleaf binder.

Computer workstations include personal computers, advanced work processors, data

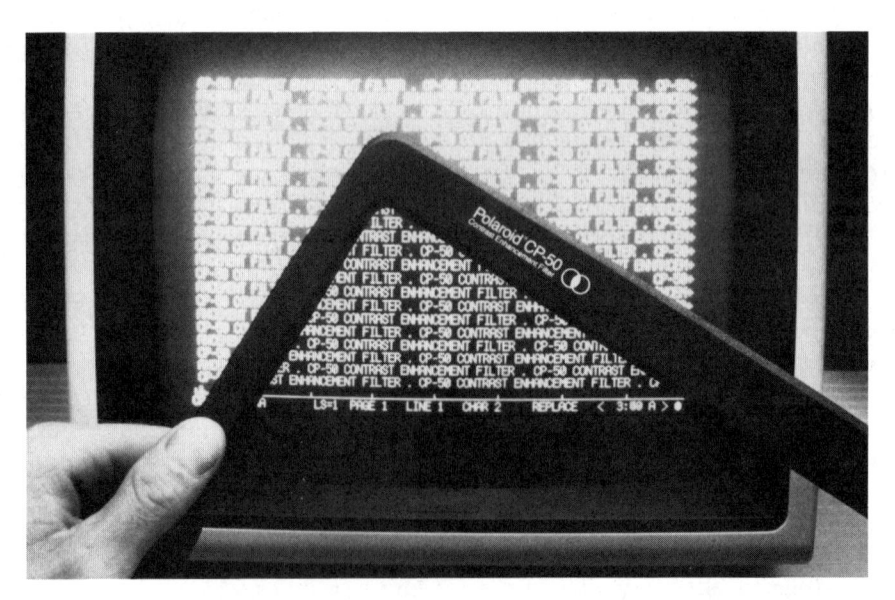

Figure 19-2. This filter improves the contrast of VDT display screens while eliminating reflections and glare.

Source: Courtesy of Polaroid Corporation.

communication terminals, and network devices. Most of today's workstations perform multifunctional tasks and are physically designed to conform to ergonomic standards.

Now, detachable keyboards, display screens with tilt, swivel adjustments, antiglare screens, and compact footprints are becoming the rule in workstation design. Ergonomists are continually working to create future workstation design that will be safer, healthier, and more comfortable for office workers.

Ergonomic Furnishings

The information work force has moved into sleek, modern towers and has changed the landscape of the American city. As workers spend more and more time sitting in front of computers, new furnishing are being developed to ease the backache, neckache, and general fatigue that have accompanied the new technology.

Corporations that are occupying new office towers and suburban centers are as concerned with interior furnishings as they are with the buildings' exterior appearance. The offices that occupy the new buildings are showcases for ergonomic furnishings. Tables tilt for better computer-screen viewing; chairs give ample support to the lower back; lighting is designed to reduce glare on computer screens; and modular office partitions mute computer sounds, hide cables, and offer privacy.

Vendors of office furnishings are designing products that respond to the worker's needs. Vendors now realize that support equipment for the new technology can only improve a person's job.

Of the $6.6 billion spent nationally for office furnishing in 1985, 90 percent went for ergonomic furnishings. Almost $1 billion of that was spent on chairs.

Chairs

Ergonomic chairs are being designed to provide aesthetic quality, user comfort, and increased productivity. Designers of office furniture, orthopedic surgeons, and physical therapists suggest that if you improve body posture, you improve job productivity. Proper seating involves total support of shoulders, neck, and back. Also, it reduces stress and fatigue by improving circulation.

Furniture manufacturers have converted skeletal-muscular pains into a strong selling point. Their chairs may look much like those of the past, but they are easier to adjust to the operator's weight and dimensions (Figure 19-3). Furthermore, they do not have to be turned upside down to be adjusted.

Office seating can be described as having two primary functions: work compatibility and comfort. Compatibility refers to the diversity of motions and the ability of the chair to adapt to and support each motion. Comfort refers to the state of mind that allows the seated worker to concentrate and work effectively.

Adjustability is another important factor in chair comfort. Tasks performed at a terminal cause changes in working posture. The office worker must consider the following guidelines for correct seating:

- Distance between seat and work surface
- Angle between upper and lower arm and wrist
- Screen height combined with viewing distance

Figure 19-3 illustrates correct seating by an office worker.

The Ergonomic Chair

The high-performance ergonomic chair provides comfort and compatibility to allow the

Figure 19-3. The right chair encourages good posture and decreases fatigue and muscle pain.

Source: Courtesy of Haworth, Inc.

office worker to perform many diverse tasks and to move about in the course of doing them. Even those whose jobs focus on a single function do not sit still: they change position, shift their weight, and stretch. Basically, the ergonomic chair adapts to the individual worker, providing support in any position the worker may take while performing a task.

Many ergonomists feel that the chair is the most important element in the office and should be selected with great care. The following is a checklist of features to look for in a good ergonomic chair:

1. *Pneumatic technology.* Pneumatic technology provides fingertip adjustability in a wide range of settings. When a chair is used by several workers, a wide range of easy adjustments are needed.

2. *Human-engineered cushions.* Seat and back cushions are molded to fit body contours. *Seat* cushions should comfortably support the legs without hindering circulation under the knees. Front edges of the seat should curve downward, and the rear of the seat should curve upward to cradle the hips and pelvis. Correct cushion support helps reduce worker fatigue during extended sitting periods.

3. *Backrest.* The backrest should be adjustable to accommodate the height of various users.

4. *Base.* The chair should have a stable base, equipped with casters. Some chairs have five-star bases with either casters or glides. These chairs offer excellent stability and safety.

5. *Fabric and coverings.* The chair should be covered with a knitted durable fabric that permits air circulation and offers a non-slip surface.

6. *Arm supports.* Cantilevered arms should allow nonrestricted movement from side to side. The arms should be strong, yet they should be softly rounded for user comfort.

Posture at the Workstation

Selecting an ergonomic chair does not ensure that a worker will use it correctly at the workstation. The information handler, naturally, is the flexible point of the workstation. The posture he or she assumes is determined by several fixed points within the workstation.

The best ergonomic posture for a seated VDT operator (Figure 19-4) is the following:

- Feet should rest directly on the floor or footrest with the thighs and lower legs at a 90° angle.

- The forearms should be parallel to the floor.

- The head and eyes should be aligned on the screen, and the copy should be 20° below the horizon.

Figure 19-4 illustrates correct ergonomic posture. Obviously, workers differ in size and weight so trying to get the tallest and the shortest person in the office into these recommended positions will be difficult. Therefore a good deal of adjustability in furniture and terminal design is needed.

The worker's posture is further affected by the "correct" placement of *copy* (documents, book, and materials) from which the worker keyboards onto the screen. Ergonomists recommend that copy should be placed upright on a *copy holder* the same distance from the eyes as the screen. That distance itself (16–24 inches) depends on the size and form of the screen's characters and the operator's eyesight.

The Desk (Work Surface)

The work surface on which the computer is placed must support the keyboard at twenty-six inches above the floor. This has been found to be the most comfortable height for continued arm and wrist action.

The position of the computer terminal can affect a worker's posture, eyesight, and efficiency. Placing a VDT on an existing desk will not accommodate the difference in body size of the operators. Traditional desks fit few people well and require workers to adapt in order to perform paper handling tasks.

Perhaps the most prevalent application of sophisticated information processing is the dedicated terminal workstation, where the operator is intensively involved in keyboarding. In fact, the need for ergonomically correct furniture is greater for the data entry or word processing operator than for those higher up in the corporate hierarchy, doing less demanding work.

In lower-level positions, ergonomic considerations of desks and work surfaces are critical to ensure comfort and improve operator performance. The full *adjustability* of work surface height and terminal position is an important factor the space planner must consider.

Adjustable furniture is important to accommodate the differences in size between users, and it is necessary to allow an individual to adjust his or her work environment throughout the day.

Some of the major factors that must be considered in determining the right desk work surface are the following:

Height. Although the height of the ideal work surface is considered to be 26 inches above the floor, workers must decide their most comfortable work postures according to their own body size (Figure 19-5). Therefore, it is vital that each person can adjust his or her

A. Synchronized seat back movement. Tilt range: back 16°, seat 8°. Seat automatically tilts 1° for each 2° the back tilts.

B. Front edge pitch adjustment range, 3°.

C. Seat height adjustment, 17"-21".

D. Normal eye-to-screen viewing range, 18"-22".

E. Vertical viewing angle comfort range, to minimize eye, neck and shoulder fatigue, 30°. VDT height should be adjusted so viewing range starts 10° below the "horizon line" level.

F. VDT screen height adjustment, 22³⁄₈" to 32¾".

G. Screen angle adjustment, ±7°.

H. Keyboard-to-screen adjustment, 0 to 7½".

I. Keyboard height adjustment, 22⁷⁄₈" to 32¼".

J. Keyboard angle adjustment, 0° to 15°.

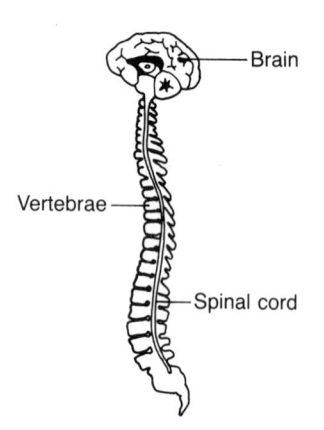

Figure 19-4. Ergonomic posture for working at a terminal.

Source: Reprinted from the June issue of MODERN OFFICE TECHNOLOGY, and copyrighted 1984 by Penton/IPC, subsidiary of Pittway Corporation.

work surface height. If this is not possible, the work surface should be established for the largest worker, and footrests must be supplied for the smallest workers.

Adjustable Desk Components. Adjustable keyboard pads can be fitted to any standard work surface or terminal stand. With standard work surfaces, most pads can be stored be-

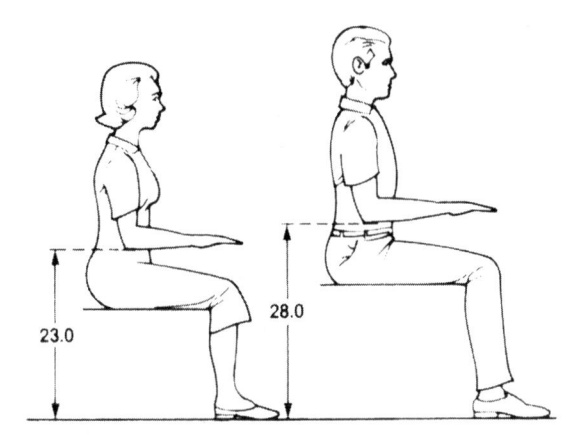

Figure 19-5. Range of work surface heights.

neath the work surface, then pulled out when needed.

A work surface extender is another example of a desk component that stores "breadboard" style beneath a work surface. Other work station extenders fold at the sides. These additional surfaces provide convenient areas for writing and reference materials as well as putting calculators and telephones within easy reach.

Elbow and Leg Room. Desks and workstations must provide adequate clearance underneath the surface for legs, thighs, and knees. It is also desirable to work with the arms fairly close to the body and the forearms approximately parallel to the floor to reduce fatigue. Research indicates that users prefer a variety of postures when they are seated at one place for a long period of time. The workstation should provide room for the worker to stretch out to relieve some of the stiffness and fatigue that result from sitting too long in one position.

The work surface should be located slightly above elbow height to prevent leaning forward too far. For keyboarding, the surface must be low enough to allow space for the keyboard. Most workers prefer a keyboard height slightly above elbow level. A separate adjustable keyboard platform can provide comfortable height for operators.

High-Quality Construction. Ergonomically designed computer work surfaces should include features such as curved legs, rounded edges, sturdy leg designs, and nonglare durable tops.

A Variety of Sizes and Shapes. Size is determined by the tasks required and the equipment used. Furniture manufacturers provide a full range of choices, from simple data tables to a complex cluster of multitask surfaces and heights. Special modular designs enable work stations to be connected in a clustered pattern.

Workstation Accessories and Support Components

Accessories to complement desk and work stations are designed to meet specific requirements for the information worker. Some of these components were mentioned in the previous section and are defined as follows.

Turntable. This is a rotating platform that provides for shared or casual use of a video display terminal among two or more users. The turntable rotates for easy accessibility.

Machine Support Trays. These trays are ideal holders for work-in-process documents. They may be designed for use on flat top terminals, or they may be swivel-based.

Adjustable Footrests. Footrests provide proper leg support enabling the VDT operator to work comfortably. This accessory elevates the operator's feet to overcome any potential ache in the lower back and prevents the legs from falling asleep. By pushing the footrest forward and backward, the operator can compensate for his or her height in relation to the work surface height.

Copy Holder. Copy holders are used for placing data in the best position for an operator's efficiency. Also, they keep work surfaces clear. Copy holders have a clamp-on design and an adjustable arm tension feature that holds up to several pounds of material.

Wrist Rest. A wrist support maintains operator productivity by reducing fatigue. This convenient support is a little platform that slips under the keyboard. It provides a slanted deck extension to support the palms and helps to relieve any wrist tension.

Keyboard Drawer. A keyboard drawer is designed to fit under any work surface. It has wire access holes or a corner work surface to position a detached keyboard. *Suspension drawers* hold letter and legal size hanging folders or standard computer printouts.

Other accessories and components support the ergonomic office:

1. *Printer support table*—a freestanding table on casters. It provides support and mobility for access by one or more users of a bottom feed printer.
2. *Vertical storage closet*—designed to hold coats, attache cases, and has a top shelf providing storage for personal items.
3. *Overhead storage unit*—a freestanding unit (Figure 19-6) that provides for horizontal and vertical separation of paper documents within an office. This unit may contain a retractable door with a lock.
4. *Marker board*—a panel-hung vertical surface upon which users can record and display written information using various colored erasable markers. The marker

Figure 19-6. A workstation organizer that fits over the work surface is a convenient means for keeping a multitude of records, reports, printouts, and reference materials.

Source: Courtesy of Acme Visible Records.

board can also serve as a convenient projection board for presenting visual information to small groups.

5. *Media storage bar*—a storage component (Figure 19-7) that supplements remote filing systems attached to panels. The media storage bar meets user needs for storing data processing forms, printouts, manuals, microfiche, and diskettes.

These information processing support components provide an *expanding system* for workers. Because the office environment is constantly changing in response to automation, new and additional components will be introduced to complement the ergonomic office and provide comfort and productivity for the worker.

Environmental Controls in the Workplace

The automated workplace creates changes in the environment. Many experts feel that office environments support efficiency and productivity at the expense of creature comfort and the needs of the human spirit. In rooms without windows, people feel isolated from their coworkers and cut off from the natural world of weather and landscapes. In spaces without walls, they feel vulnerable. In offices that do not respond to individual personalities and working styles, people tend to feel unappreciated and undervalued. Without planning for environmental factors and personal touches, workers feel that they are working in a "high-tech ghetto." Working in such an environ-

Figure 19-7. A media storage bar attaches to end panels that are hung from office panels to store data processing forms, printouts, manuals, microfiche, and diskettes.

Source: Courtesy of Haworth, Inc.

ment fosters feelings of alienation and worthlessness.

Ethospace: Designing the Workplace for Human Spirit

It is more difficult to design an office today than it used to be. In this age of information, when computers are changing work processes and relationships, people need workplaces that are comfortable—even comforting—and functional.

The science of ergonomics has been continually refined and expanded to accommodate changing organizational requirements and work patterns. Now *ethospace interiors* are being designed. They are interiors that work, but still make room for the human spirit. *Ethos* refers to the characteristic spirit of a particular group of people. The ethospace concept in office planning combines furniture, interior architecture, and an understanding of human needs that go beyond ergonomics. Through ergonomics and ethospace planning, office systems can now be designed to tailor environments individually at every level for maximum working efficiency and comfort.

The new environment will enable workers to function most effectively and productively through:

1. Improved lighting, acoustics, furnishings, and privacy
2. Better management of wiring and power needs
3. Improved design in space management where walls, partitions, and aesthetic surfaces can be changed quickly
4. Proper ventilation, humidity, and temperature controls
5. A variety of choices in color, texture, and aesthetic expression

Landscape, Layout Design, and Partitions

The first dramatic move away from the conventional office layout, characterized by rows of neatly arranged desks was the open landscape design (Figure 19-8). This design, introduced in the early 1970s, made use of modular-type furniture instead of desks. It featured movable partitions, screens, and other amenities such as plants, paintings, and carpeting. It was an improvement over rigid, conventional offices as it brought flexibility and expandability to the workplace. The open-plan landscaped office, with its movable partitions and modular furniture, seemed a promising approach. Heavily promoted by office systems vendors, eager to sell walls along with desks and chairs, the open-landscape office offered management the economic advantage of flexibility. Meanwhile, there has been a renewed leaning toward privacy and the tranquility of private offices in the workplace (Figure 19-9). In spite of vendor claims of inexpensive, easy-to-move walls and partitions, once they are in place they are rarely moved. Information workers are rediscovering the quiet and status of closed doors and ceiling-high walls. The major office systems vendors have responded by offering full-length walls that give their own sense of space and aesthetic character. Several manufacturers have introduced retractable floor-to-ceiling partitions. These partitions furnish the flexibility of the open plan office, yet permit privacy. Facility and design managers should look for a smooth transition between these new partitions and open office systems.

Wire Management

The wiring and power needs of terminals dramatically increase the requirements of power cords, data cables, and networks that must be dealt with in each workstation. Where should

Figure 19-8. The open landscape design.
Source: Courtesy of Haworth, Inc.

Figure 19-9. Private offices are preferred by many managers and executives.
Source: Courtesy of Haworth, Inc.

one place all the wiring that links telephones and computer terminals? All major office equipment makers have addressed this problem. They have built *wire management* features into their systems, the components that organize, conceal, and direct telecommunication in the office.

Power communication cables can enter from the floor or ceiling through a core and can feed into individual workstations. Cables can also be run through the end panels. Each workstation can be independent, and installation or modification can take place without affecting other workstations.

Technology, moreover, has a way of solving some of the problems it creates. Wire problems may soon disappear as such methods as fiber optics rapidly advance.

Climate Conditions of the Office

The environment people work in should help them feel comfortable while they do their work. The climate conditions of the office are known as *ambient* conditions. These conditions include temperature, air quality, lighting, and noise. Workers themselves are not waiting for facility planners to make improvements. They are taking a harder look at their work places. Moreover, social problems have been recognized pertaining to such issues as safety and indoor pollution. Workers complain about cigarette smoke, windows that do not open, and other quality-of-life issues.

Lighting

At the turn of the century, daylight was the prime source of illumination, supplemented by oil and gas lamps (Figure 19-10). Today, office lighting has reached new levels of sophistication and office planners are placing a new priority on proper lighting. Many electronic components are themselves light sources.

Proper lighting depends on two factors: quantity of light provided and the quality of light. *Quantity* of light refers to the distance and strength of the light source to the work object. Illumination is provided by lighting fixtures, sunlight, and reflection. The *quality* of light refers to the absence of glare, flicker, shadows, and the even distribution and color of light. *Glare* is usually caused by bright windows or other lights. The degree of glare depends upon the intensity of the source, the size of the source, and the distance from the line of sight. Most lighting problems result from too much light, which causes glare on written documents and reflections on the equipment. Windows seem to create glare and reflection problems, especially for those using screens, reading, or typing. Other lighting problems are excessive contrast and harsh shadows.

Video display terminal screens must be free of reflections and glare to assure visual acuity for the user. At the same time, copying documents and performing manual tasks require a sufficient quality and quantity of light.

However, direct light on source documents can result in eye strain from too great a contrast between the copy and the screen. The best solution to this circular problem is to provide the user with as much control as possible for adjusting the position and brightness of the light source.

Task (direct) lighting fixtures are placed directly above the work surface to provide enough light for specific tasks such as writing, reading, and keyboarding. They are similar to desk lamps in concept; however, there are some important differences. The light rays are filtered through a lens that diffuses the light so that the work surface is evenly illuminated. With a regular desk lamp, there is usually a band of light across the desk, although the remaining area of the desk is in shadow. With task lighting, both glare and shadows are eliminated, enabling users to see what they are

Figure 19-10. Business office circa 1900.
Source: Reproduced with permission of AT&T Corporate Archive.

reading or typing without eye strain. Fixed task lights, standard for most furniture systems, will not completely solve the needs of most VDT users. Instead, adjustable swivel-arm task lighting provides ideal light by permitting users to control their use. General task lighting, mounted under shelves, evenly distributes low-glare illumination over work surfaces wherever required.

Ambient lighting uses indirect fixtures that direct light upward to be reflected off the ceiling onto other surfaces that surround the work area. Fixtures are strategically located so that the light is distributed to areas where it is most needed, areas where people work.

Noise Control

Noise can become a mental distraction in a busy office. The sounds of business machines humming, beeping, clacking, whirring, people talking, and phones ringing can create a disturbing environment. Noise consists of those sounds that are unwanted or irrelevant. However, information workers cannot work in a totally noise-free environment. In the work environment, the acoustic problem is essentially the relationship between sound that one feels comfortable with and sound that is distracting or disturbing. As noise levels increase, job satisfaction and productivity decrease. To provide workers with a comfortable working environment, there must be a plan to control noise. The following are some specific ways to control noise.

Acoustical Ceiling and Wall Panels. Machine noise, such as printer noise, radiates upward. Good acoustical ceiling material prevents noise from being reflected back into the work environment. Wall panels and partitions should have the ability to absorb sound. This ability is known as a *noise reduction coefficient* (NRC) and pertains to how well the panels reduce noise reflections.

Automated Equipment. Electronic keyboards make little noise other than the sound of the key clicks or cooling fan noises. Printers, on the other hand, are becoming almost as prevalent as terminals and are a major source of noise.

Acoustical covers or sound enclosures are designed to reduce the noise levels of printers. These sound-absorbing devices consist of foam-lined cabinets. They are usually made with particle boards on the sides and back. They also contain an acrylic shield in the front for visibility. Some covers are now made of plastic.

Nonimpact printers, such as ink jet or laser printers, may make sound covers unnecessary in the future.

Unfortunately, the high cost of these printers will prevent them from overtaking impact printers in the foreseeable future.

Isolating Noisy Equipment. The isolation of noisy equipment such as printers and copiers may help solve the noise problem in certain work groups. People who perform similar tasks will probably have similar needs for acoustical privacy; therefore, grouping them together may solve some of the noise problems. Reducing the distances that people have to go to confer with one another can also reduce noise levels.

Background Sound. Background sound can be used to improve acoustical design. This concept involves adding proper sound frequencies at an appropriate volume to cancel out the sound (noise) transmitted from one open office to another. Human ears tend to hear everything equally. If a person hears two sounds that are essentially equal, the ear mixes them together. This is what added background sound essentially tries to do. It closes the gap between the residual background sound level and the level of intrusive sound. Thus, if the intruding noise is lower in level than the background sound, it will not be heard. If it is at or near the background sound level, it will cause little annoyance. Masking sound systems are installed in open offices to provide a uniform level of unobtrusive background sound that significantly improves the acoustical privacy.

Open office systems (with strategically placed panels) provide a less noisy work environment than a completely open office; however, noise levels may still be considered too high.

Acoustical panels, when combined properly with carpet, ceiling, and window treatments, can reduce noise that reaches the operator and the surrounding work areas.

Color

Color and texture humanize the work environment by lowering stress, battling fatigue, and lifting spirits. Ergonomic offices with modular components allow work groups and departments to project different images. Now, organizations recognize the increasing need for personalization of the individual work areas through colors, textures, and artistic displays.

The people who work in offices and those who design them want variety and choice in color, texture, and aesthetic expression. Furniture vendors now offer a full range of colors, fabrics, and finishes that can be combined in an endless variety of ways. A basic color can be used singly or in combinations to signify a variety of environments. Color can express a workplace with a light, airy, casual feeling to one that is dark, rich, and formal.

The fabric colors bring a wealth of textures to the workplace, and they can contribute significantly to a particular character or corporate image.

Intelligently combined, the color, fabric, and finish choices of the ergonomic office give it an aesthetic appearance that is uniquely different from the conventional offices. Advertising agencies do not have to look like law offices, and people do not have to suffer under the antiseptic monotony that characterizes offices without color.

Heating and Ventilation

Heating and ventilation also affect the quality of the environment in today's automated office. Temperature and air quality refer to the proper temperature, the extent that it fluctuates, and the clean aroma of air in the office.

Computers, as well as people generate heat. An organization that provides each employee with a computer terminal is doubling the heating load of its environment. Additional heat is created by energy-efficient buildings, which are windowless and tightly sealed. Although these new buildings save money by recycling a fixed volume of inside air, they also can produce dangerous buildups of toxic substances and bacteria.

Toxic agents, cigarette smoke, and other indoor air pollutants can cause health problems for office workers (Figure 19-11). Other indoor air pollutants include formaldehyde, asbestos,

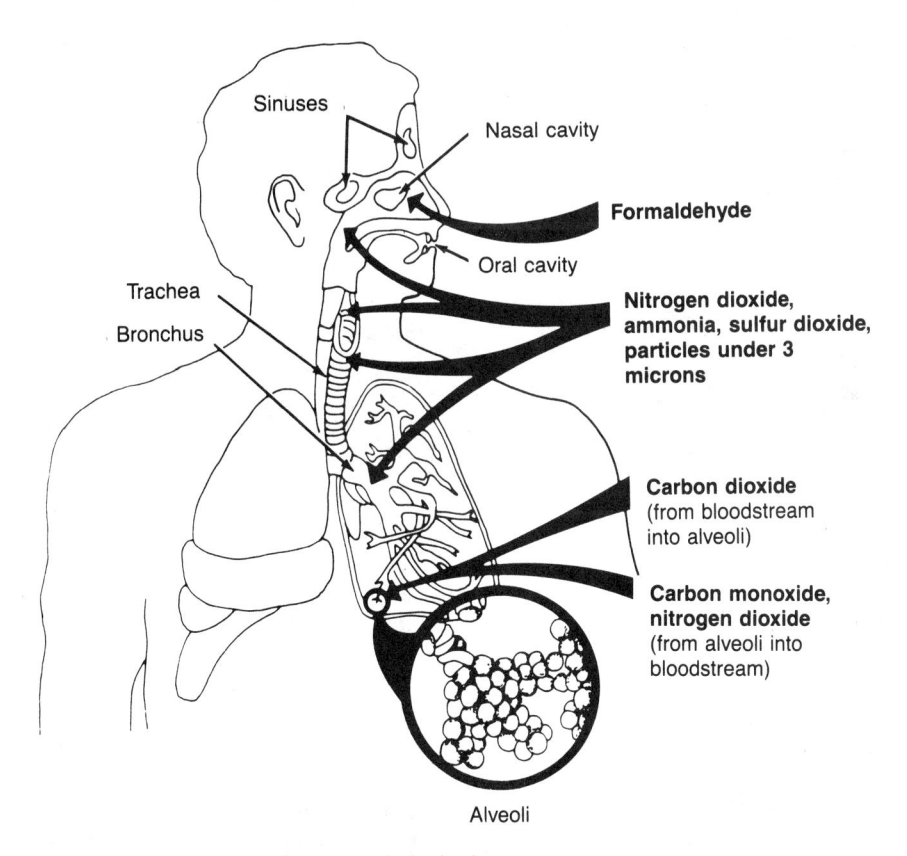

Figure 19-11. How pollutants attack the body.

Source: Reprinted with permission of *OFFICE SYSTEMS '86*.

carbon monoxide, carbon dioxide, bacteria, and viruses. The sources of these contaminants are linked to cleaning products, office furnishings, equipment, fabrics, appliances, and cigarettes.

Two conditions have to be met to assure indoor air quality is safe and healthy: (1) air cleaning equipment should keep pollutants at acceptable levels, and (2) ventilation must be adequate to provide air movement in concentrated areas, where workers spend most of their time.

A study by the Buffalo Organization for Social and Technological Innovation, Inc., found that thermal problems affect about 50 percent of office workers, and that air quality problems affect about 10 percent. Both temperature and air quality affect job and environmental satisfaction.

Improved building designs are beginning to address the problems of heat, ventilation, and air conditioning.

Intelligent (or smart) buildings (Chapter 12) use computers to control and monitor energy sources. Energy-cost control is a major consideration in "smart" buildings. These buildings offer a vast improvement over wasteful building systems that use light, heat, and air conditioning zones to cover whole floors or even groups of floors. In an older building, thousands of square feet must be heated or lit if one worker stays late. However, newer buildings have lost a feature since the advent of sealed windows: they do not have the ability to circulate unheated, uncooled air from the outside, in spring and fall.

Ergonomists—A Career Area Emerges

The discipline of ergonomics originated in the 1940s, when scientists began to relate the efficiency of products to their design. Ergonomic importance was fully realized when organizations began to equate computers with productivity, and productivity with a competitive advantage.

Interest has surged in this field during the decade of the 1980s. Managers perceived that the design of the tools that workers use and the workplace itself is a major factor in increasing the productivity of business. Today, human-factors scientists are focusing not only on consumer and industrial products, but also on the laboratory, the manufacturing line, and the office in their quest to improve productivity.

Individuals working in the field of ergonomics are known as *ergonomists,* or *human-factors engineers.* They engage in the measurement of human needs, limitations, and abilities. These measurements are factored into the design and engineering of tools, machines, and other synthetic objects that are used in work environments.

The field has expanded into scientific areas and uses the skills of behavioral and experimental psychologists, physiologists, and engineers.

A growing number of colleges and universities have set up graduate courses in ergonomics. Most of the 3600 members of the Human Factors Society, the U.S. professional association for ergonomists, work for government, universities, or as consultants.

Professional ergonomists work in areas such as consumer products, environmental design, safety engineering, and computer systems. They are involved in the design and development of objects as simple as can openers and scissors and as complex as network security systems and aircraft ejection seats.

The employment outlook for this new breed of scientist is excellent. Kodak has forty ergonomists, who do everything from helping to make its copiers and cameras user-responsive to studying the work practices in its factories. Automakers, appliance manufacturers, and office equipment/furniture designers employ teams of ergonomists. However, nowhere are the ergonomists currently more numerous and influential than in the production of desktop computers and their *display screens.*

Young people are finding an ergonomics career attractive and challenging. Experienced professionals are also moving into the ergonomics field along with sales managers and even high-level executives. For computer makers, ergonomic thinking will continue to make terminals more friendly to the average user.

Because of the attention ergonomics is receiving, this chapter serves only as an overview of an ever-changing field. The ambiance of the working environment has finally reached a level or prominence among corporate management teams. Ergonomics is really about balancing the needs of machines and people. Also, ergonomists, or human-factors engineers, should never lose sight of the fact that the machine should be shaped by the needs of its users. After all, people make the machines work.

Summary

Ergonomics is the study of human beings at work. Its goal is to make the things people use and the places in which they use them as safe, healthy, and comfortable as possible. Ergonomists are people engaged in the design of furniture, machines, and environment.

There has been a growing concern that video display terminal (VDT) operators suffer from eye irritation and other health factors. New screen designs and modular workstations have helped to improve the comfort and safety level for workers.

An *ethospace* interior is a new concept in office planning, combining furniture, interior architecture, and an understanding of human needs that goes beyond ergonomics. Through ergonomics and ethospace planning, office systems can now be designed to tailor environments at every level individually for maximum working efficiency and comfort.

Ambient conditions relate to the climate within the office environment. They include air quality, lighting, and noise. Achieving a satisfactory ambience within a balanced ergonomic setting has finally received a level of prominence among corporate planners.

Review Questions

1. Define ergonomics. Describe the origin of the word.
2. What are the basic reasons that motivate management to include ergonomics in strategic planning?
3. List the attributes of a well-designed keyboard.
4. Describe some of the ailments that workers may suffer as a result of prolonged VDT use.
5. What are pixels? How does pixel size affect improved working conditions for VDT operators?
6. Why is glare troublesome to operators? What can be done to reduce it?
7. What is a *footprint?* How is this a factor in workstation design?
8. Briefly describe how ergonomic chairs can be adjusted to accommodate differences in worker size.
9. What devices can be used to adjust desk and work surfaces? How do adjustable desk components benefit the office worker?
10. What is the purpose of a wrist rest? a copy holder? a keyboard drawer?
11. Briefly, describe some of the methods used to manage the maze of wires and cables used within an electronic office.
12. Differentiate between ambient lighting and task lighting.
13. Which computer system component makes the most noise? How can this noise be reduced?
14. How can color and texture contribute to a pleasant work environment?

15. Glass-enclosed modern buildings are tightly sealed to conserve energy. How does this design affect air quality and temperature within the work group?
16. Who are ergonomists? What do they do?

Projects

1. From a historical perspective, trace the early beginning of the science of ergonomics. Write a paper describing the origin of ergonomics, the early scientists who worked in the field, and the variety of applications.
2. Write to office furniture vendors. Request brochures and materials on their seating systems. Design a chart or matrix comparing several key vendors' ergonomic chairs. Include headings that describe various categories of chairs and special features that respond to specific tasks and user comfort needs.
3. The text refers to noise as a distracting influence that impedes concentration and productivity. From a health standpoint, can noise be dangerous? Can it affect hearing, or loss of hearing? Research the topic of noise as it pertains to health and hearing loss. Write a research paper on this topic and describe the impact of noise and its effects on one's word.
4. Observe your college computer lab or word processing applications class. From the standpoint of ergonomic design, suggest ways of improving the following areas:
   a. Lighting
   b. Air control and temperature
   c. Workstation layout
   d. Noise control
   e. Color
   f. Terminal design
   g. Seating
   h. Wire management

PART SIX
Personnel and Training

CHAPTER 20

Personnel Selection, Job Descriptions, and Career Options for Information Workers

After reading this chapter, you will understand

1. The change in the demographics of the workforce
2. The functions and differences between public, private, and temporary employment services
3. Flextime, jobsharing, and other alternate work styles
4. Job descriptions and career opportunities within a changing information work force

Personnel policies remained relatively stable in U.S. offices throughout the first half of this century. Well-tested approaches in the selection, evaluation, training, and promotions varied little among organizations. Things began to change, however, with the advent of the information age. Instant access to extensive amounts of information dramatically altered the way decisions were made. Telecommunication and computer networks transformed the office into a beehive of vitality. Word processors and personal computers changed the way information processors did their jobs. Managers began to alter their approach to job selection and training. As a result of these new technologies, new job titles, descriptions, and career paths emerged. Those that embraced the new technologies thrived. From the employee's point of view, the major advantage of the electronic office is the promised relief from some of the routine work. These technological devices and the new ones that followed helped to make office work more creative and interesting.

The Changing Work Force

Just as technology has changed, so has the makeup of the people who use the systems. Environmental changes and other external forces are dramatically reshaping personnel policies. Dataquest, a California consulting firm, has identified four major trends in the composition of the work force.

- A higher degree of computer literacy among business people
- More direct interaction with computers by more individuals
- A trend in the U.S. economy toward more jobs in the highly information-intensive service industries, with fewer jobs or a leveling off of jobs in the manufacturing sectors
- A slower growth rate of the 20–34-year-olds entering the work force, which may increase the need for investment in labor-saving devices such as computer systems

The last trend pertains to *demographics,* the changing age distribution of the population. The U.S. Bureau of the Census identified a sharp decline in the population aged 20–29 years in 1985 compared with 1968 and forecast a continuation of this trend (see Figures 20-1 and 20-2). Such a dramatic change in the work force demographics has ramifications for the selection and recruiting of information processing personnel:

- The younger, less experienced workers are usually given those jobs that are most routine, repetitive, and easily computerized. With fewer of these workers available, companies will look to computers to fill the gaps.
- The growth in output is still expected and needed, despite a slowdown in the growth of labor. Computers will be expected to make up for this shortfall in available human labor.

In addition to these trends, a large influx of women into the work force has changed the recruiting and selection processes of organizations.

These were the kinds of changes that employers had to recognize and deal with. Success in adapting to these changes is imperative if an organization is to attract skilled and talented people. In addition, society has established new laws to deal with the new work force in an information-intensive world. This chapter will discuss these changes and how organizations are embarking on human resources management strategies to serve everyone's needs better.

Sources for Employment

Information processing employees may be recruited from a variety of sources. The objec-

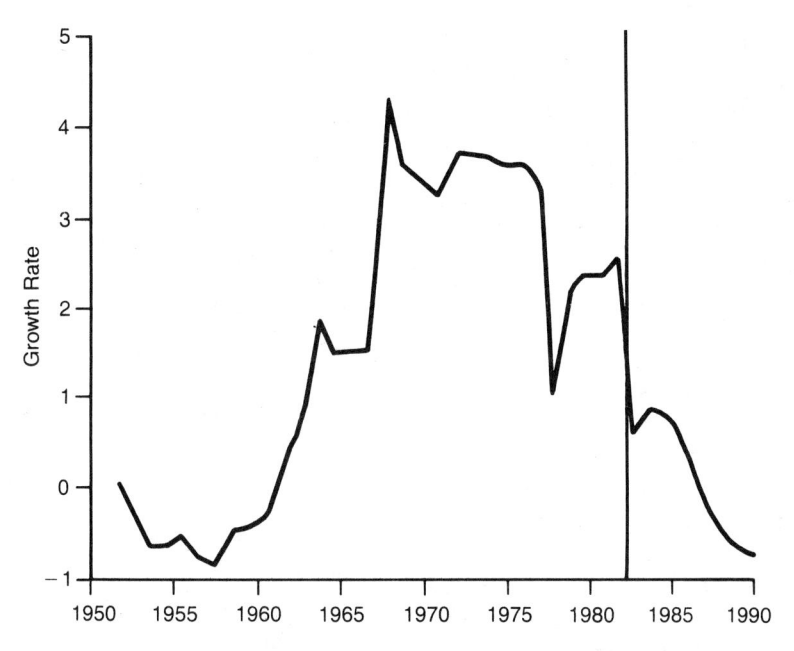

Figure 20-1. Growth in the young adult population aged 20–34.

Source: Courtesy of Dataquest Incorporated.

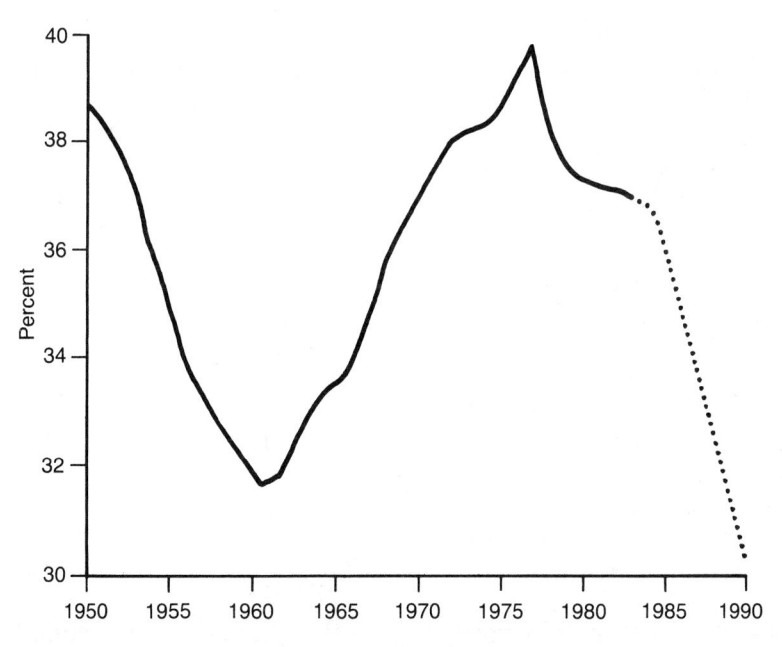

Figure 20-2. Ratio of population aged 20–29 to total labor force.

Source: Courtesy of Dataquest Incorporated.

tive of any selection process is to recruit and hire the best and brightest candidates. Enticing the best to join your organization is only one phase of the recruiting effort; keeping them can be more difficult. Personnel managers involved in the selection process must come to grips with the technological explosion in the information processing field.

The first phase of the selection process is to determine the best way of finding potential employees. Selection of a source varies from company to company. Some companies use employment agencies; others recruit through college placement bureaus or through newspaper classified ads. Organizations must determine the best source or combination of sources, hiring without conflicting with the laws relating to age, race, and sex discrimination. The following sections discuss sources for selecting information processing personnel.

Referrals and Recommendations

Referrals and recommendations are internal sources of recruiting. A friend tells a friend of an opening in the firm. In many instances, this may be a more desirable means for recruiting than an external source. However, great care must be taken when using such referrals. A recommendation from a worker's friend does not necessarily indicate that the candidate is the best person for the job.

Schools and Colleges

Schools and colleges are excellent sources for finding office information systems candidates. College curriculums are constantly keeping abreast of the trends in integrated systems. They are adding courses and curriculums to train students for the skills and concepts necessary to function in an information-intensive world. Special career days are devoted to bringing together students and personnel directors who conduct interviews on campus.

Newspaper Ads

There is a growing trend to become more selective in the use of advertisements. New job descriptions and titles are given to fit new office information systems categories. However, there is still much confusion in recognizing job categories by both the applicant and the employer. The line between office automation (word processing) personnel and data processing personnel is still hazy. It has not been clearly defined within the industry or within organizations. Employers who use advertisements as a source of recruiting must be explicit in their job descriptions. In addition, care must be exercised when choosing the wording and placement of the help-wanted advertisements.

Professional Organizations

Professional organizations specializing in office information systems, data processing, and office automation are developing and operating placement services for their members. Trade shows and conventions are used by some organizations to recruit candidates. Local chapters of professional organizations often offer employment advertisements in their local newsletters.

In recent years, several professional associations have changed their names to appeal to a broader base of technology professionals. The International Word Processing Association, for example, changed its name to the Association of Information Systems Professionals (AISP). They had the foresight to look beyond technology that applies to word processing alone and focused on a broader objective that encompasses a cross section of information professionals.

Employment Services

There are two types of employment services—public and private. These services put an employer in contact with qualified job candidates.

Public employment agencies are branches of state, county, and federal government. There is no charge for services provided by these groups. Any person seeking employment may register without charge with one or more of these agencies. The largest public employment service is the U.S. Employment Service (USES), which is under the direction of the Department of Labor. Public employment agencies offer a variety of services for the employer as well as the job seeker. Registration by the applicant usually includes a comprehensive interview and a skills test so that the applicant can be properly classified according to his or her abilities, personality traits, training, and experience.

Private employment agencies differ from public agencies in one important aspect: they charge a fee for their services. In most cases, however, the employer pays the fee for the employee. Private employment agencies perform valuable services for both the employee and the employer. The agency acts as an agent for the job seeker, as a recruiter for the employer, and as a job market information center for both parties.

Temporary personnel services have an important role in the marketplace. According to the National Association of Temporary Services, half a million companies, or nearly nine out of ten, use temporary workers in one way or another (Figure 20-3).

About 700,000 jobs in the United States are now filled each day by temporaries. The number of temporary workers is growing at a rate of about 20 percent annually. Nonetheless, demand still outstrips supply. Managers of some of the nation's more than 2000 temporary employment service companies say their growth is limited only by a shortage of skilled candidates.

Temporary help can play an important role in virtually any business. Their services are essential to those companies needing skilled workers on less than a full-time basis. Temporaries are available to alleviate peak work loads, cover for vacations or illness, handle specialized tasks, cover during unpopular work shifts, or serve as interim replacements. Temporary workers do not usually replace the

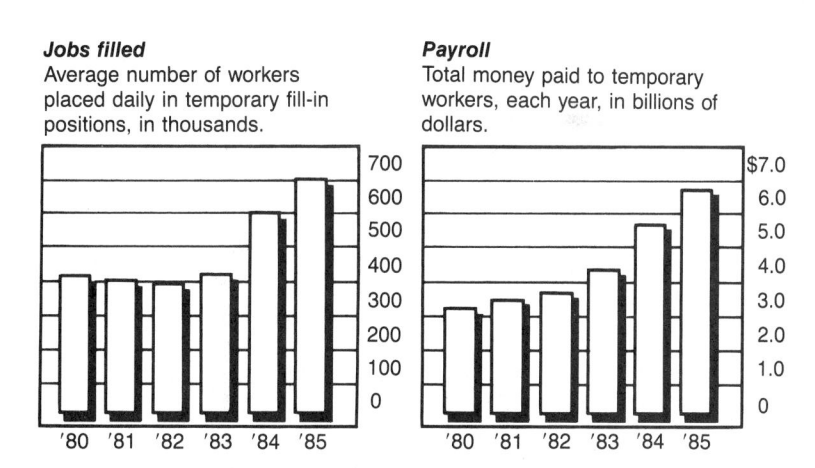

Jobs filled
Average number of workers placed daily in temporary fill-in positions, in thousands.

Payroll
Total money paid to temporary workers, each year, in billions of dollars.

Figure 20-3. Temporary employment is a growth industry.

Source: Courtesy of the National Association of Temporary Services.

regular workers; they augment them. The expansion of computerization and high technology into the office is ideal for temporary help.

However, organized labor officials are opposed to expanding the use of temporaries because it allows companies to cut their hiring of permanent employees. These critics say the presence of temporaries undercuts efforts toward collective bargaining by office employees.

Organizations use temporary services to staff entire operations, such as newly formed work groups and information centers; and during office relocation periods, until formal jobs can be established, temporary workers are also valuable.

Some organizations are staffing certain facilities only with temporary help year round. This hiring practice is called *facility planning*. Employees staff certain facilities such as mailrooms that have a high turnover or fill peak periods with temporary help so that businesses do not have to keep permanent employees on the payroll.

For job seekers, especially those reentering the job market, temporary services provide an opportunity to readjust gradually to the business world. Employees work where they want, for as long as they want, and whenever they want. Some workers want to work only on a temporary basis. They are part of a new trend called the *career temporary*. They find this type of work highly suitable because they can acquire different skills and have flexible hours.

Temporary service organizations provide a full range of services for their client firms. The worker is interviewed and paid by the temporary service. Elaborate training facilities are provided to train applicants in the latest hands-on skills and to give them computer literacy. Qualified temporaries can enter an office and be productive on their first day. These specialists understand office automation concepts and are skilled in a wide range of word processors and personal computers.

Executive Search Firms

Most employment services are structured to place applicants at the operator level. The applicants may be document production specialists, also called *word processing operators, transcriptionists,* and *secretaries.* They primarily perform keyboarding functions for a variety of documents. *Executive search firms,* also referred to as *headhunters,* serve the needs of companies looking for supervisory and management positions.

These positions deal with the daily handling of people and operations. Management positions involve more planning and decision making and less daily coordination of people, equipment, and procedures on an operational basis.

Most candidates have a bachelor's degree and/or several years of appropriate experience, including supervision and knowledge of the area to be managed. The positions include regional sales directors, business development directors, consultants, and marketing management positions.

Executive search firms have provided candidates for the small computer companies that have emerged during the past several years. Some of these small firms offer candidates equity positions that provide profit-sharing plans rather than salaries.

One of the major problems facing many small high-tech companies is that they have outgrown current management. There is a need for a new breed of managers who can think and act with an open mind. Many companies are looking for "generalists," who, they feel, will fit better into the newly emerging, flexible work groups. Executive search firms dealing with office information systems managers usually advertise in newspapers or special journals.

Electronic Recruiting

The competition for jobs and job candidates has prompted organizations to use electronic recruiting to hire from the high-tech labor pool. The use of electronic mail enables job seekers to send their resumes via their personal computers directly to an organization. This in-house electronic recruiting system also details available jobs by their titles and descriptions.

To read the newly arrived resumes, the recruiters flip on their screens and review their electronic incoming mail. When they spot a good job match, they call the respondent immediately. Electronic recruiting has the advantage of enhancing the speed with which people can get in touch with each other. Personnel agencies can call and leave a resume, and candidates can find out what positions are open.

Another high-tech recruiting system is a satellite communication network. Through this means, corporations are able to transmit their recruiting messages to campuses all over the country instantly.

Personnel tracking software is available to allow personnel managers to sort through an almost infinite number of files to find people with specific skills and to track their progress. Depending upon the size of the memory of the hardware, personnel planning and executive recruiting software can profile the people of even the largest companies. One program, called *CAPTURS* (an acronym for *Complete Automated Personnel Tracking and Utilization Review Systems*) contains information such as salary, education, and previous positions. It also has a built-in past job performance rating system, so that employees can also be screened for the qualities a company views as necessary for a particular job.

Electronic recruiting is not intended to replace the traditional method of recruiting—the face-to-face meeting. That is still the most important method of recruitment. Electronic recruiting provides an alternative means of narrowing the search. It is a new technology that offers a cost-effective way for employers to present themselves to graduate students and other job seekers.

Alternate Work Styles

Along with the changing work force and demographics that arrived with the information age, alternate styles of work have appeared in the workplace. An increasing number of employees are taking advantage of new job options and alternate work styles. As a result, many people are no longer working at the same company, eight hours a day, five days a week. Resourceful workers are selecting non-traditional work styles. They are changing their working hours, sharing jobs, limiting themselves to part-time work, and even working on the job while staying at home. The following sections summarize alternate work styles and schedules.

Telecommuting

Telecommuting is an alternate work style that allows people to work at home. Usually, telecommuting involves work performed on terminals and computers connected to a mainframe, as well as work done on personal computers, which is stored on disks and carried to the office when necessary. Telecommuters may either be employed directly by the company or work as subcontractors or freelancers on special projects. The implications of the "electronic cottage" and the neighborhood workcenter are discussed in more detail in Chapter 12. Advocates of telecommuting predict that by 1990, 10 million people could be earning a living in their own homes.

Flextime

Flextime enables jobholders to decide for themselves when they come to work and when they leave—within limits set by management. Usually, flextime requires that employees be at work during the peak or "core time" of the business day. They can work the remaining time that they owe the company at their convenience, within a broad period. Employees can stagger their work hours to do errands or to avoid commuter traffic. Flextime can ease the commuter crush as thousands of workers pour out of offices at 5 P.M. rushing for subways, trains, and buses. Flextime can also be referred to as *flexitour, gliding time, variable day,* or *maxflex.* All plans have the major advantage of allowing workers to have some control over their working hours.

Jobsharing

Jobsharing involves two half-time workers sharing one job. This work is based on the premise that two halves equal more than one whole. Two segments of the population are attracted to job sharing: (1) parents of small children who do not want to drop out of the work force completely and (2) men and women who are phasing into retirement. Jobsharing saves the company money because workers who work less than half a normal work week do not receive fringe benefits.

As with flextime, employees who select jobsharing enjoy it because they have control over their working conditions and environment.

Changed conditions in the economy, in technology, and in society are creating changes in job time and place. These alternate work styles offer a greater opportunity for job fulfillment and satisfaction for a growing portion of the work population. Progressive organizations are now restructuring their personnel policies to accommodate employees who no longer choose to work "at the office" between 9 A.M. and 5 P.M. Monday through Friday.

Information occupations entail the creation, processing, and distribution of knowledge. Over the past two decades, automation and the computer have changed the description and title of these jobs. More than 65 percent of us work with some form of information; for example, teachers, insurance people, accountants, lawyers, and in office-related jobs. Those born in the industrial age have witnessed several social and technological changes within a single decade. Almost overnight, the industrial age has moved into the space age, the technological age; today we are only at the beginning of the information age. Computers, electronics, and automation have created a variety of new career opportunities for the office worker.

The role of the office worker is changing and evolving as the office itself evolves. One of the most significant changes taking place is in the number of career options available. The door has opened to a myriad of choices not previously available for office information specialists.

New technology has created new jobs and new career opportunities. Reorganization of office personnel and procedures has produced increased specialization. Integrated information systems and the merging of word processing and data processing have created an almost overwhelming variety of job titles related to information processing subsystems. Information processing can be a career in itself or an opportunity to branch out into a lateral promotion or a spin-off career. A *lateral promotion* is a horizontal advancement across an organizational chart into another department that may lead to *vertical* (upward) career mobility.

Job Titles and Descriptions

A job description is a tool that serves as a blueprint for identifying where and how jobs fit into a company.

A job description helps managers in the following ways:

- It helps organize work functions.

- It defines relationships between jobs.
- It details job responsibilities.
- It provides an equitable basis to determine salary levels.

As the definition implies, a job description provides a "word picture" of a job, explaining its basic responsibilities. The description is intended to define the duties of a job as opposed to explaining the work activities of a particular employee.

Once developed, the job descriptions must be updated as needed, in order to reflect the dynamic environment in which most office personnel operate.

Proper job descriptions must be established to provide a personnel manager with information on the type of person needed to fill positions from the entry level (the trainee) to the upper-echelon level (supervisor/manager).

Career versus Job

Those entering the world of work should contemplate whether they want a job or a career. The diversity of job titles and opportunities increases the importance of *career planning*. A career provides enjoyment and job satisfaction; a job is mainly a way to earn a living. A person plans for a career, choosing jobs and training carefully, in order to make choices that will advance that career. People with successful careers in information processing are those who have realistic expectations and who plan to grow professionally instead of taking a job that offers immediate satisfaction.

Job titles and descriptions within information processing range from administrative support to office automation, data processing, supervisory, and management-level positions. Each category and level requires special skills and traits. The following section discusses the career options and the special skills and traits required to function successfully.

Career Options for Office Information Systems

People working in information occupations possess a common set of skills, attitudes, and personal characteristics that set them apart from those in other occupations. Specific jobs within information occupations, however, have a set of unique skill requirements in addition to the general requirements for all information workers.

General Qualities

The Bureau of Labor Statistics forecasts that within the next ten years, 70,000 information workers will be entering the market annually. The impact of office technology will certainly change the role and specific tasks of office occupations, but what qualities should the ideal candidate possess? Will computers and other automated office equipment drastically change the human relations, technical skills, and personal qualities needed by information systems personnel? The successful candidate should possess the following characteristics:

- Interpersonal skills
  Sense of teamwork
  Willingness to accept constructive criticism
- Technical skills
  Keyboarding
  Computer literacy
  Mechanical ability
- Communication skills
  Language arts skills
  Desire to communicate effectively
- Personal Skills
  Creativity
  Responsibility and dependability
  Motivation
  Loyalty
  Knowledge of the company
  Ethics

Willingness to learn and listen
Flexibility

There is still some uncertainty in the minds of job applicants about the requirements of information systems jobs. The merging of word processing and data processing scares a lot of people away from newly created jobs because of misconceptions about the amount of technical and mathematical ability required. What it does take is common sense, organizational and communication skills, and analytical capabilities. Computers are a necessary component of these jobs, but they are only tools to help solve problems quickly and efficiently.

The Secretary

Of all the information jobs subject to sweeping automation, secretarial positions represent the front line.

The U.S. Department of Labor defines a *secretary* as someone who "maintains a close and highly responsive relationship to the day-to-day activities of the supervisor and staff." Duties include varied clerical and secretarial skills requiring a knowledge of office routine.

The word *secretary* has been replaced by *word processing operator, office technician, executive assistant,* or *administrative, correspondence,* or *document production specialist.* Because there is a variety of job titles, estimates on the number of secretaries in the United States vary from 2.5 million to more than 3 million.

Women still make up 99 percent of the secretarial ranks, but today's secretary has rejected the role of the "office wife," who fetched coffee and remembered anniversaries, as well as the concept that traditional wifely duties mean an end to her career. If she is trading in her steno pad or typewriter for anything, it is for a word processor or personal computer.

In the early stages of word processing, organizations sought to specialize the work group into a division of labor for secretaries. They described the job of correspondence secretary as operating (keyboarding) the word processor. The administrative support tasks generally included most of the nontyping functions previously performed by the traditional secretary. There are still organizations that adhere to this strict job description, but they are in the minority. With the proliferation of small work groups and integrated information systems, the "new secretary" will perform a combination of tasks and duties. Workstations and terminals will be on each desk. They will have an abundance of software programs and be tied into a computerized network.

The new secretary will be a professional. Selection, pay, and promotion will be based on ability, technical skills, and proficiency in language arts skills for composing communications, proofreading, and editing. Many secretaries will be working as team members within small work groups or in a decentralized structure. These individuals should be people oriented, well organized, poised, mature, and willing to take on new and challenging tasks. Today's secretaries have more opportunities to participate in the management process and have more potential influence on their organization than ever before. They want and need to participate in decision making. And the advent of office technology makes lower-level decision making not only feasible, but actually desirable, productive, and logical.

For those who aspire to move up the career ladder, professional growth and self-improvement are positive ways to move into positions with more responsibility.

Computer Specialist (Data Processing)

Jobs within the data processing field include those of programmers, data entry input operators, computer operators, systems analysts, and others with a wide variety of computer-related skills.

Programming is actually more closely related to the creative arts than to mathematical and engineering skills. Both verbal and written skills are essential because computer professionals do not function in a vacuum, but rather serve several departments in a company.

Personal characteristics of the computer specialist that are typically sought vary widely, but most organizations look for a logical mind and a demonstrated interest in problem solving. Many organizations believe that *communication barriers* between users and computer professionals account for a large number of failures in computerization. As a result, dealing with people is an integral part of most data processing jobs. Being a good listener and being articulate, sensitive, and able to communicate are key characteristics.

The data processor programmer must also be methodical and meticulous, for thousands of individual instructions might be required to program a computer. If something goes wrong, every one of them may have to be rechecked.

Understanding of computer systems and knowledge of computer languages such as BASIC, FORTRAN, COBOL, and Pascal are important. Computer specialists also need extreme patience and a sense of responsibility. Working in a computer environment means getting paid to get work done—on schedule.

Today's data processing specialists no longer work in isolation. They are not satisfied toiling in daily solitude behind a desk. The most satisfied employees are people who find their jobs energizing and derive pleasure from handling clear and understandable programs.

The demand for computer professionals has always been great, and the need exceeds the supply in almost all areas of the United States. Estimates by the Bureau of Labor Statistics reveal that the number of computer professionals employed in 1990 will be 50 percent greater than in 1980. During the same period, overall employment growth is expected to increase only 20 percent. Because computers are so pervasive and will continue to have an impact upon society at so many levels, the computer field will be virtually recession-proof.

Telecommunications

The great divestiture of the Bell system brought about drastic changes in providing equipment and service to thousands of corporate users. Equipment companies that used to manufacture telephone sets turned to building sophisticated electronic switchboards that can remember numbers and route calls over the least expensive circuit. Long-distance calling, which was the exclusive preserve of AT&T, is now brimming with competitors large and small. Corporations now have separate *Telecommunications Departments* to handle all of the organization's telecommunication needs and applications.

Organizations are struggling to handle a rapid growth in communication applications with a limited pool of specialists. Although new telecommunication technology has, in some organizations, reduced staffing needs, the growth in demand for telecommunications services is causing shortages in a number of highly technical job categories. These include radio engineers, data communications specialists, computer programmers, and network and fiber optic specialists. The main reason for the shortage of qualified workers is the rapid emergence of the technology. The technology is so new that few workers have received enough on-the-job training to make them effective telecommunications specialists.

Telecommunications includes much more than it did only a few years ago. Newer technologies such as fiber optic networks, modern PBXs, satellites, microwave transmission, videotex, artificial intelligence, security software, integrated voice and data devices, and mobile cellular devices are finding a growing acceptance by organizations.

The forecasts by leading consultants indicate that the demand for technical people in communications will far outstrip the supply. However, openings also exist for numerous categories of business people, such as marketing and financial specialists, strategic planners and experts in the area of government regulation.

Thousands of new job titles and descriptions are emerging in this field. However, telecommunication careers can be divided within two major categories: technical and nontechnical.

For the *technical* job seeker, the most common preparation for a career in telecommunications is to earn an engineering degree. Suitable degrees will be those of electrical engineers, physicists, chemists, computer designers, and programmers. Engineers design communication equipment such as solid-state switchboards and big switches for telephone company central offices. They help build the long-distance networks of the new carrier companies.

Nontechnical applicants require a business degree or M.B.A. degree. This area of telecommunication offers opportunities for traditional business people to use their skills in telecommunication applications. Marketing specialists, for example, may sell discount long-distance service or telephone switching systems. Government specialists deal with federal agencies, such as the Federal Communications Commission (FCC), that oversee phone companies' operations.

As technology expands, the concept of the world as a global village will be realized more fully through the increased use of international communications. Because of these changes, opportunities for people with broad knowledge in marketing and finance and inventive skills will be needed.

NYNEX Business Information Systems President Richard J. Santagati estimates that there are 100,000 jobs available now for college graduates skilled in telecommunications.

And by the year 2001, a great percentage of the work force will need basic telecommunication skills to perform various job-related tasks.

Spin-off Careers

Career opportunities in the fields that support the office are increasing. Within the next two decades nearly all industrial communication will take place through electronic means. Computers will speak, listen, recognize voices, translate words into foreign languages, send and take messages, schedule appointments, and even make telephone calls. Those who have entered these newly emerging information occupations on the ground floor have the opportunity to build a solid foundation and a wealth of experience for career growth.

An emerging work force will be employed in "spin-off careers" as an outgrowth of providing services and information to society. Some of the new computerized technologies will include industrial robots, genetic engineering, electronic medical devices, and scores of applications not yet known.

A relatively short background in a new industry can serve as a springboard for breaking into related careers.

Sales Representative. If one has a unique flair for selling and a background in systems and applications, selling hardware or software may be a promising career. Manufacturers with new products in telephone systems, long-distance service, computer, software, and supplies are always in need of energetic sales people. As these markets continue to expand worldwide, greater demands for sales people will be made.

Trainers. Closely related to sales are marketing support personnel. These positions require the person to demonstrate and train a user who has purchased or leased the vendor's product. The training may take place on the user's company premises or at the vendor's branch office. Both sales representatives and trainers may be required to do some traveling.

Journalism. Writing about office automation systems can be another career choice. Many trade journals are devoted to information systems, computers, and telecommunication. They require people with knowledge of equipment, applications, and management issues. Reporters and editors must be computer literate and possess excellent writing skills.

Consulting. Opportunities in the consulting field have grown as a result of the implementation of office automation, communication networks, and the need and desire to redesign office systems. Consultants may be independent freelancers or a part of a consulting team.

Small Business Entrepreneurs. Resourceful employees strike out on their own to form small businesses such as personnel agencies, executive search firms, retail computer shops, service bureaus, and freelance typing and proofreading services. There are hundreds of case studies about former employees' running their own companies or serving as independent contractors.

Professionals. Then, too, there are opportunities for licensed professionals. For example, nearly all emerging computer and telecommunication companies, phone companies, computer makers, and long-distance service companies hire lawyers and accountants. Some even have architects on staff for the design of computer and communication systems in buildings.

A new breed of worker will be needed to work in and manage predominantly information-intensive corporations. Changes are taking place in the once tradition-bound world of work.

Summary

Environmental changes and other external forces are dramatically reshaping personnel policies. The computer and office automation have influenced the requirements and skill level of workers entering the work force. In addition, more young people and women are selecting information processing careers. These new groups have prompted employers to initiate changes in recruiting, hiring, and promotion policies.

Workers are recruited from a variety of sources. They include referrals, colleges, advertisements, professional employment services, and electronic data banks.

An increasing number of employees are taking advantage of new job options and alternate work styles. Nontraditional work styles comprise such plans as telecommuting, flextime, and jobsharing.

The computer society has brought with it a new breed of workers. They are more motivated, computer literate, and challenged than the generation of workers who preceded them. Success in adapting to a changing work force is imperative if an organization is to attract and retain skilled and talented people.

Review Questions

1. Describe the change in the work force age group.
2. What disadvantage may result from recommending a friend for employment?
3. What is the key difference between a public employment agency and a private employment agency?
4. Briefly describe the function of a temporary personnel service. Under what circumstances would companies require the services of temporary help?

5. What is electronic recruiting? Why does it have an advantage over traditional recruiting methods?
6. Compare flextime with jobsharing. Why do employees select these alternate work plans?
7. What is a job description? How is it useful to managers?
8. What is the difference between a job and a career?
9. Why has the field of telecommunications opened up so many job opportunities since the breakup of AT&T?
10. Describe the opportunities for a nontechnical person in the field of telecommunications.
11. Compare the job descriptions of a sales representative and a trainer. What are the similarities?
12. Describe how a freelance typist or word processor can supplement his or her income.
13. The text mentions that some computer and telecommunication companies hire professionals. What services, for example, do staff lawyers perform?

Projects

1. You are the manager of a temporary employment service in your local community. You specialize in placing temporary office workers and word processing specialists. You receive a request from an insurance company to select six operators skilled in operating a specific brand of personal computer. The client firm specifies that it wishes only operators with experience on its type of equipment.

   As an employment specialist, what resources would you use? After several weeks of searching, you have been unable to match an operator with some experience on the specific piece of equipment for the job. What suggestions can you make to your client to fill these openings?

2. Your organization has decided to set up a telecommuting program. In the initial phase, it has selected twenty employees to perform work on terminals in their homes connected to a mainframe in the office.

   You are selected to set up a work-at-home program for your company. A substantial amount of hardware, especially computer and communication equipment, is required to equip each employee's home office adequately. You are to conduct further research into this area and create a list of guidelines for an effective telecommuting program. Include some advantages for your company as well as for the participating employees.

3. You are asked to develop a job description for the ideal temporary employee for a new temporary employment service. Write a generic job description of the ideal temporary candidate, listing positive attributes, skills, education, and personal traits. Include the following five key sections in your job description:
   a. Job title
   b. Job summary
   c. Reporting relationship
   d. Qualifications
   e. Typical duties

CHAPTER 21

Training, Salary Administration, and Measuring Job Performance

After reading this chapter, you will understand

1. How today's information workers compare with workers of a decade ago
2. The information manager's role in training
3. The variety of training methods and resources available to organizations
4. Special devices and resources available for disabled workers
5. Aspects of various compensation and employee performance measurement plans

Advanced office technology and microcomputers are part of the landscape and tools that comprise most modern organizations. Although the rate of change and introduction of newer, more powerful systems has subsided, upgrading and add-ons present new challenges for information workers. When office automation was new, a vast population of office workers had to be trained on word processors, microcomputers, and assorted devices. A generation has passed. New products are introduced with less frequency, and most of today's product introductions are evolutionary rather than revolutionary.

Today's information workers and business users bear little resemblance to the neophytes of a decade ago. They are knowledgeable, sophisticated, and "computer literate." The new worker entering the information work force has graduated from high school or college with some level of understanding of the concepts and skills of computer systems.

This does not mean that we have gone as far as we can go. There will always be improvements and new frontiers in office information systems.

Newer supercomputers, image processors, artificial intelligence applications, and a vast array of software will open up whole new avenues for the information user. High-technology training will be more critical than ever for the intelligent use of these emerging technologies.

Objectives of Training

Although the majority of information workers are knowledgeable and skilled in computer systems, training workers is a vital function of any organization. *Training* provides employees with directed experience and the means to achieve desired learning goals. Even if employees have a level of computer literacy and operational skills, good training provides the bridge between prior experience and desired performance. Effective training goes beyond achieving a level of competence. It boosts workers' morale and lessens turnover for the company. With proper training, errors are reduced, employee self-confidence is increased, and attitudes toward the organization are enhanced.

On the opposite end of the computer literate work force are entry-level workers who do not possess good communication skills. These workers need to be brought up to a competency level through remediation in English, language arts, and computational skills. This level of basic skills training must be achieved before the worker moves into a computer training program. This need is urgent. All entry-level workers who do not understand the automated office must be brought into the mainstream of computer literacy. If they do not understand the concepts, applications, and functions of these systems, they will become almost obsolete. Workers not only need training in the operational "hands-on" aspect of automation; they also must understand the logic and concepts behind these systems.

Changing Nature of Office Automation Training

In the early days of word processing, vendors assumed all of the training functions. No other source was available. Word processing vendors set up training sessions on-site and sat with operators until they became efficient, skilled, and confident. They "held their hands" and even retrained the company's operators when they did not perform as promised. Many organizations purchased word processing equipment because the training was included in the purchase price. When the evolution of dedicated word processors moved into microcomputers, WP vendors dropped their prices and "unbundled" their training. Training was priced as a separate item, rather than an integral part of the equipment cost. Organizations began scampering around to do their own training, something with which they had little or no experience.

As prices for word processors and personal computers dropped, vendors simply could not afford to include training as part of the purchase price. The information manager was suddenly thrust into an additional role with responsibilities for training. Vendors provided instructional manuals and self-paced training packages; however, it was clear that a comprehensive training program with human interaction was needed. As departments grew and interactive computing spread among users, an intelligent strategy for training end users, operators, and middle managers had to be implemented. Organizations soon realized the value of properly trained operators. All the hoped-for productivity gains of acquiring expensive equipment would be meaningless if workers could not operate the equipment.

Establishing a Training Program

When corporations saw the need to establish in-house training programs, their initial budgets for these ventures were modest. Traditionally, the office has never had large training budgets. Before formal training programs, organizations relied upon on-the-job training, with one worker teaching the other in informal sessions. These informal methods proved ineffective.

Today, organizations cannot overlook the importance of training. Teaching operators and end users to use their tools helps an organization to obtain a return on its investment. A growing number of enlightened companies are now providing computer training programs. Only through a willingness to commit more capital for high-quality training as an investment in human resources will improved office productivity be achieved.

Who Needs Training?

To get started on a training program, the organization must determine who needs training. This can be done through a survey to better understand the organization's state of computer literacy and to find out who lacks needed skills. A *job analysis* can identify specific skills, equipment used, applications, authority, responsibility, and performance standards. Managers who use microcomputers, for example, need training that includes an overview of the applications and benefits of microcomputers. They also need to be aware of application operations, such as backup of data files, and data security. Operators, on the other hand, perform task-oriented functions such as keyboarding, inputting data, and copying and printing reports.

Who Should Train?

Trainers may be employed within the personnel, human resources, office systems, or information services department.

Some organizations have set up microcomputer trainers. Although they are not called micro managers, they manage all the personal computers for the department. This job includes selecting software and requesting equipment. As the needs vary, so do the positions. One microcomputer manager may be familiar with all aspects of this technology, whereas a coordinating manager may use administrative skills, rather than technical skills, to manage certain aspects of training.

Some managers delegate training to other employees. Yet few employees have the necessary knowledge of techniques or sufficient time to perform a top-quality training job.

It is important that information managers play a role in training. Managers understand and appreciate the means by which their departments carry out the firm's overall objectives. Managers come into contact with their employees on a day-to-day basis. They should be involved in training to ensure the success of the work group in the initial stages of training and on a follow-up basis. The information manager may not be well versed in the fine points of classroom-type instruction and may have little formal training in the field of education. Ideally, the very nature of the managerial role should indicate that the information manager has leadership ability and enjoys dealing with people. If this is not the situation, a change should be made.

Software specialists and hardware support people may be brought in to teach intensive training. They may train the manager or in-house training specialist, who in turn may train the employees.

Sequence of Training

Training designed to teach how the machine works, rather than how *people* use the machine for their work, is likely to fail. This type of training is not designed with a sensitivity to the trainees' concerns, worries, or needs.

1. Concepts of Office Automation. Before "hands-on" keyboarding takes place, trainees should be oriented to understand the entire picture of office automation. They must be made aware that the main role of the office is to handle information. They should know what their role is and how their office machine will help in the total flow of information within the organization. They should be given an explanation of the main components of their machine and its essential features and functions. This orientation will help to overcome computer fear. It will give them an overall picture of the logical purpose of using these devices. It will assure them that they can learn to use these machines. The learning process should be gradual and presented in a logical sequence.

2. Features and Functions of Computers. The next sequence of training should present the main features and functions of computers. It includes storage media, floppy disks, hard disks, and the use of software. Software capabilities for the word processor and personal computers should be presented in this sequence. Communication capabilities can also be introduced. The trainees should understand how information stored on a PC or word processor can be transmitted through telephone wires to another terminal. The explanation of communications can be expanded to include micro-to-mainframe links, LANs, and information retrieval. Each phase of learning brings success and motivates the learner toward the next learning sequence.

3. Advanced Applications. Records processing, graphics, desktop publishing, spreadsheets, calendaring/scheduling programs, and integrated software are examples of topics in the applications training phase. The list of applications is long. The best approach to introduce this phase is to apply the software programs for the trainee's particular application to illustrate the way this training can help on the job.

4. Hands-on Keyboarding. The final sequence of training should be hands-on keyboarding experience. Use a logical sequence that includes turning the system on, loading software, keyboarding practice material, simple editing, storage, and printing operations.

Comprehensive training that includes the first three phases of conceptual training builds an awareness and foundation for understanding the overall picture of office automation. Although trainees may become restless during these phases and may become eager to get their hands on the equipment, this preliminary learning is vital. In the long run, the learner will have a level of computer literacy and develop a solid foundation in office automation and computer systems.

The following is a summary of contents of a comprehensive computer training program.

Comprehensive Training Program Course Outline

- Phase I—General Concepts
  Principles of office information systems
  Information flow
  Components of a computer system
  Computer operation

- Phase II—Features and Functions of Computers
  Storage media
  External storage and memory
  Floppy disks
  Hard disks
  Internal storage and memory
  Random access memory/read only memory
  Operating systems

- Phase III—Advanced Applications
  Forms
  Graphics
  Integrated software applications
  Word processing/desktop publishing
  Financial and spreadsheet applications
  System security
  Data-base management systems
  Calendaring/scheduling programs

- Phase IV—"Hands-on" Applications
  Turning on the computer system
  Loading diskettes
  Caring for diskettes
  Using the keyboard
    Cursor controls/mouse/touch screen
    Automatic word wraparound
  Keyboarding practice material
  Simple editing
    Backspace/strikeover
    Insertion
    Deletion
  Storing
  Printing
  Removing diskettes and turning off system

The first three phases should present a brief overview of these topics to enable trainees to understand the process of computer operations and to allay their fears of computer systems. Time and resources dictate the amount of time spent on these phases. One should not go into extensive detail about every nut, bolt, circuit, and quirk of computer systems.

Formal training on Phase IV should move from the classroom to the job as the trainee becomes confident in the basic startup procedures.

Training Techniques

The training sequence described here must be concise, present usable information one step at a time, and it should also be cost effective. The following training techniques should be practiced:

1. *Do not teach too much at one session.* Learners can absorb and retain just so much at one time. Do not overwhelm them with many new topics. Break new topics into small and logical segments.

2. *Reinforce and repeat.* Repetition and reinforcement of previously learned materials are basic rules of instruction. Oral instructions must be repeated several times in varying forms for some learners.

3. *Set realistic goals.* Goals provide direction for the employees' work. If the trainees have realistic goals at the outset they are likely to reach desired levels of performance. Sometimes trainees have the skills and intellect but do not achieve the desired levels because they really do not know what is expected. Goals should be specific, attainable, and clearly understood by each learner.

4. *Demonstrate.* Demonstrating the sequence of steps should be followed by let-

ting trainees repeat the tasks on their own workstations.

5. *Criticize constructively.* Constructive criticism, when applied judiciously, can be an effective learning technique. It is impossible to learn without specific correction and clear direction for improvement.

6. *Give positive reinforcement.* Employees learning something new need generous encouragement. Acquisition of skills or knowledge is a form of change, and change can be intimidating. Sincere praise and reassurance of a job well done are needed to bolster confidence.

7. *Create a comfortable learning environment.* The trainer can set the stage for a comfortable learning environment. A flexible and unstructured approach may create a relaxed atmosphere for learning. Trainees need less warning about how to avoid breaking the machine and more assurance that they cannot harm it.

    Allow students to interrupt your lecture if they have a question. Answering questions helps "cement" information in the trainees' minds. Inject humor when appropriate, to provide a break in the tension.

8. *Do not teach everything.* Do not try to "spoon-feed" trainees every step of the way. If they come across a function that has them stumped, ask them to try to figure it out themselves. By looking it up in the manual, they experience a sense of independent discovery. Whenever trainees have the experience of discovering information independently, their self-confidence builds.

9. *Evaluate, follow-up, and revise.* Once trainees are sent into their corporate workstations, establish a periodic check to make sure that they are meeting expectations. This can be done by conferring with department supervisors or information managers. Successful training results in good performance on the job. Immediate evaluation and subsequent follow-up are needed to measure the training program's effectiveness accurately. Training must be revised as new software and systems are introduced into the workplace.

Training Methods

There are a variety of approaches to training. By choosing the right computer training method, a company can make the most of its investment in equipment. One approach may not work for all purposes, so it is a good idea to know about the pros and cons of each. The most common training approaches usually involve one of the following categories.

Group Instruction

Teaching by the group instruction method involves setting up organized classes. Group instruction can combine logical, structured presentations in concepts as well as hands-on practice. The organized class approach teaches a group of people how to use the same hardware or software at the same time. This method assures that all employees in a particular area receive the same information.

One-on-One Tutoring

In one-on-one training, a skilled user/trainer works with a new computer user, in several intensive sessions extending over several weeks. Many companies use this approach when training executives with little or no computer expertise. They perceive that these executives expect (and need) individual attention. Given the right tutor for the executive's personality and style, this method can be effective.

On-the-Job Training (OJT)

On-the-job training occurs in the real work environment. It is a highly individualized approach to teaching new users how to work with equipment. OJT allows a new office worker to learn how to use the equipment by working at the job under the direction of a co-worker, a supervisor, or a work group manager. The trainee performs actual tasks and can be productive immediately. However, the value that is gained from OJT depends upon the teaching or coaching ability of the trainer to whom the new worker is assigned. Some in-house experts are not necessarily good teachers.

Self-Paced Training Materials (Programmed Instruction)

The materials used in programmed instruction are specially produced for each hardware and software product. Self-paced training materials or programmed instructions help employees work individually through much of the training program. In this method of instruction, training material is presented in small bits of structured information. The learner proceeds in a step-by-step sequence from the basic elements of a skill or concept to more difficult material. Programmed instructional materials can be used to teach information system concepts, hardware training, and telecommunication.

Programmed instruction is especially well suited for training in machine operation. These skills are relatively clear and uniform. Keyboard sequencing is a procedure that has few variations.

Corporate Information Center

In Chapter 4 you read about the information center concept. The information center is designed to meet two very pressing needs. It is geared to help data processing personnel reduce the massive applications backlog by giving users the resources and the expertise needed to build some systems of their own. It is also intended to provide a training and learning environment for a growing population of microcomputer users.

The corporate information center can be a viable training approach, with a staff of data processing and microcomputer personnel who are available to provide assistance to users. Individualized training occurs as it is needed, and the personnel assigned to the information center can control and standardize training.

Computer-Based Training (CBT)

Computer-based training (CBT) or computer-assisted instruction (CAI) is a one-to-one interactive learning experience between the student and the computer. Lessons consist of explanations or lectures, and the computer provides the instruction and feedback. CBT training is only as good as the quality of instruction programmed. CBT, like other computer-assisted instruction programs, lacks the human element.

User Manuals and Documentation

Documentation is a written record detailing the design, function, and operating procedure for a computer system. Early user manuals that were provided with first-generation word processors and personal computers were cumbersome and difficult to understand. They were written for well-trained engineers, programmers, and technicians. When the typical operator/user is a business professional or office worker, documentation must be nontechnical and easy to read.

User manuals must have comprehensive *indexes* so that users can find what they need. They must be able to read and understand the material.

A user's manual should be the *primary source* of reference. User reference documentation continues after the training is over and the system is installed and operating. User manuals and documentation should provide quick, clear answers. The user should be as comfortable with using the resource as he or she would be with using a dictionary to check the spelling or meaning of a word.

In the rush to acquire the latest computer system, organizations have not devoted the attention, money, and resources to establishing an effective training program. Without proper training, hardware will be underused. Proper training and continuing support can help make the most of an organization's investment in computer systems. The training can also make the lives of information workers more pleasant and more productive.

Training and Career Opportunities for the Disabled

The Vocational Rehabilitation Act of 1973 has stimulated opportunities for qualified physically and mentally handicapped individuals. Now they can consider information systems as a new career. Employers with federal contracts have the obligation not only to prevent discrimination against handicapped individuals but also to take positive steps to hire and promote people with handicaps.

Computers are creating new lives for more than 20,000 handicapped workers across the country (Figure 21-1). Once forced to rely on others to read, write, or speak for them, this group now uses specially adapted personal computers to help make up for their disabilities.

New advances in computer technology will soon provide the opportunity for about 4 million additional handicapped workers to compete in the U.S. work force.

The Visually Handicapped

Equipment is being modified to enable blind operators to become more productive. The Optacon, manufactured by Telesensory Systems, Inc., of Palo Alto, California, is a read-

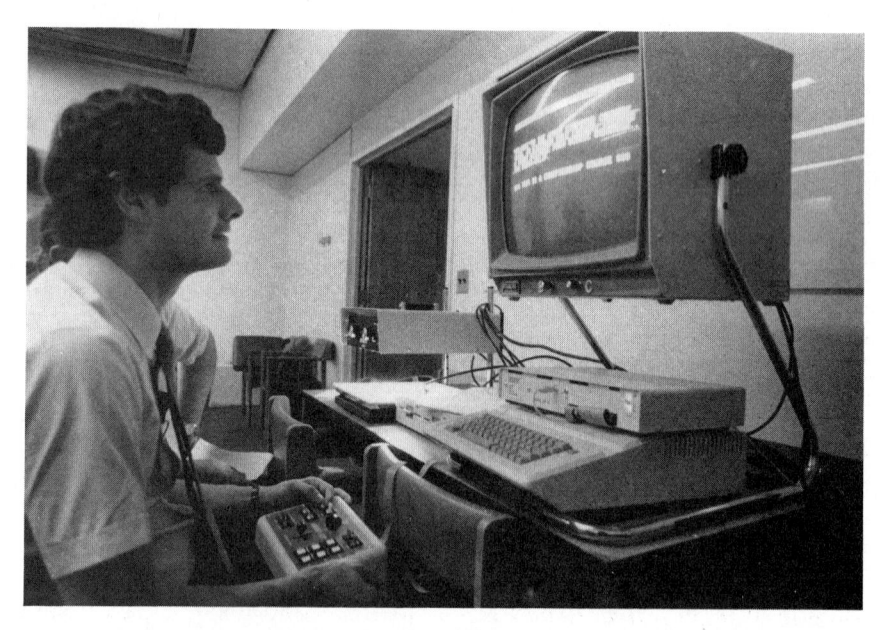

Figure 21-1. Visually impaired himself, Professor Lawrence Gardner, director of Columbia University's Program for the Education for the Visually Impaired, works with a large-screen computer designed for people with visual handicaps.

Source: Courtesy of Teachers College, Columbia University.

ing system for the blind. It is equipped with a specially designed, light-sensitive lens that reads the copy displayed on the CRT and transmits what it reads in electronic impulses to the index finger of the operator's left hand. What the operator feels with his or her fingers are the fully formed alphabetic characters.

A *voice synthesizer* can be attached to a computer to read words that appear on a display screen. A similar device can be programmed to speak for people who cannot talk.

Software can translate the text appearing on a computer screen into printed Braille. The same text can be displayed on a VersBraille, which has plastic pins that can be raised and lowered to form a line of Braille.

Workers with Limited Mobility

Computer systems with oversized keys or devices that can replace keyboards completely are some of the enhancements designed for people with limited mobility. Other designs include plastic tubes. They can be operated by what is known as "sip and puff," a system for translating inhaled and exhaled air into text.

New Technology for the Deaf

For the deaf, new advances in computer technology that can replace office telephones are now available. These include such techniques as computer mail and printed information that can be transmitted instantly to and from other computers. The advances in computer technology are creating new jobs for the deaf.

Technological Advances for Workers with Handicaps

Technology is providing immense opportunities for a large source of previously untapped capable employees. Specially designed computer systems not only will change the lives of the handicapped worker but will provide a productive way of life for millions of people unable to enter the job market. For the first time in history, the handicapped worker and the computer will be able to do almost anything anyone else can. If the worker cannot see, the machine can. If the worker cannot go to and from work, telecommunication technology will transmit data and information from the worker's home to the office.

Computers and computer-assisted instructions are already playing a major role in training and education. Computers are now helping children with disabilities in school. A deaf child can participate in class, can answer a teacher's questions, and can even joke with other students in the classroom.

Special computer schools for the disabled are providing education and training so that candidates can become productive members of society. One such school is the Computer Training and Evaluation Center, C-TEC, in Palo Alto, California. It offers information and training on computers for blind users.

To the credit of computer vendors, research and development in computer systems for the handicapped are continuing. Voice-recognition machines that would allow people to talk to a computer and have their words appear on a screen are now being developed by IBM, Kurzweil Applied Intelligence Inc., and AT&T Bell Labs. These devices would be used as electronic interpreters for the hearing-impaired and eventually as a replacement for dictation and keyboarding devices in the office.

Vendors, educational institutions, and dedicated professionals are working with computer technology to create new opportunities in the professional and personal lives of our handicapped citizens. It is through their combined efforts that the positive benefits of the ever changing technology will aid the handicapped worker.

Salary Administration

The cost of salaries represents a significant part of the total operating expenses for most organizations. It is therefore vital that adequate attention is given to this area. Then a business can operate in a highly competitive society, attracting and keeping skilled and competent workers. Top executives, information managers, and other officers of an organization may share in the responsibility of establishing and administering policies for a salary administration program.

In larger companies salary and payroll specialists create and administer a salary program.

New Approaches to Salary Programs

Benefits + Direct Salary = Compensation Package. Compensation in the form of direct salaries has long been perceived as the total compensation package for employees. However, the benefits paid or provided to an employee are clearly a part of any compensation package. Benefit coverage is just as much a cost to the employer as salary and wages and should be considered in the value of an employee's worth.

Enlightened employees are redefining compensation packages. Both employees and employers are including benefits and direct salary payments as items in compensation packages. However, the task of putting a dollar value on benefits coverage is difficult and involves consideration of a variety of standards.

Compensation packages have different values for employees compared to what employers must pay. The dollar value of what an employee receives in benefits, for example, is quite different from its cost to the company because of tax advantages and other factors. Some benefits such as health insurance, education plans, and child care have different values to different employees.

Flexible Benefits. Some companies have adopted a flexible benefit program to meet the needs of a changing work force as well as to manage rising costs of employee benefits.

A flexible benefit program allows employees to choose nontaxable benefits, taxable benefits, or cash.

The advantages of a flexible benefit program are

1. Meeting the needs of employees by allowing them to select their own packages
2. Helping manage rising costs of benefits
3. Aiding in redesigning a new and more effective program.

The various benefits employees can select include medical insurance, dental insurance, life insurance, hearing care, and long-term disability.

Another benefit that is receiving increasing employer attention is child care. There are now about 1800 U.S. companies that provide a wide variety of aid to help their employees meet child care needs. Corporate efforts range from operating on-site day care centers to making contributions to community-based child care programs; however, the vast majority of employers are providing direct financial assistance to employees. Provisions for child care benefit both the employer and employee. For working parents, child care has become a critical problem. It now trails only food, housing, and taxes as a budget expense. Changes in U.S. tax laws have made child care a nontaxable benefit for employees and a tax deduction for employers.

Rewarding High-Quality Workers. Workers in the information sector are no different from factory or other workers in their desire for adequate compensation for their time, energy, and mental discipline. The worker, whether functioning as a keyboard operator, a computer programmer, or a supervisor, gives up the best hours of the day in the interests of the company. Employers buy this time much as they buy commodities and other services. Skilled, experienced, and dependable information systems workers are a valuable and scarce commodity in many organizations. As a result, new merit compensation systems are being added to organizations to reward performance and creativity.

To be really effective, a compensation system should not treat all employees the same. It should not reward workers only for longevity. Creativity and excellence in performance should be recognized and rewarded as well, regardless of how long a worker is on the job. This philosophy is not without controversy. Merit pay for teachers has been debated for many years. Too many compensation systems today do not offer a significant difference in pay between the better performers and those who are not as good. The difference needs to be great enough to provide an *incentive* for high performance. In order for any organization to grow and thrive, it must tap and channel the creativity of its workers. Establishing an environment in which ideas and creativity can flourish is an essential prerequisite for this to occur.

Work group managers should have the freedom to encourage and reward extraordinary performance on the job. Creativity and merit pay (reward systems) can be achieved in a variety of ways. Employee suggestion programs, performance appraisals, and development and use of new methods by employees are examples. Whatever form the reward takes, meaningful recognition should be part of a management philosophy that recognizes high-quality workers.

Salary Surveys

In order for organizations to attract and keep good people, they must be aware of the salary structure of competitive firms in their area. Similarly, employees should be aware of the going rate for their job titles, experience, and credentials. Professional organizations such as the Association of Information Systems Professionals (AISP), Data Processing Management Association (DPMA), Association of Records Managers and Administrators (ARMA), and the Association for Systems Management (ASM) provide timely salary surveys. These reports are usually prepared annually and provide a basic idea of what other firms in the area pay for office automation jobs.

These surveys measure numbers, spot trends, list new job titles, and analyze comparative measurements for the years ahead. Management can use these surveys to plan recruiting, training, and compensation strategies more effectively.

Job Performance: Measurement and Review

Measuring Job Performance in the Office

Automating office operations has not only helped enhance productivity but has also brought about the need to change the way organizations measure employee productivity. There has always been difficulty in applying work measurement to office workers. *Work measurements* determine how much work is completed and how effectively it is completed. Office work consists of a series of complex and varied tasks. Repetitive information processing operations do not exist in as many phases of office work as they do in assembly-line factory work.

In the early days of centralized word processing, work was specialized and operator tasks were clearly defined. A modified approach of measuring output was practiced. Problems such as inequities in pay, work loads, opportunities for advancement, and other discrimination associated with this organizational structure soon surfaced. Measuring operators *only* soon became controversial and workers voiced their objections because of pressure and stress. As a result, many firms modified or discontinued measuring operator output and performance.

However, organizations now realize that some degree of measurement of office operations is needed because the investments in technology and human resources are very high. In addition to evaluating how effectively office technology enhances productivity, a system of measurement is needed. Productivity gains can be measured in tangible terms such as units produced, time saved, money saved, quality, accuracy, and turnaround time.

Traditional methods for measuring the efficiency of operators, such as keystroke and document counting, are still being used. But determining the degree of productivity for management is a much more complicated task.

Managers, for example, with the help of personal computers and integrated software, can take notes, draft reports, create graphics, access data bases, and transmit documents to a branch office. Office information systems have redefined the evaluation of white-collar workers and have made the word processing centers obsolete. Automation has placed personal computers and multifunctional terminals on all workers' desks regardless of position. Office automation has changed the nature of work at all levels. Managers and workers now perform tasks such as calendaring and networking. Automation has altered the process of measurement. It has created an environment where workers must adapt their unstructured and diverse work activity in a highly structured electronic work place.

The Rationale of Measurement

Today, measuring office tasks has taken on a positive image. The office systems analyst uses measurement to

1. *Justify equipment needs* by determining activity levels of potential users
2. *Determine* the application programs accessible to the user
3. *Justify the cost* of software and hardware purchases
4. *Verify the effectiveness* and/or performance of a system or program
5. *Perform* quality control testing
6. *Anticipate* future needs

Although methods vary, at least measuring office performance has become democratic. It applies to all levels and is no longer personally threatening to one's job security or status.

Summary

Training provides employees with directed experience and the means to achieve desired learning goals. Good training provides the bridge between prior experience and desired performance. The most common training methods are group instruction, one-on-one tutoring, on-the-job training, self-paced training materials, computer-based training, and documentation (user manuals).

Computers are creating excellent employment opportunities for handicapped workers. Specially adapted computers and workstations provide them a safe and comfortable environment.

Effective compensation plans should not reward workers simply on the basis of longevity. Creativity and excellence in performance should be recognized and rewarded as well. Creativity and merit pay can be achieved through employee suggestions programs, performance appraisals, and development and use of new methods by employees. Measuring employees' performance should be democratic. It should apply to all levels and should not be used to threaten job security or status.

Review Questions

1. How does today's information worker compare to the worker of a decade ago?
2. Describe how workers were trained in the early days of word processing systems.

3. What does the term *unbundling* mean? What factors led to this practice?
4. How does a job analysis determine who needs training?
5. Briefly describe the role, if any, an information manager or department manager plays in training.
6. The suggested sequence of training in this chapter emphasizes the importance of explaining the concept of office automation prior to hands-on training. Why is this knowledge necessary for workers who will only be involved in machine operations?
7. List some advantages of each of the following training methods:
   a. Group instruction
   b. One-on-one tutoring
   c. On-the-job training
8. Describe how the corporate information center (IC) serves as a resource for training.
9. Briefly describe computer-based training (CBT).
10. What were some of the early problems in documentation as a means of training resource? What improvements have been made in training manuals to help users understand the hardware?
11. What special devices and adaptions are now available for blind computer operators?
12. Why do employees place different values on the same benefits in a compensation package?
13. What are flexible benefits?
14. A merit compensation plan seeks to reward employees on the basis of outstanding performance. yet some employees oppose this compensation plan. Why do you suppose they do?
15. What were some of the deficiencies of early employee performance measurement plans? What effect did office automation, a new corporate structure, and an abundance of personal computers have on performance measurement plans?

Projects

1. Microcomputers have been added to the desks of many executives. Some organizations are not willing or equipped to handle the task of training. As a result, these organizations have contracted with specialized training companies devoted to custom-designed courses in microcomputer applications for managers. Describe the services that one specific company provides.
2. Collect data and information on a training program in an organization in your community. Write a summary report describing its training methods, resource material, and type of computer used. Include a course outline, the levels of trainees, and the assignment of responsibility for training.
3. Analyze the user manual provided with the computer used in your college class. Write a critique indicating good points and bad points in terms of organization ease (or difficulty) in finding operations and features, index, writing style, and language.
4. The widespread availability of the computer has become a concern to executives and professionals. The high-level executives who now use personal computers are not data processors, programmers, or typists. They resist learning to develop keyboarding skills and feel that their hunt and peck typing skills of five words per minute are adequate.

    As a director of training, you feel that it is important that executives learn to keyboard using the traditional touch typing method. Prepare a report for top management to justify your position. What reasons can you cite to support your position that managers should learn to type?

PART SEVEN

The Future

CHAPTER 22

Toward the Twenty-first Century

After reading this chapter, you will understand
1. The allure, curiosity, and excitement of perceiving an advanced age of information technology
2. Artificial intelligence and its application to office tasks
3. Alternate methods of text entry and their potential effect on our traditional method of keyboarding
4. Some of the exotic future technology of robotics, smart cards, and futuristic telephones
5. The way that individuals can prepare for technological change

> The mind of man is capable of anything—because everything is in it, all the past as well as all the future.
>
> —*Joseph Conrad*

Human beings have risen above the other species that inhabit the earth because they were endowed with an ability to learn that vastly exceeds their need to survive. Not only have human beings struggled to feed, clothe, and protect their brood, but they have persevered to understand the world. Morever, human beings are capable of logical analysis and able to communicate complex ideas.

According to Greek mythology, Prometheus gave to humanity the gift of fire—the symbol of humankind's relentless, questing intellect. Since then, an age of discovery and enlightenment has thrived. History has vividly demonstrated that the more innovative and revolutionary the science, the more profound and far-reaching are its practical applications.

And as Thomas J. Watson, former president of IBM, once wrote: "An invention is the product of imagination and human aspiration achieved through hard work. Its purpose is to improve the way of life, both physical and spiritual."

The Allure of the Future

Throughout the course of human history, the future has always had an allure and fascination. As early as 600 B.C., Greek philosophers and storytellers weaved magical tales based upon past events and their own intuition. Later, scientists used forecasting and predictions to help them create and invent new tools.

As we approach the twenty-first century, we are still drawing on our presumptions. However, in the area of computers and office automation, modern-day soothsayers may be working with a formidable handicap—the unknown. To predict the future is difficult. To predict the future of something that has almost no past is impossible.

Despite the handicaps, predicting what the office of the future will be like is as compelling to modern forecasters as it was to the sixteenth century soothsayers who tried to guess what was on the other side of the ocean.

Change: A Constant and Predictable Ingredient

The one thing that is certain about the future of information technology is change. Change is rarely easy or comfortable. Like other implementations of high technology, office automation is an agent of change. We try to understand it, predict it, and control it. However, what happens when the agent of change is itself changing? This final chapter will try to detail some of the forces that have triggered the evolution in information management and computing. The "offices of the future" are still evolving, and some exciting new capabilities are just around the corner. Let us look at projected trends, developing technologies and the effect they will have on the office environment by the end of this century.

Artificial Intelligence (AI)

Artificial intelligence (AI) was mentioned briefly in Chapter 14 as a part of knowledge-based software. It has grown into a vastly intriguing subject and deserves a more detailed explanation in this section. The simplistic definition of artificial intelligence is the capability of a machine to imitate intelligent human behavior. As with any evolving and overused buzzword, the acronym *AI* currently has expanded to have a variety of meanings, focused on computer systems that are capable of performing "intelligent" human activities, such as language, understanding, reasoning, learning, and problem solving. The predominant type of AI system allows people to communicate with their computers using free-form English. Through artifical intelligence, information workers may soon find themselves actively engaged in conversation with a computer, just as they now sit down to chat with a fellow worker.

Personal computers, no matter how fast they become or how much memory they can store, can only serve humans as productivity tools, turning paper-based functions into electronically automated procedures. Using conventional software, computer systems will ultimately compute faster and become more memory intensive. However, they will not be able to provide answers or solve problems. This is the area in which artificial intelligence combined with *expert systems* can offer change. They can give advice based on knowledge of a particular field, the user's goals, and a system of rules for making judgments. Expert systems can store specialized information about a particular aspect of the world and then use the information to draw conclusions about a problem. Perhaps the best known expert system is one that doctors use to help diagnose illnesses.

Commercial applications, using artificial intelligence in the office, are beginning to appear. They will aid managers in a variety of business and human resource situations, such as advice on handling difficult business and employee situations, advice on developing training and sales skills, and advice to office workers, on all levels, who need a system to perform scheduling.

Natural language is another form of artificial intelligence, which differs from the expert systems just described. Natural language systems allow a personal computer user to interact with it in more comfortable ways. A user can type in words and statements that look more like English and less like "computerese." Natural language systems either replace or supplement the command and menu-driven systems.

The realm of artificial intelligence will ultimately transfer the burden of understanding from the computer user to the computer. Natural language will be a key component in the next generation of computer software. Conversational interaction between people and computers is the sphere in which future breakthroughs will come.

Voice Communications Between People and Machines

Major shakeups in the human-machine relationship will occur in advanced voice technologies, including speech synthesis, speech recognition, and speech verification. Innovative research and imaginative approaches to artificial intelligence have placed these technologies on the leading edge of the human-machine connection with the spoken word. Controlling machines directly by voice would link us to them in a most basic way. The realization of this potential has come to symbolize a major breakthrough. These concepts, first presented in Chapter 7, provide a radical alternative to keyboard use.

Text-Entry Systems

Other text-entry systems, including touch-sensitive screens, the mouse, and tracking balls, are drawing executives, who have traditionally resisted typing, closer to the computer. These alternate methods are image-oriented and are geared to letting the user visualize the tasks. Icons or pictures on the screen, for example, quickly identify tasks, and "windows" allow a user to see multiple tasks simultaneously.

Workstations

On the threshold of the next century, the typical workstation will be dramatically changed. Analysts predict that workstations will feature 2–5M bytes of memory in a system designed around a 64-bit processor. The bulky CRT will be replaced by slim, high-resolution, flat-panel displays. A typical workstation will include a 10–50-M-byte hard disk. Future workstations will operate in a "seamless partitioned" mode, where communication, at all levels, will be completely transparent to users.

Tomorrow's workstation will be smaller,

designed for the single user; and the workstation will listen, talk, and see.

The workstation will perform almost every conceivable office application and will have every peripheral available. All of these features will be linked into one unit, with a desktop footprint smaller than the smallest personal computer, but one hundred times more powerful. This single-unit workstation will replace just about everything that is now on top of one's desk—calendar, in-out basket, telephone, index file, and calculator. It will perform all of these functions more accurately and more speedily as it helps to create a better document. Many of these multifunctional systems are available today and vendors are working on creating still more powerful and versatile tools to help the workers of the future.

Advanced Storage

The first computers occupied large rooms, cost a fortune, and had limited storage capacity. Today, computers have grown smaller, cost less, and are capable of storing tremendous quantities of information. Miniature random access memory (RAM) chips hold massive amounts of data, and their capacity keeps growing. Hard disks with billions of bytes will be the norm for the future.

Soon, we will have at our disposal a full-scale computer embedded in a credit card. We will be able to carry it in a purse or pocket and throw it away after a few years of use. These pocket marvels, known as smart cards, were developed in Europe. Smart cards will soon replace credit cards with programmable microchips that can store and forward information electronically to thousands of customers. These applications will end the problems of counterfeiting, stealing, and other credit card abuses. Future smart card applications will also protect software from unauthorized access and duplications. Department stores, health groups, and other organizations will use smart card technology. They will be able to distribute smart cards to each of their customers or subscribers. Each card can hold the equivalent of 800 pages of information. Medical subscribers, for example, can have smart cards that include a digitalized photograph of the carrier, a fascimile of his or her signature, the extent of one's health insurance, an electrocardiogram, a chest X ray, a list of medicines being taken, the names of physicians who have provided treatment, and other elements.

The smart card of the future could very well include a keyboard and display. The device would look like a solar-powered credit card calculator, increasing simplicity and security, while permitting the user to check his or her checkbook balance immediately.

Robotics

Although robotics are used mainly in factories (Figure 22-1), they have the potential for becoming an integral part of education, medicine, the home, and the office. Today's robots are mostly used in repetitive tasks that can be preprogrammed, such as spot welding, grinding, spray painting, stacking, loading, and unloading materials to and from machines. Robots are ideal for these unpopular tasks since they are untiring, performing their duties flawlessly and uncomplainingly. Doctors even use computerized robotic arms to help with surgery on the human brain. These medical robots calculate the angles, hold and direct a surgical drill, and position a biopsy needle while the doctor applies the pressure on the instruments to penetrate the skull and brain. In the future, surgical robots may be used to drain abscesses, implant radioactive pellets directly into tumors, repair blood vessels, or guide laser beams to tumors.

The real value of robots for office applications will be apparent when they can utilize artificial intelligence concepts. The objective is to create machines with human attributes

Figure 22-1. Four computer-controlled robots weld the underbody of full-size luxury cars at the GM-Orion assembly plant. These are among 138 robots used to weld bodies at the plant to achieve structural soundness as part of building a high quality car.

Source: Courtesy of GM Corporation.

such as sight, speech, perception, movement, and decision-making ability.

Telecommunications

No single development is playing a more significant role in reshaping the information processing industry than the telecommunication revolution. The convergence of data processing, office automation, and communication technology is rapidly creating a whole new range of possibilities for communicating and handling information. Sparked by divestiture, the communications market has grown during the decade of the 1980s in terms of dollar volume and product offerings (Figure 22-2). Because no one company will be able to satisfy the demands for information processing systems and products, merges between computer and communications companies will continue to proliferate.

Some new applications for digital technology are global networks of simultaneous voice, data, and image communications. These sophisticated networks hold the promise of communications through the use of intelligence terminals so that information can be organized, stored, accessed, and retrieved from anywhere in the world. Today, the world is on the threshold of a new communications era. Some of these communications wonders will be available before the end of the century.

Supernetworks will link the globe through high-capacity optical cables and sophisticated computers. Such a network will allow a variety of reciprocal banking and retailing services, open up access to a wide range of specialized information, provide home-security alarm services, and even deliver newspapers and magazines to subscribers through a computer terminal and printer.

Cellular car phones will expand in terms of subscribers and quality of service. Within a few years, telephones will be everywhere—in

Figure 22-2. Airfone is the first air-to-ground telephone system designed for the commercial airline passenger.

Source: Courtesy of Airfone, Inc.

boats, cars, airplanes, and coat pockets (Figure 22-3). Autos will not only have telephones as standard equipment but also satellite navigational devices to pinpoint a vehicle's location and guide the driver to any destination.

Telephones will evolve into computer terminals. Text and pictures will be viewed on a display screen attached to the phone, and additional data will be delivered as electronically synthesized speech. A caller's identity can be learned before the receiver is lifted from the hook, thus permitting the screening of calls. A microprocessor-controlled transmitter will send the calling party's number to the user's display screen during the silent period, between the first and second rings.

These developments give only a small sampling of the marvels of telecommunications that lie ahead. Some of the technologies are in limited use today; others are in the drawing-board stage and are considered feasible within this decade.

How Will Computer Technology Shape Our Society?

There has been much speculation about how the future of the information age will shape our society. Social scientists, sociologists, and futurists have probed these issues. Will automation and computers be used to create a world in which individual actions and thoughts are monitored and controlled? Will the information processing capabilities of computers create an environment of stress and strain? Or will the world become a technological wonderland making possible a rich and better life for all, with freedom for individual thought and actions everywhere?

As we move toward the next century, it seems unlikely that either of these extremes will prevail. If we are to approach the brighter outcome, it will require a thoughtful plan in order to break through some formidable barriers. Resistance to change has been dominant in limiting progress in information processing. Holdouts of another era still feel uncomfortable or intimidated by computers. They argue that using a machine is difficult or time-consuming. Through education and gradual phase-in, companies are striving to counteract this resistance.

Economic and technological barriers have also slowed the acceptance of information technology. However, these barriers are also rapidly failing.

Other breakthroughs will be achieved when resistance by organizations, institutions, and governments can be altered in favor of technology. Old laws that no longer serve us well must be changed. The work force needs an in-

How an idea in yesterday's funny papers can become tomorrow's front page headlines.

Dick Tracy's wrist radio used to belong strictly in the Funnies.

But a revolution in electronics is moving ideas like these onto Page One.

Thanks to a semiconducting compound called gallium arsenide that's being used to make super microchips by the people at ITT.

These miniature integrated circuits work ten times faster than conventional silicon chips.

And at higher frequencies in smaller spaces.

Which could make possible satellite phone calls from personal wrist phones.

And night vision devices for crime detection that are thousands of times more efficient than the human eye.

To find other ways gallium arsenide can help advance the state of the art in electronics, ITT is building a multimillion dollar research center.

There, ITT engineers will be able to use this revolutionary technology to turn the ideas of yesterday into the news of tomorrow.

Figure 22-3. Yesterday's comic strip can become tomorrow's telecommunication norm.

Source: Courtesy of ITT Corporation.

fusion of younger executives who have grown up with personal computers, are comfortable with machines, and have a thorough computer literacy. To enter the twenty-first century on a positive course, there must be a willingness to reorganize existing work patterns in order to blend the best of technology and human resources.

Learning from the Past

Throughout this chapter, we have discussed some of the anticipated problems and opportunities of the future. This knowledge may help us to understand and appreciate how these technological changes may affect our personal and professional lives. Contemplating the future can offer promise and hope, or the future can be perceived as threatening, with no room for freedom of thought or deed or individuality.

In today's culture we must live with change. We must learn to adapt to it and harness it into constructive, productive, and profitable channels if we are going to survive. Every society undergoes change. When Alfred Lord Tennyson said a century ago, "Let the great world spin forever down the ringing grooves of change," he could have been talking about today. He would be amazed at the changes taking place during the remaining years of the twentieth century.

The Best Is Yet to Come

Information technology has been a relatively recent phenomenon. Anyone who still believes the proverb "The more things change, the more they stay the same" after observing three decades of technological upheaval must have been asleep or well insulated from the events taking place in this information age. They will be in for more surprises in the next decade and the century ahead. New information technologies, developing for the past thirty years, are about to burst upon the office environment:

- Intelligent humanoid machines in the form of robots
- Computers that see, hear, speak, and reason
- Knowledge systems that are nourished by the input of a warehouse of genius and are able to hold and use that knowledge forever
- Dick Tracy–like watches that can integrate a computer, and a telephone, and have an automatic translation capability
- A work force that is more individualistic and creative, using tools and technologies in a comfortable, safe environment

These are glorious times to be part of the information society. They can be the best of times for aggressive, knowledgeable, computer literate professionals. If these individuals are willing to use the technology to their advantage, they can reap the fruits of career advancement and economic gain.

Summary

Throughout the course of history, there has always been an allure and fascination with the future. The gadgets and devices that await us in the area of computers and office automation are spectacular to behold. Consider artificial intelligence that will be capable of imitating and responding to human behavior; expert systems software that will solve problems, offer advice, diagnose illness, and render verdicts; voice communication computers that will recognize speech commands and talk back to people; smart cards that will be carried by consumers and hold thousands of pages of information; and robots that will have human attributes such as sight, speech, perception, movement, and decision-making ability.

We may view these phenomena as dreadful and frightening or as positive forces that will improve the quality of life. No one can accurately predict whether or not all of these ideas will become commercially successful. Humankind must have the foresight and wisdom to use what will benefit society and to abandon those processes that make little or no contribution to our well-being. Beethoven said, "Bring all the parts together in magnificent harmony." Managing information takes more than computers and devices alone. To orchestrate the modern office, we need both—machines and people, working together as a unified whole.

Review Questions

1. The text suggests that people have always had a curiosity about forecasting the future. Why is forecasting in the age of information technology particularly difficult?
2. What is the meaning of artificial intelligence, and how can it be applied to the office?
3. Compare natural language software with expert system software.
4. What alternative methods of text entry can be expected in the future?
5. Briefly describe some of the characteristics of the future workstations.
6. What is a smart card? How can it be used in medicine? in banking? in shopping?
7. How can robotic technology be used in the office?
8. Briefly describe some of the exotic features of the telephone of the future.
9. What are some of the barriers that may impede progress in advanced information technology?
10. What can individuals do to prepare for technological change? How can we use change as a positive force within our personal and professional lives?

Projects

1. There has been confusion and concern about the future and practical application of artificial intelligence. Research some recent articles in this area and write a summary of the present and future direction of artificial intelligence. Be prepared to give an oral report to the class.
2. Who are the year's five top inventors or scientists who have contributed the most toward information technology? Write a report on the current year's group of men and women and include their names, backgrounds, and a description of their contribution in computers, office automation, or information management.
3. What will it be like in the year 2000? What will you be doing? Will you be worried about health, crime, war, pollution, or just plain survial—or will these issues be behind us? Will people be more friendly and concerned, or will they be more scared and defensive?

   There is no right or wrong answer to this project. Create a scenario of how you think life could be in the year 2000. Describe your home, your job, your community, and your work environment.
4. Prepare a report on speech technology. Include in your report answers to the following:
   - How does speech recognition work?
   - What are the different types of speech recognition?
   - What is involved in making a computer recognize human speech?
   - In what applications do you think voice systems can be used?

GLOSSARY

Acoustics. An ergonomic consideration relating to the level of noise within an office and workstation. Noise can be controlled through the engineering and/or the architecture of the space.

Affirmative action. A law designed for employers to set numerical goals and take positive steps to guarantee equal employment opportunities for everyone.

Ambient conditions. The climate conditions of the office such as temperature, air quality, lighting, and noise.

Analog transmission. In telecommunications, a transmission technique of varying (rising and falling) magnitudes of frequencies.

Applications software. Software designed to instruct the computer to perform specific tasks, such as electronic spreadsheets, word processing, or inventory or payroll programs.

Artificial intelligence. Computers and software that could replicate human abilities, that could understand simple conversation, see, and in some cases, reason.

ASCII (American Standard Code for Information Interchange). A standard set of codes used to represent characters within a computer.

Asynchronous. A data transmission method that uses "start" and "stop" bits before and after each character, allowing the time interval between characters to vary.

Audiotex. A computerized home information service that provides data and information to subscribers via the telephone.

Authority. The right to invoke compliance by subordinates on the basis of formal position and control over rewards and sanctions.

Bandwidth. The difference between the highest and lowest frequencies of a transmission channel.

Bar code. A series of lines that identifies items and prices for computers.

Bar code reader. A scanning device that translates black and white bar codes of different widths into electrical impulses.

Baud rate. A standard measure, in bits per second, of the speed by which computers transfer data from one place to another. The higher the baud rate, the faster the data transmission.

Bit. The basic unit of information in a computer. A bit indicates a single value: 0 or 1. Bits can be added together to form larger words called *bytes*.

Boot. The initial starting-up of a PC. The opening system is brought into main memory and takes over control.

Bus. A group of electronic paths in a computer that permits data to flow from one place to another.

Bypass. The use of one or more transmission media to link end users that excludes the use of local telephone company exchange plants.

Byte. Each combination of bits, representing a letter or a number, is called a *byte*. There are 8 bits in a byte.

CAD (computer-aided design). The use of computers and advanced graphics hardware and software to provide interactive design assistance for engineering and architectural design.

CAI (computer-aided instruction). The use of computers to aid in the instruction process. Almost any topic or subject can be taught with computer-aided instruction.

CAM (computer-aided manufacturing). The use of computers to automate the operational system of a manufacturing plant. Also see MAP and TOP.

CAR (computer-assisted retrieval). The use of the computer to index and retrieve COM (computer output microfilm) files of randomly microfilmed information and documents.

CBT (computer-based training). See CAI.

Channel. An electronic transmission path between two or more stations. Channels may be furnished by wire, radio, or a combination of both.

Chip. A microprocessor that is a complete computer on a single chip of silicon.

COM (computer output microfilm). A micrographic output method in which data are placed on microfilm rather than on paper.

Communicating word processor. A system that can send text from another communicating system or in exchange with a computer, over phone lines or via other electronic hookups.

Communication. The imparting or interchange of thoughts, opinions, or information by speech, writing, and electronic means.

Communications terminal. Any device that generates electrical or tone signals that can be transmitted over a communications channel.

Company store. A facility within an organization where employees with an interest in microcomputers can come to select hardware and software and receive advice and training.

Compatibility. The ability to run the same software programs and connect the same peripherals and add-on equipment (boards, printers, modems) as another PC.

Computer. A machine designed to receive and store instructions and execute them with the capacity for the input and output of data, as well as processing and storing various forms of communications.

Computer crime. The unlawful use of a computer to steal information.

Copyboards. An electronic presentation board that can produce copies of material written on the board's surface (also referred to as a *whiteboard* or *electronic blackboard*).

CRT (cathode ray tube). A televisionlike screen that shows information as it is entered into the computer.

Cursor. The movable dot on a CRT screen that shows the place on a displayed document for entering new text or making editing changes.

Data. Facts or statistics entered into or taken out of a computer.

Data bank (data base). A collection of information in a computer, arranged by means of various indexing procedures so that the data are readily accessible.

Data communication. The transmission of data from one computer to another.

Data processing. The execution of a programmed sequence of operations upon data. A generic term for computing in business situations and other applications with machines such as microcomputers.

DBMS (data-base management system). A toolbox of programs, utilities, and processes that actually "manage" the physical contents of the data base.

Dedicated word processor. A computer system used primarily for word processing.

Delegation. A function of supervisors who provide for the assignment of meaningful tasks to subordinates.

Desktop publishing. A process of in-house publishing that combines a personal computer, page layout software, and a laser printer to create professionally typeset-quality documents.

Digital. Pertaining to data in the form of digits.

Digital transmission. Data represented in discrete, discontinuous form, as contrasted with analog data represented in continuous form.

Disaster recovery plan. A plan to assure that an organization's operations, employees, and assets effectively survive the impact and consequences of a catastrophic event.

Disk. See Floppy disk.

Divestiture. Breaking up. Occurs when one or more companies are split from a parent corporation.

Documentation. A collection of information and instruction (usually contained in a manual) that describes a computer program or sequence of operation.

Download. A process of calling up (retrieving) information from a large computer onto a microcomputer.

DSS (decision support system). Interactive computer-based services, including data-base management, modeling, report generation, and electronic spreadsheets.

Duplexing. A copier feature that makes two-sided copies.

Electronic banking. A means of electronically handling the request, production, and delivery of banking service.

Electronic mail. Communication of nonvoice information (the message) from one location to another, accomplished electronically (without the physical movement of paper) either wholly or partly.

Electronic typewriter (ET). A typewriter that looks and operates like a standard electromechanical typewriter but contains a microprocessor that enables it to perform word processing and other advanced functions.

Emulation. A process by which one computer system is made to function like (imitate) another, in order to accept the same data, execute the same programs, and achieve the same results as the imitated system.

Encryption. A system in telecommunications security that provides a special code so that only users with the appropriate terminals and decoding program can interpret the scrambled message.

EPROM (erasable programmable read only memory). Memory chips that can be erased.

Equal access. Allows phone customers to use any of the long-distance carriers without dialing extra numbers.

Ergonomics. The study of people at work. Ergonomics emphasizes the safety, comfort, and ease of use of human-operated machines, such as computers. The goal of ergonomics is to produce systems that are safe, comfortable, and easy to use.

Ergonomists. Professionals in the field of planning and designing ergonomic workstations, peripherals, and environments (also referred to a *human-factors engineers*).

Executive search firm. A personnel service that provides organizations with candidates for supervisory- and management-level positions.

Expert systems. A software program to aid in the decision-making process that guides the user in making the right decision. This program differs from other software programs in that it has the ability to reach conclusions that are not programmed into it.

Facsimile. Process of transmitting textual and graphic material electronically, usually via telephone lines.

Fault-tolerance system. A back-up system to prevent a computer from "going down."

Fiber optics. A technology that uses light-conducting glass or plastic rods to transmit information at high data rates.

File. A collection of related records.

Filing system. A way to store documents or other materials in an organized and standardized fashion, so that they can be retrieved by following some logical procedure.

Flat-panel display screens. Non-CRT display screens that are small, thin (less than three inches thick), lightweight, and designed for portable computers.

Flextime. A work schedule that allows employees to decide for themselves when they go to work and when they leave.

Floppy disk. A flexible piece of mylar plastic contained in a protective envelope that stores data. The data are recorded magnetically in a number of concentric tracks.

Floppy disk drive. A disk drive that records data onto a rapidly moving magnetic disk.

Footprint. The amount of space a device takes up on a desktop surface.

Gateway. A form of network that allows different products and devices to communicate with each other.

Gigabytes. A unit of measure in which 1 gigabyte (Gbyte) is equal to 1024 megabytes, roughly a billion bytes of information.

Glossary. A word processing function that enables the user to store frequently used terms or phrases and display them on the screen anywhere in a document.

Graphics. A form of image processing that presents information in the form of charts, lines, curves, and other visual displays.

Hacker. A computer enthusiast who attempts to gain access to computer networks (also known as an *electronic trespasser*).

Hard card. A miniature circuit board that can be inserted into an expansion slot of a PC that provides the storage capability of a hard disk.

Hard disk. An aluminum platter coated with a magnetic surface. It is rigid and mounted in a sealed box. Hard disks are capable of storing large amounts of data (25 or more megabytes).

Hardware. The equipment that makes up the computer system.

Header/footer. A word or series of words, and/or page numbers, that appears consistently at the top (header) or bottom (footer) of all pages of a document. This could include copyright notices, company logos or names, and so on.

Hierarchy. The number of levels of management within the organization arranged by degree of authority.

Hypertext. An electronic outline software program.

Icon. A pictorial representation of a command to the PC. An example would be a picture of a garbage can to ask the computer to erase a file.

Image processing. Occurs when data are converted into visual and graphic representation.

Index generator. A word processing feature that allows the user to generate a reference index as part of a records processing function.

Information center. A concept and place within an organization designed to offer support, aid, and resources to end users of computer systems.

Information processing. A concept that covers both the traditional concept of processing numeric and alphabetic data and the processing of text, images, and voices.

Information resource management (IRM). A management concept that views data, information, and computer resources (computer hardware, software, and personnel) as valuable organizational resources that should be efficiently, economically, and effectively managed for the benefit of the entire organization.

Information resource planning (IRP). The process of identifying the fundamental structure of information available to an organization.

Information workers. Professionals who create, process, and distribute information (also referred to as *knowledge workers* and *white-collar workers*).

Input. The process of entering information into a computer.

Input/output (I/O). Software routines or hardware architectures that receive or transmit data with peripherals external to the computer.

Integrated services digital network (ISDN). A future network that would provide a common standard for computers and other devices to send and share information.

Integrated software. Software that can perform more than one function, such as word processing, spreadsheet, graphics, data-base management, and telecommuncations.

Intelligent building. A building that includes computer-controlled energy management systems, electronic security, and advanced telecommunication services, usually controlled by a central computer and communication system (also referred to as *smart buildings, shared tenant services,* or *technology-enhanced commercial buildings*).

Interactive function. A two-way communication with the capability of manipulating information, resulting in a finely honed message.

IVDT (integrated voice/data terminal). A device that combines voice, data communication, and a range of functions in a single desktop unit (also referred to as *integrated voice/data workstation, universal workstation, executive workstation,* or *computer phone*).

Jobsharing. An alternate work style that involves two half-time workers' sharing one job.

Kilobyte. 1024 bytes or units of information.

Lap-size computers. Portable computers that generally weigh less than 10 pounds and can be carried in a standard-size briefcase.

Light pen. A stylus-shaped device that may be used to enter data on a display screen.

List processing. A feature of a word processing program that provides the user with creating, retrieving, and sorting lists in desired formats.

Local area network (LAN). A network that interconnects devices by using nonpublic conductors within a small geographical area.

Machine transcription. The preparation of a typewritten document from dictated material by means of a transcribing machine. The operator transcribes directly from sound rather than from visual materials, such as shorthand symbols or longhand notes.

Macros. Defining keys on the keyboard to be commands, words, or a specified sequence of machine instruction.

Mailbox. A word processing feature that provides for sending and receiving short messages to other word processing terminals hooked into a network.

Mainframe. The largest type of computer system.

Maintenance agreement. An insurance policy for a computer system.

MAP (manufacturing automation protocol). A set of evolving and established standards designed to allow factory equipment made by a variety of vendors to communicate over a single network.

Menu. A type of user interface that lists the possible commands that the user can perform on the PC.

Merge. A word processing function that combines text from two independent documents. The main document is referred to as the *primary document* and the variables for the main document are referred to as the *secondary documents* (also referred to as *list/merge*).

Microcomputer. A relatively small-size computer system (also called a *home computer* or a *personal computer*).

Micrographics. A process of photographing paper records for the purpose of recording and storing data and information in greatly reduced form.

Micro-mainframe link. An interorganization network that links remote microcomputers to corporate data bases stored on mainframe computers.

Microprocessor. A CPU contained on a single chip, designed to perform a specific job.

Minicomputer. A medium-size computer system.

Modem. A device that converts electronic signals from the computer into sounds that can be carried over telephone line and then converts the sounds back into electronic signals. The term is a combination of two words: *modulator* and *demodulator*.

Monochrome. A one-color screen.

Mouse. A small box with a large button on it, attached to the PC with a cord. The user pushes ("clicks") the button to issue commands to the PC.

Multiplexer. A device that collects low-speed data from terminals and combines them into a high-speed stream for transmission over a single channel.

Nepotism. Favoritism in the employment of relatives.

Network. A method of linking several communicating terminals through electronic transmission at various locations.

Nonvolatile. A condition whereby a computer will not lose its data when the power is shut off.

OCR (optical character reader). A machine that can read printed or typed characters and then digitally convert them into input to a data or word processor.

OCR (optical character recognition). A process that scans text images and stores the scanned characters in digital form.

Off-the-shelf. Standard software package that is not customized for any particular vertical or industry market, such as medical or legal.

On-line. Computer equipment and other information systems that are available to interact with a central processing unit (CPU).

On-line data bases. Warehouses of stored information, accessible by microcomputers over conventional phone lines.

Open-landscape design. Office layout design with minimal enclosed areas, using movable wall panels, screens, and modular furniture.

Operating system. The set of rules that control the computer. PC DOS is the most common operating system found on PCs, although there are many others.

Optical disk. A circular computer disk that can store large amounts of information in digital form (also referred to as *laser disks* or *digital optical disks*).

Optoelectronics. A technology that combines light, using lasers and electronics, to provide storage, processing, and communcation for telephone and computer applications.

PABX (private automatic branch exchange). A telephone system that switches calls between the public telephone network and inside extensions.

Password. Any combination of alphanumeric characters assigned to a computer that the user must supply (to meet security requirements) before gaining access to data.

PBX (private branch exchange). A telephone switching system used primarily for voice applications.

Phototypesetting. A computerized photographic printing process that uses film to produce an image (camera-ready copy).

Plain-paper copier. Copier machine that produces copies on plain bond paper.

Port. An outlet that can be perceived as a communication gateway for signals to and from the computer.

Printers. *Dot matrix printers* print a series of dots formed together to create a character or graphic. *Letter-quality printers* print fully formed characters on a wheel or thimble. *Nonimpact printers* print without a hammer or character wheel striking the page.

PROM (programmable read-only memory). Chips that have their programs written on them after they have been made at the factory.

Proprietary. Usually, equipment, programs, or technologies that belong exclusively to one company.

Protocol. A formal set of conventions governing the format of how devices will communicate (baud rate, duplex, parity, character length, and so on).

RAM (random-access memory). The temporary memory of a computer system in which each element has its own location and from which any element can be easily and conveniently retrieved.

Record. Any form of recorded information.

Records management. A process that includes the creation, storage, retrieval, retention, protection, and disposal of all vital records.

Reprographics. The reproduction of hard copies (printed paper) by photocopy, printing, and other office duplicating methods.

Resellers. Organizations that buy long-distance services in bulk and in turn sell the services to others for less than common carrier rates. Resellers also offer other special telecommunication services.

Retrieval function. A one-way communication with the capability of finding information rapidly and then communicating it on a screen.

ROM (read-only memory). The permanent memory of a computer system wherein a computer chip resides in the computer and cannot be reprogrammed by the user.

RS232. An industry standard for a 25-pin serial interface that connects various peripheral devices to computers.

Satellite. An orbiting vehicle above the earth that reflects or transmits communication signals.

Scrolling. A word processing feature that permits horizontal or vertical movement of a cursor in order to access more characters than can be shown on the display screen at one time.

Search and replace. A word processing function that searches through a page and/or entire document for a specific word or phrase. The word or phrase can then be automatically replaced.

Security system. A special option designed for companies requiring maximum security for their computer systems and telecommunication network, designed to prevent unauthorized personnel from seeing confidential information.

Sexual harrassment. Unwelcome sexual advances, requests for sexual favors, and other verbal or physical conduct of a sexual nature.

Shared-logic systems. A system in which multiple workstations and output devices simultaneously use the memory and processing powers of one computer.

Shared tenant services. A form of bypass whereby sophisticated telecommunication and office automation services are provided to the tenants of a building by a landlord at substantially reduced prices.

Silicon. An abundant element in the earth's crust used for making computer chips.

Simplex channel. A channel that permits transmission in one direction only.

Smart card. A plastic card that contains a programmable microchip that can store a variety of information and can be used for a variety of applications.

Smart modem. A modem that can perform several automatic tasks such as automatically dialing and answering calls from other computers.

Software. The set of instructions or programs given to the computer to tell it how to perform certain tasks.

Software piracy. The unauthorized copying of software programs that have been published and copyrighted.

Span of control. The number of subordinates whom a superior can supervise effectively.

Speech synthesis. An advanced voice technology that enables a computer to generate speech.

Spelling checking program. A function of word processing that electronically scans a document and identifies spelling errors (also referred to as a *dictionary program*).

Spreadsheet software. A software program that provides an electronic worksheet for calculations and other mathematical tasks.

Storage. The amount of space allotted for storing information in a computer. Storage capacity can be as small as a few hundred kilobytes and as large as more than 20 megabytes.

Streaming. A process using tape to back up data stored in memory.

Superchip. A 32-bit microprocessor able to process 32 bits of information simultaneously and execute up to 8 million tasks in a second.

Supercomputer. A category of the largest, fastest, and most powerful computers available.

Synchronous. A data transmission method used by faster modems in which each bit or group of bits is sent under precise timing conditions.

System security. A special option designed for companies requiring maximum security for their computer systems and telecommunication network, designed to prevent unauthorized personnel from learning confidential information.

Technostress. A condition resulting from the inability of an individual or organization to adapt to new technology.

Telecommunications. A technology that provides the movement of information between two locations by electronic means. Information can take the form of voice, data, or images and can be sent by telephone wire, radio signals, fiber optics, or satellite.

Telecommuting. A technology that allows information workers to work at home, using electronic communicating devices to send and receive data and information.

Telecomputing. Use of the personal computer for communications activities such as electronic mail and data bank retrieval.

Teleconference. The processing of meetings, conferences, and exchange of information through electronic transmission.

Telemarketing. The use of telecommunication to promote, sell, or market a business's or organization's product or service.

Telephony. The branch of telecommunication related to transmission and reproduction of speech and, in some cases, other sounds.

Teleport. A cluster of commercial and/or residential buildings sharing advanced telecommunication facilities.

Teletext. A one-way service delivered to the user's standard television set (equipped with a simple decoder) over broadcast or cable television systems.

Telex. An automatic dial-up teletypewriter switching service provided worldwide by various common carriers.

T-1 (Transmission-1). A high-speed telecommunication line developed by AT&T that transmits digitized voice, video, and data at 1544 megabits per second.

TOP (technical office protocol). A set of evolving and established standards designed to allow office equipment to communicate over a single network.

Touch-sensitive screen. A type of display screen that activates a computer response when a user touches a graphic symbol on the screen.

Transactional function. Goes beyond the retrieval and interactive function to result in the buying, selling, or transfer of goods, services, or funds.

Transparent. A machine that is "user-friendly." The user of a machine should hardly be aware of the existence of the hardware or the operating system that tells the computer how to carry out the commands.

TWX. A teletypewriter service provided by Western Union. Generally used to transmit brief messages.

UNIX. A trademark for a family of Bell Labs–developed, general-purpose computer operating systems that have been used extensively in the Bell System (AT&T) in a variety of ways, ranging from helping to maintain and manage a telecommunication network to word processing and document preparation.

Upload. A process of downloading and then storing the data onto a microcomputer. See Download.

UPS (uninterruptible power system). Devices that protect a computer system from the irregularities of commercial power.

User-friendly. Systems that are easy to learn and use.

USASCI. USA Standard Code for Information Interchange.

Videotex. A two-way, interactive communication service offering both information and transactions to subscribers. Videotex transmits text and graphic information via cable, radio, or microwave to a terminal or home TV set.

Voice mail (messaging). A computerized means to allow users to send and receive voice messages from any touch-tone telephone. This technology uses a voice store-and-forward system. Voice mail can be stored on a terminal, and messages are forwarded to the recipient through a mailbox or callback system.

Voice processing. A method of using voice combined with electronic devices to create, process, and transmit information and data.

Voice recognition. A technology that interprets continuous speech by converting the speech generated by sound waves into digital impulses and enables computers to recognize and respond to spoken words.

Voice synthesis. A technology whereby computers talk in human speech.

Voice verification. A technology that has the ability to store a human's unique "voice pattern" or "voice print" in a system's memory. It is used in security applications to determine whether the speaker is actually authorized to gain entry or access into a system or facility.

Volatile. A condition whereby a computer loses its data and information when the power is shut off.

WATS (wide area telephone service). A U.S. telephone company service that permits a customer to dial an unlimited number of calls in specific areas for a flat monthly charge.

Windows. A display screen divided into several areas so that several files can remain in view simultaneously (also may be referred to as *split-screen editing*).

Wire management. A system to organize and conceal wires and cables within an office setting.

Word processing. A system that combines hardware and software to aid in the composition, revision, printing, and storage of text.

Word processing software. A software program that provides automatic processing, revision, and manipulation of text.

Word wraparound. A word processing feature that automatically moves words to the next line when the right-hand margin is reached and the word does not fit within it.

Work group. Consists of a small segment of people working within an organization structure.

WYSIWYG ("what you see is what you get"). The ability to view on a computer screen exactly what will be printed, including different typefaces, the placement of graphics and scanned images, column layouts, and headers/footers.

INDEX

A

Accountability, organizational, 36–37
Accounting software, 208–9
Acquired immune deficiency syndrome (AIDS), employees with, 60
Affirmative action, 62
Age Discrimination Act of 1967, 62
Age discrimination in hiring, 62–63
Aiken, Howard, 95
Alcoholism among employees, 58–59
Allnet Communications, 176
Amdahl Corporation, 133
American Telephone and Telegraph (AT&T) divestiture, 46, 97, 172, 175
Animation, 154
Aperture cards, 226 (fig.)
Apple computers, 281
Applications, computer systems
  deciding on needs of, 135–36
  most frequently used, 168 (fig.)
  touch-sensitive screen variety in, 128
  word processing, in industry and the professions, 165–69
Applications software, 207–10
Architecture, word processing applications in, 166
Armstrong, Neil, 172
Artificial intelligence (AI) software, 215, 332–33
Association of Data Processing Service Organizations (ADPSO), 273
Association of Information Systems Professionals (AISP), 7, 54, 304, 325
Association of Records Managers and Administrators, 54, 325
Association for Systems Management (ASM), 325
AT&T. *See* American Telephone and Telegraph (AT&T) divestiture
Attitudes, and security issues, 275–76
Audioconferencing, 186
Audiographics, 186
Audio processing, 28–29
Audit trail, 271
Authority, organizational, 36–37
  information management and delegation of, 53–54
Automation. *See* Office automation (OA)
Auxiliary storage, 98, 105–7
  future trends in, 334
  universal workstation, 125

B

Backup systems, 105, 235–36, 261, 271
  built-in, 258
Bell, Alexander Graham, 80
Bell, Chichester, 80
Bell Laboratories, 95, 182, 190, 323
Bell System. *See* American Telephone and Telegraph (AT&T) divestiture
Bernoulli box, 106 (fig.)
Bits, 97, 98 (fig.), 105, 180

Boilerplate, 167
Buffalo Organization for Social and Technological Innovation, 296
Business culture, 14
Business forms/paper products, 247–49
Business plan, 12–13. *See also* Strategic planning
Bytes, 180

C

CAD/CAM software, 211, 212 (fig.)
Call-backs, 271
Camus, Albert, 58
Careers. *See also* Job(s)
 vs. jobs, 309
 in office information systems, 309–13
Carterfone court decision, 46, 172
Cathode ray tube (CRT). *See* Display screens; Terminals
Cellular mobile phones, 182, 183 (fig.), 335, 336 (fig.)
Central processing unit (CPU), 98, 104 (fig.). *See also* Microprocessors
Chairs, 283, 284 (fig.), 285, 286 (fig.)
Champion International Corporation, 8
Chief information officer (CIO), 33, 45
Chrysler Corporation, 54
Civil Rights Act of 1964, 62, 63
Clock, computer's, 105
Coated paper copiers, 142–43
Color, effects of, in the workplace, 281, 294–95
Color copiers, 143
Color terminals, 103, 281
Committee organization, 41
Communication(s). *See also* Telecommunications
 bridging gaps in, 69, 311
 electronic typewriter capabilities, 119
 information manager's tasks, 53
 microcomputer-mainframe computer, 194–200
 software, 179–80, 209–10
 universal workstation options, 125
 voice, 333
 written, 26
Communication cards, 179
Compact disks, 234
Company stores, 45–46, 71
Compatibility, and integrated systems, 68, 136
Comprehensive Crime Control Act of 1974, 269
CompuServe, 180, 209, 213 (fig.)
Computer(s), 94–116. *See also* Integrated information systems; Office automation (OA); Processors
 arithmetic/logic unit, 105
 classification of, 99–101
 clock, 105
 compatibility, 68, 136
 computer phones, 131, 132 (fig.)
 definition of, 97–98
 disasters. *See* Disaster recovery

evolution of processors, 95–97
expansion capabilities, 125
graphics. *See* Graphics
information transfer between, 194–200
interfaced with copiers, 146
maintenance. *See* Preventive maintenance
measuring computer power, 105
memory. *See* Memory
microprocessors. *See* Microprocessors
peripherals, 98, 101
personal. *See* Personal computers (PC)
portable, 130, 131 (fig.)
printers. *See* Printers
projected market 1984 to 1989, 118 (fig.)
smart cards, 334
software. *See* Software
supercomputers, 133–35
systems. *See* Computer systems
terminals. *See* Terminals
user-friendly aids, 126–33
voice technologies, 87–91
Computer-aided design/computer-aided manufacturing software (CAD/CAM), 211, 212 (fig.)
Computer-assisted instruction (CAI), 321
Computer-assisted retrieval (CAR), 229
Computer-based training (CBT), 321
Computer conferencing, 187
Computer crime, 266
 and the law, 267–69
 perpetrators of, 266–67, 268 (fig.)
Computer-output microfilm (COM), 227, 229
Computer specialist, career as, 310–11
Computer systems, 98
 applications. *See* Applications, computer systems
 connecting, 197–99. *See also* Networking
 emulation of, 198
 environmental hazards to, 255–57
 fault-tolerance, 258
 strategies for acquiring, 135–37
Computer Training and Evaluation Center, Palo Alto, CA, 323
Computerworld, 44
Connectivity issues, 31, 197–99
Conrad, Joseph, 332
Consultative Committee on International Telephony and Telegraphy (CCITT), 200
Consulting, information careers in, 313
Continuous forms, 248
Copiers, 29, 141–43
 dry, 142–43
 features, 143–46
 interfaced with computers, 146
 supplies, 249
 wet, 141–42
Copyboard. *See* Electronic presentation boards

Corporate information resources, 4–5, 13–14, 32–33, 206
   disaster recovery for, 259–63
Corporation for Open Systems (COS), 198
Costs
   computer systems, 136
   printer supplies, 112–13
   software, 218
   telephone systems, 177
Cray-1 supercomputer, 133
Cursor control, 126

D

Daisy-wheel printers, 107 (fig.), 108
Data
   backup, 235–36
   dictionary recovery, 87
   encryption, 271
   vs. information, 28, 95
   input. *See* Input
   integration, 196–97
   output. *See* Output
   uploading/downloading, 195, 209–10
Data base(s), 114
   financial data, 212, 213 (fig.), 214
   graphics, 152
   terminology, 222 (fig.)
Data base management software, 168 (fig.), 210–11
Data communication, 30
Data processing, 5, 6
   careers in, 310–11
   as information processing, 28
Data Processing Management Association (DPMA), 54, 267, 325
Datasources, 215
Deaf workers, 323
Decentralization in the information age, 8–9
   of office and information systems, 42–44, 123
Decision making
   on acquiring computer systems, 135–36
   role of information in, 36
Decision processing in word processing systems, 162
Decision support systems (DSS), 71
Delegation of authority, 37
   and information management, 53–54
Departmentalization, organizational, 37
Desk (work surface), 285, 286 (fig.), 287 (fig.)
Desktop publishing, 148, 164
Devanna, Mary Ann, 60
Dictionary recovery of computerized data, 87
Digital Equipment Corporation computers, 281
Disaster recovery, 259–63
   plan, 259, 260 (fig.)
   strategies for implementing, 260–62
   telecommunicating data, 262–63

Discipline in information management, 53
Disk drive, 106
   cleaning, 254
Diskettes. *See* Floppy diskettes
Display screens. *See also* Terminals
   dictation/transcription machine, 85
   electronic typewriter, 119 (fig.)
   ergonomic design, 280–83, 297
   etching, 254
   flat panel, 126
   macro screen protection, 216
   maintenance, 254
   personal computer, 102, 103 (fig.)
   touch-sensitive, 126–28
Distributed data processing (DDP), merged with office automation, 44
Distribution of information, 29–31
Divestiture of AT&T and the Bell system, 46, 97, 172, 175
Document(s)
   defined, 26
   life cycle management of, 229–30, 236, 237 (fig.)
   source, 227
Documentation
   computer system, 321–22
   software, 217–18
Dot-matrix printers, 107 (fig.), 108, 109 (fig.)
   maintenance, 255
Dow Jones Retrieval Service, 179, 212, 213 (fig.)
Downloading data, 195, 209–10
Drug abuse among employees, 58–59

E

Eckert, J. Presper, 95
Edison, Thomas Alva, 79 (fig.)
Editing
   electronic, on copiers, 145
   split screen, 163 (fig.)
   typeset copy, 147
   word processing, 28, 159–61, 163, 167
Education
   for office automation. *See* Training
   on security issues, 275–76
Electronic calendar software, 211
Electronic employee recruiting, 306–7
Electronic imaging systems, 146
Electronic mail, 30, 165, 178–81. *See also* Voice mail
Electronic Numerical Integrator and Computer (ENIAC), 95, 96 (fig.)
Electronic outline software, 210
Electronic presentation boards, 154, 155 (fig.)
Electronic voice messaging. *See* Voice mail
Electrosensitive printers, 108
Employee(s). *See also* Information workers; Job(s)
   benefits/salaries for, 324–25

Employee(s) (*continued*)
  external forces affecting personnel policies, 60–61
  laws/regulations affecting hiring of, 61–64
  measuring job performance, 325–26
  motivating, 54–57
  orientation/training, 19, 45–46
  smoking among, 63, 64 (fig.), 256, 295 (fig.)
  sources of, 302–7
  strategic planning participation, 17
  stress/alcoholism/personal problems among, 57–60
  training. *See* Training
  unions and, 61
  women, 60–61
Employment. *See* Job(s)
Employment services, 304–6
Encryption of data, 271
End-user computing, 71
Engineering, and word processing applications, 165 (fig.), 166
Entrepreneurship, careers in small business, 313
Equal Employment Opportunity, 62
Equal Employment Opportunity Commission (EEOC), 63
Equipment. *See* Computer(s); Office equipment
Erasable programmable read-only memory (EPROM), 104 (fig.)
Ergonomics, 278–98
  definition, 279
  display terminals, 280–83
  furnishings, 283–89
  office climate conditions, 292–96
  workplace environmental controls, 289–92
  workstation design, 279
Ergonomists, 101, 279
  careers as, 296–97
Ethospace, 290
Executive search firms, 306
Expansion slots, 179
Expert systems, 215, 333
Eye strain, 281, 282 (fig.)

F

Facility planning, 306
Facsimile, 28, 155
  transmission, 178, 179 (fig.)
Fairness in information management, 53
Fiber optics technology, 190
  copiers, 143
Fielden, Rosemary, 158
File-by-file backup, 236
Filing systems, 223
  equipment and supplies for, 223–25
Financial data software, 212–14
Firmware, 207
Flexibility in information management, 54

Flextime, 307–8
Floppy diskettes, 105 (fig.), 106
  care/protection of, 244–46, 271
  information storage/retrieval, 231, 232 (fig.), 235
  word processing management of, 161
Florida Computer Crimes Act of 1981, 269
Footprint space, 132, 282
Forecasting, 114–15
Format of information, 26
Forms tractor, 248 (fig.)
Freeze-frame teleconferencing, 186–87
Full-motion videoconferencing, 187
Functional organization chart, 40 (fig.), 41
Furnishings, ergonomic, 283–89
Fuse-box security, 273, 274 (fig.)

G

Gallium arsenide chips, 235
Gardner, Lawrence, 322 (fig.)
Gas plasma display, 126
Gateways, 31, 199 (fig.)
General Motors Corporation, 41, 198
Getting the Most Out of Your Word Processor, 158
Gigabytes, 134–35
Glare, effect on workers, 281–82, 292
Glossary, word processing, 162
Goals and objectives
  of integrated systems, 69–70
  office information, 16–17
  organizational, 15, 36
Graphics, 114, 150–55
  data bases, 152
  hardware, 110, 152–54
  integrating with text, 163, 164 (fig.)
  software, 150–52, 211
Group instruction, 320
GTE Corporation, 8, 176

H

Hackers, 267
Handicapped employees, training, 322–23
Hard cards, 231, 232 (fig.), 233
Hard disks, 106–7
  information storage/retrieval, 231, 235–36
Hardware. *See also* Office equipment
  auxiliary storage, 105–7
  graphics, 152–54
  microcomputer, 101–3. *See also* Microprocessors
Harvard Medical School, 281
Health hazards
  and furnishings, 283–89
  of video screens, 281–82
Heat
  hazards of, to computers, 255–56
  and ventilation in workplaces, 295–96

Hewlett Packard, 96, 127
Hierarchy, organizational, 37
Hierarchy of needs, 55, 56 (fig.)
Hiring practices, 62–64
Hollerith, Herman, 95
Human Factors Society, 296
Human resources, 14, 33
    management of, 57–60. *See also* Personnel management
Humidity, hazards of insufficient, 255
Hypertext, 210

I

Iacocca, Lee, 54
IBM, 133, 323
    electronic typewriter, 95, 120
    Scanmaster I, 146
    telecommunications activities, 172 n, 176
    word processor, 96
Icons, 71, 129 (fig.), 130
Image processing, 28, 140–56
    computer-interfaced copiers, 146
    copiers, 29, 141–46
    desktop publishing, 148
    developing visual literacy, 154
    electronic presentation boards, 154–55
    graphics, 150–54
    optical character recognition, 148–50
    other technologies of, 155
    phototypesetting, 146–48
Impact printers, 107
Indexing
    dictation/transcription tapes, 84
    generated by word processor, 164–65
    micrographics, 229
Information, 4
    backup, 235–36
    controlling access to, 196, 270–71
    as a corporate resource, 4–5, 13–14, 32–33, 206
    vs. data, 28, 95
    defined, 26
    factor of, in office automation systems, 13–14
    management. *See* Information management
    processing. *See* Information processing
    security. *See* Security of information and computerized systems
    stages, 26, 27 (fig.)
    storage/retrieval. *See* Information storage and retrieval
    transfer of, between computers, 194–200
Information age, 3–10
    beginning of, 4–5
    concept of information systems, 5–6
    workplace/workforce in, 6–9
Information center(s), 44–45, 71, 321

Information float, 32
Information management, 31–33
    personnel management, 54–64
Information manager, 51–54
Information processing, 26–31. *See also* Computer(s); Image processing; Processors; Telecommunications; Voice processing; Word processing
    distribution of information, 29–31
    input phase, 26–27. *See also* Input
    output phase, 29. *See also* Output
    processing phase, 27–29
Information resource(s), expanding, 71–72
Information resource management (IRM), 33. *See also* Information management
Information resource managers, 33. *See also* Information manager
Information resource planning (IRP), 33
Information retrieval services, 114, 179, 180
    financial data, 212, 213 (fig.)
Information storage and retrieval, 29, 221–40
    automated systems, 225
    computers and micrographics, 228–31
    data/information backup, 235–36
    filing systems, 223–25
    implementing systems, 236–38
    micrographics, 225–28
    new technologies in, 231–35
Information systems
    concept of, 5–6
    integrated. *See* Integrated information systems
    office. *See* Office information systems (OIS)
Information utilities. *See* Information retrieval services
Information workers, 6–7, 8, 14. *See also* Employee(s)
    alternate work styles, 307–8
    careers as, 309–13
    handicapped, 322–23
    health hazards to. *See* Health hazards
    professional organizations, 304
    telecommuting, 188
    unionization of, 61
Ink-jet printers, 107 (fig.), 109
Input, 26–27, 98
    icons, 129 (fig.), 130
    keyboard. *See* Keyboard(s)
    light pens, 128
    mouse, 128, 129 (fig.)
    terminals. *See* Terminals
    text-entry systems, 333
    touch-sensitive screens, 126–28
    voice processing, 78
In Search of Excellence, 45
Integrated information systems, 18, 67–74
    developing a strategic plan for, 68–73
    electronic connectivity and, 31, 197–99
    guidelines for an effective, 72–73

Integrated information systems (*continued*)
   personal computers and, 113–14
   software for, 71, 216–17
Integrated Services Digital Network (ISDN), 191
Integrated voice/data workstation (IVDW), 131, 132 (fig.)
Intel Corporation, 96
Intelligent buildings, 189, 296
Intelligent copiers, 29, 143
Intelligent text, 271
Interface, computer-printer, 112
International Communications Association, 54
International Word Processing Association, 304
Investment advice program, 214

J

Jackson, Jesse, 62
Japanese management, 54
Job(s)
   analysis, 317
   career vs., 309
   in computer servicing, 258
   market in the information age, 6–7
   in office information systems, 309–13
   sources for finding, 302–7
Job performance, measuring and evaluating, 325–26
Job sharing, 308
Job titles and descriptions, 7, 33, 113, 308–9
Journalism, and information careers, 313

K

Keyboard(s)
   ergonomic design, 280–81
   maintenance, 254–55
   numeric keypad, 102 (fig.)
   personal computer, 101, 102 (fig.), 126
   phototypesetting, 146–48
   redefiner, 216
   word processing, 28
Key ring security system, 273
Knowledge-based software systems, 214–15
Kodak Ektaprint Electronic Publishing System (KEEPS), 229
Kodak Image Management System (KIMS), 146, 229, 230 (fig.)
Kurzweil Applied Intelligence Inc., 323

L

Labor Department, U.S., 62
Labor unions, 61
Laser printers, 107 (fig.), 109, 110 (fig.), 111 (fig.)
Law, word processing applications in, 166
Laws/regulations in personnel management, 61–64
Leadership, and information management, 52–53

Letter-quality printers. *See* Daisy-wheel printers
Library services feature of word processors, 165
Light-emitting diode (LED) display, 85 (fig.), 126
Lighting, workplace, 292–93
Light pens, 128
Line and staff organization, 38, 39 (fig.), 40
Line departments, 39
Line organization, 38, 39 (fig.)
Liquid crystal display (LCD), 126
List processing on word processors, 162–63
Local area networks (LAN), 30, 183, 184 (fig.), 185
   configurations, 185 (fig.), 186
   gateway connectors for, 199 (fig.)
Long-distance service, 175–77
   bypassing telephone companies, 176–77
   choosing a, 177
   equal access, 176
Lotus 1-2-3 software, 214, 216

M

McGregor, Douglas, 55
Machine dictation/transcription systems, 28, 78–87
   beginnings of recorded sound, 79–80
   benefits of, 79
   central recording management systems, 86–87
   dictation management/control, 85–86
   dictation media, 83–84
   document flow, 78
   equipment, 80–83, 84
   remote dictation, 83
   special system features, 84–85
Macros, 208, 216
Magnetic bubble memory, 235
Magnetic media
   conversion, 236
   dictation, 83–84
   storage, 245–46, 289 (fig.)
Mailbox, word processor, 165. *See also* Electronic mail
Mainframe computers, 99
   information transfer from, 194–200
Maintenance. *See* Preventive maintenance
Management
   of information. *See* Information management; Information manager
   integrated systems, and issues of, 69
   participation in, 37
   personnel, 51–64
   principles of, and organization structure, 36–37
Management information systems (MIS), 33. *See also* Information systems; Office information systems (OIS)
   centralized, 42, 43 (fig.)
   and integrated systems, 70
   and personal computers, 44

Index

Manufacturing, word processing applications in, 166
MAP (manufacturing automation protocol), 198
Maslow, A. H., 55–56
Math capabilities on word processors, 163
Matrix organization, 42 (fig.)
Mauchly, John, 95
MCI long-distance service, 176
MCI Mail, 180, 200
Medicine, word processing applications in, 166–67
Meese, Edward, 62
Megabytes, 135
Memory
  bubble, 235
  electronic typewriter, 119
  erasable PROM (EPROM), 104 (fig.)
  nonvolatile, 235
  programmable read-only memory (PROM), 104
  random access (RAM), 103, 105, 334
  read-only (ROM), 104
Microcenters. *See* Company stores
Microcomputers, 100 (fig.), 101. *See also* Personal computers (PC)
  information transfer from mainframes, 194–200
  supermicrocomputer, 134
Microfiche, 226 (fig.), 227
Microfilm, 155, 225, 226 (fig.)
  for computer-copier interface, 146
  computer-output, 227
Microforms, 225, 226 (fig.)
  readers/printers, 227, 228 (fig.)
Micrographics, 225–28
  applications of, 230–31
  benefits of, 229–30
  computers and, 228–29
  conversion process, 227
  processing, 227
Micro-mainframe link, 194–200
  applications/advantages of, 194–95
  controlling mainframe use, 195–96
  data integration, 196–97
  degree of accessing files, 195
  gateways, 199
  servers, 199–200
  standards for, 197–99
  technologies available for, 197
  X.400 plan, 200
Microprocessors, 96, 103–5. *See also* Central processing unit (CPU); Processors
  measuring the power of, 105
  superchips, 234–35
Minicomputers, 99, 100 (fig.)
Mirror-image backups, 235–36
Mitchell Field project, Nassau County, New York, 8
Mobil files, 224 (fig.), 225
Modem, 178, 179 (fig.)

Morse, Samuel, 172
Motivating employees, 54–57
Mouse/mice input, 71, 128, 129 (fig.)

N

National Association of Temporary Services, 305
National Computer Network, 213 (fig.)
National Institute for Occupational Safety and Health (NIOSH), 281, 282
National Office Products Association (NOPA), 244
Natural language systems, 333
Needs assessment, 16
Nepotism, 63–64
Networking, 30–31, 114
  local area, 30, 183–85, 199
  network configurations, 185 (fig.), 186
  standards issues, 197–99
  supernetworks, 335
Noise control in offices, 293–94
Nonimpact printers, 107, 108–9, 110 (fig.), 111 (fig.)

O

Office. *See also* Workplace in the information age
  climate conditions, 292–96
  development of, and telecommunications, 189–90
  ergonomic design, 289–92
  function of, 13
  landscape, 290, 291 (fig.)
Office automation (OA), 5–6. *See also* Integrated information systems
  business plan for, 12–13
  change and, 12, 13
  evolution of, 31 (fig.)
  future of, 332–38
  implementation of, 17–19
  information resources and, 13–14
  merged with distributed data processing, 44
  moving into, 12
  people resources and, 14
  preimplementation of, 17
  strategic planning for, 14–17
  technology resource and, 14
  training. *See* Training
  word processing's role in, 168 (fig.), 169
Office equipment. *See also* Computer(s); Hardware
  copiers, 141–46
  dictation/transcription, 80–83
  electronic presentation boards, 154–55
  filing, 223–25
  furnishings, 283–89
  graphics, 152–54
  optical character recognition, 148–50
  printers. *See* Printers
  telephone, 173–77

Index

Office equipment (*continued*)
   typewriters, 95, 118–20
   workstation. *See* Workstation
Office of Federal Contract Compliance Programs (OFCCP), 62
Office information systems (OIS), 6, 12. *See also* Integrated information systems
   careers in, 309–13
   elements of, 13–14
   organizing for, 42
   planning, 14–17
Off-line storage, 29, 105
One-on-one tutoring, 320
On-line systems, 71, 105
On-the-job training (OJT), 320–21
Operating system software, 207
Optacon, 322
Optical character recognition, 148–50
Optical disk storage/retrieval, 233–34 (figs.)
Organization(s)
   automation and structure of, 18
   goals/objectives, 15
   impact of technology on the structure of, 42–46
   management principles affecting the structure of, 36–37
   structure of, 36, 37–42
Organization charts, 38–42
Original equipment manufacturer (OEM), 243
Output, 98. *See also* Image processing; Voice processing; Word processing
   hard copy, 29, 107, 141
   soft copy, 141

P

Pagers, 181 (fig.)
Page size, and printer selection, 112 (fig.)
Paper products/business forms, 247–49
Parallel computer-printer interface, 112
Parks, Sandy, 281
Participative management, 37
Passwords, 270–71
PBX technologies, 173–74
Pen plotters, 153
People. *See* Employee(s); Human resources; Information workers
Personal computers (PC), 8, 71, 101–3. *See also* Microcomputers
   acceptance in business, 113–15
   decentralization and, 44
   impact of, on software, 206–7
   system components, 101 (fig.)
   word processing integration, 164
Personal problems among employees, managing, 59–60
Personal signature recognition security, 271

Personnel management, 50–64
   forces affecting, 60–61
   human resource management, 57–60
   laws/regulations affecting, 61–64
   measuring job performance, 325–26
   motivating employees, 54–57
   salary administration, 63, 323–25
   sources of employees, 302–7
   training. *See* Training
Peters, Thomas, 45
Phonograph, 79–80, 81 (fig.)
Phosphor, 102
Photocomposition, 29
Photocopying. *See* Copiers
Photographs, integrating into documents, 154
Phototypesetting, 146–48
Pitney Bowes Inc., 8
Plain paper copiers, 143
Planning. *See* Business plan; Strategic planning
Plus Development Corporation, 231
Policy, organizational, 36
Portable computers, 130, 131 (fig.)
Portfolio management program, 214
Post security, 272
Power, and information management, 32
Power failures and surges, 256–57
Powers, James, 95
Preventive maintenance, 254–55
   checklist for computer, 259
   hazards to computer systems, 255–57
   planning for disaster recovery, 259–63
   service plans/contracts, 257–58
Printers, 29, 107–13
   criteria for selecting, 110–13
   graphics, 110, 152–53
   maintenance, 255
   server links for, 199–200
   supplies, 112–13, 246–49
   types of, 107–9
Print quality, 110
   letter-quality, 108
Printwheels, 247 (fig.)
Processor modules, 258
Processors, 117–39. *See also* Computer(s)
   evolution of computer, 95–97
Product cycles, 9
Product organization chart, 40 (fig.), 41
Professional growth for information managers, 54
Programmed instruction, 321
Programmable read-only memory (PROM), 104
Project, and matrix organization, 42
Project management software, 211–12
Protocols, communication, 30 (fig.), 31, 180
   converters, 199
Publishing, word processing applications in, 148, 166

Q

Qyx electronic typewriter, 118

R

Random access memory (RAM), 103, 105
Read-only memory (ROM), 104
Read/write head, 106
Reagan Administration, 62
Recordings. *See also* Machine dictation/transcription systems
  beginnings of sound, 79–80 (figs.)
Records management, 222–38
Remote diagnosis, 258
Remote networking, 31
Reprographics, 141
Requirements analysis, 16
Responsibility, organizational, 36–37
Retrieval. *See* Information storage and retrieval
Ribbons, 246–47
Robotics, 334, 335 (fig.)
Rolm Corporation, 172 n, 176
Rosen, Arnold, 158

S

Salary administration, 63, 323–25
Salary surveys, 325
Sales, and information careers, 312
SAS software, 152 (fig.)
Satellite Business Systems, 176
Satellite communications, 190–91
Scanning terminals, 146
Science and research, word processing applications, 167
Secretary, career as a, 310
Security of information and computerized systems
  computer crime, 266
  digitized voice processing, 88, 89–90
  ethics and education, 275–76
  improved, with microforms, 230
  keyboard enhancers, 216
  law and computer crime, 267–69
  methods to insure, 196, 269–71
  perpetrators of computer crime, 266–67, 268 (fig.)
  software piracy, 266, 272–73
  terrorism and sabotage, 273–75
  in word processing systems, 164
Self-paced training materials, 321
Serial interface, 112, 179
Serial printers, 107
Servers, 199–200
Service and support. *See also* Vendors
  computer system, 136–37
  plans and maintenance contracts, 257–58
  printer, 112
  software, 218

Sexual harassment, 63
Shared-logic systems, 121, 122 (fig.)
Shared-resource systems, 121, 122 (fig.)
Sheet feeder, 248
Signal software, 212, 214 (fig.)
Silicon chips, 96, 105. *See also* Microprocessors
Singer Company, 8
Site selection/space planning for automation, 18–19
Slide making, and graphics capabilities, 154
Slow-scan teleconferencing, 186–87
Smart cards, 334
Smoking among employees, 63, 295 (fig.)
  hazards of, to computers, 256
  incentives to stop, 64 (fig.)
Society, future effects of computer technology on, 33–38
Software, 98, 205–20
  bundled, 206
  communications, 179–80, 209–10
  customized, 215–16
  data base. *See* Data base management software
  data integration, 196–97
  emergence of, 206
  graphics, 150–52
  illegal copying of, 266, 272–73
  integrated, 71, 216–17
  interactive, 207
  personal computer impact on, 206–7
  personnel tracking, 307
  printer driver, 148
  security programs, 271
  selection of, 217–18
  spreadsheets. *See* Spreadsheets
  types of microcomputer, 207–16
  vertical markets for, 215
  for visually handicapped, 323
  word processing. *See* Word processing
Software Catalog, The, 215–16
Sort feature
  copiers, 144
  word processing, 162–63
Source, The, 179, 180, 209, 213 (fig.)
Source documents, 227
Span of control, organizational, 37, 38 (fig.)
Speech synthesis, 90
Speed
  clock, 105
  printer, 112
Spelling checking programs, 162
Split screen editing, 163 (fig.)
Spreadsheets, 154, 168 (fig.), 208
Sprint communications, 176
Staff departments, 39
Stamford, Connecticut, 8
Static and dust, hazards of, 255

360 *Index*

Storage. *See also* Auxiliary storage; Information storage and retrieval
   diskette, 245–46
   filing systems, 223–25
   workstation, 288–89
Strategic planning
   for integrated systems, 68–73
   for office automation, 14–17
Strategy, organizational, and technology, 42
Streaming-tape subsystems, 236
Stress, 57–58
Supercomputers, 133, 134 (fig.), 135
Supplies, 241–50
   cleaning, 247
   costs, for printer, 112–13
   diskette care/protection, 244–46
   filing systems, 223–25
   guidelines for purchasing, 244
   managing/controlling, 242
   protecting/storing, 246–49
   selecting a vendor for, 243–44
   shelf life, 247
Support. *See* Service and support
Systems software, 207

T

Tainter, Charles Sumner, 80
Technology advances
   in the future, 332–38
   for the handicapped, 323
   image processing, 155
   impact on organizational structure, 42–46
   information transfer, 197
   integrated systems and, 70–71
   office, 7–8, 14
   telecommunications, 172–73, 181–91
   voice, 89–91
   workstation, 126
Technostress, 57
"Technotrends '85: Inside the American Office," 44
Telecommunications, 30, 171–92. *See also* Communication(s)
   careers in, 311–12
   effect on organizational structure, 46
   electronic mail, 178–81
   future trends in, 335, 336 (fig.)
   image, 155, 178, 179 (fig.), 186–87
   office development and, 189–90
   software for, 179–80, 209–10
   telephone systems, 173–77
   umbrella of technologies, 172–73, 181–91
   and workplace changes, 8–9
Telecommuting, 188, 307

Teleconferencing, 30, 186 (fig.), 187
Telephone systems, 30, 172, 173–77. *See also* Telecommunications
   cellular mobile, 182, 183 (fig.), 335–36
   computer, 131, 132 (fig.)
   future trends in, 335–36
   long-distance service, 175–77
   PBX technologies, 173–74
   selection guidelines, 177
   system divestiture, 46, 97, 172, 175
   system features, 174–75
Teleports, 189
Telesensory Systems, Inc., 322
Teletext, 187
Temporary personnel services, 305 (fig.), 306
Tennyson, Alfred Lord, 42, 338
Terminals. *See also* Display screens
   color, 103, 154
   dumb, 133
   ergonomic design of, 280–83
   graphics, 153–54
   input, 132, 133 (fig.)
   monochrome, 103
   personal computer, 101–3
   scanning, 146
Terrorism and sabotage protection, 273–75
Texas Instruments, 96
Text entry systems, 333
Theory X and Theory Y of motivation, 55
Thermal printers, 108
Third-party services, telecommunications, 180–81
Third Wave, The, 188
Throughput, 27
Time management software, 211–12
Time schedule for office automation, 18
Toffler, Alvin, 188
Touch-sensitive screens, 71, 126, 127 (fig.), 128
Training
   careers in, 312–13
   careers for the handicapped and, 322–23
   in company stores, 45–46
   establishing a program of, 317–20
   for information managers, 54
   methods of, 320–22
   objectives, 316
   for office automation, 19, 316–20
   techniques, 319–20
Transfer files, 225 (fig.)
Transmission lines, telecommunications, 180
Trust in information management, 53
Typewriters
   electric, 95
   electronic, 118–20

U

Ultrafiche, 227
Uninterruptible power systems (UPSs), 256
Unit sets, 248 (fig.)
UNIVAC computer, 96
Universal Product Code (UPC), 90
Universal workstation, 124–25 (figs.)
Uploading data, 195, 209–10
U.S. Employment Service, 305
User-friendliness
    aids for, 126–33
    in universal workstations, 124–25
U.S. Teleco Data Communications Co., 176
U.S. Telecom, 176
Utility software, 207

V

Vacuum tubes, 97 (fig.)
Vendors. *See also* Service and support
    selecting, for computer systems, 136–37
    selecting, for office supplies, 243–44
VersBraille software, 323
Videotex, 155, 187 (fig.)
Visual displays. *See* Display screens; Terminals
Visual literacy, 154
Visually handicapped workers, 322 (fig.), 323
Vocational Rehabilitation Act of 1973, 322
Voice communications, 333
Voice mail, 30, 88–89, 182–83
Voice processing, 77–93
    advanced technologies, 89–91
    automating input, 78
    digital voice processing, 87–88
    machine dictation/transcription, 78–87
    synthesizer, 323
    voice mail. *See* Voice mail
Voice recognition, 89, 90 (fig.), 91 (fig.)
    security systems of, 272
Voice verification, 89–90

W

Wang Professional Image Computer (PIC), 146
Warner Computer Systems, 213 (fig.)
Washington Business Group on Health, 63, 64 (fig.)
Waterman, Robert, 45
Watson, Thomas J., 332
Western Union, 180
Whiteboard. *See* Electronic presentation boards
Wire management in offices, 290–92
Women, in the workforce, 60–61
Word Processing, 158
Word processing, 5, 6, 7 n, 96, 157–70
    advanced features, 161–65
    benefits of, 158–59
    defined, 158
    features, 159–61
    industry/professional applications, 165–69
    as information processing, 28
    microcomputer software for, 208
    multifunction systems, 121
    multiuser systems, 121, 122 (fig.)
    office automation role of, 169
    stand-alone (dedicated) systems, 120, 121 (fig.)
Workforce. *See also* Employee(s); Information workers
    changing structure of, 5 (fig.), 6–9, 302, 303 (fig.)
    women in, 60–61
Workplace in the information age, 6–9. *See also* Office
    climate conditions, 292–96
    environmental controls in, 289–92
Workstation, 7 n, 8, 101, 121–26, 270 (fig.)
    accessories and support components, 287, 288–89 (figs.)
    computer phone systems, 131–32
    ergonomic design, 279, 280–89
    in the future, 333–34
    menu of options, 123 (fig.)
    as a path to new technology, 126
    server links for, 199–200
    universal, 124–25
Work styles, alternate, 307–8
Write-protected disks, 106
Writing, and word processing, 167–68

X

X.400 Plan, 200
Xerox Corporation, 8

ABOUT THE AUTHOR

Arnold Rosen received his B.S. degree in business administration from Ohio State University and his M.S. degree in business education from Hunter College. He is a professor at Nassau Community College in Garden City, New York. He previously taught at New York City Community College and Hunter College and in the New York City school system. Professor Rosen has authored numerous journal articles on aspects of word processing and office automation and has written the following books:

AVT Machine Transcription
Word Processing, with Rosemary Fielden
Word Processing Keyboarding Applications and Exercises, with William Hubbard
Administrative Procedures for the Electronic Office, with Eileen Feretic and Margaret Bahniuk
Language Skills for Transcription
Getting the Most Out of Your Word Processor
Information Processing: Keyboarding Applications and Exercises, with William Hubbard
Text Editing: Keyboarding Applications and Exercises, with William Hubbard

Professor Rosen has presented seminars and is a speaker and consultant on office automation, education, and training. He organized and served as chairperson for the Association for Information Systems Professionals (AISP) of the Educators' Advisory Council. He also served on the board of directors, as vice president, and was elected international president in 1979.

Professor Rosen edited the "Information Processing" section of *Business Education Forum,* the official magazine of the National Business Education Association. He has directed three VEA Grant proposals for Nassau Community College.

WE VALUE YOUR OPINION—PLEASE SHARE IT WITH US

Merrill Publishing and our authors are most interested in your reactions to this textbook. Did it serve you well in the course? If it did, what aspects of the text were most helpful? If not, what didn't you like about it? Your comments will help us to write and develop better textbooks. We value your opinions and thank you for your help.

Text Title _____ Edition _____

Author(s) _____

Your Name (optional) _____

Address _____

City _____ State _____ Zip _____

School _____

Course Title _____

Instructor's Name _____

Your Major _____

Your Class Rank _____ Freshman _____ Sophomore _____ Junior _____ Senior

_____ Graduate Student

Were you required to take this course? _____ Required _____ Elective

Length of Course? _____ Quarter _____ Semester

1. Overall, how does this text compare to other texts you've used?

   _____ Superior _____ Better Than Most _____ Average _____ Poor

2. Please rate the text in the following areas:

	Superior	Better Than Most	Average	Poor
Author's Writing Style	_____	_____	_____	_____
Readability	_____	_____	_____	_____
Organization	_____	_____	_____	_____
Accuracy	_____	_____	_____	_____
Layout and Design	_____	_____	_____	_____
Illustrations/Photos/Tables	_____	_____	_____	_____
Examples	_____	_____	_____	_____
Problems/Exercises	_____	_____	_____	_____
Topic Selection	_____	_____	_____	_____
Currentness of Coverage	_____	_____	_____	_____
Explanation of Difficult Concepts	_____	_____	_____	_____
Match-up with Course Coverage	_____	_____	_____	_____
Applications to Real Life	_____	_____	_____	_____

3. Circle those chapters you especially liked:
   1 2 3 4 5 6 7 8 9 10 11 12 13 14 15 16 17 18 19 20
   What was your favorite chapter? _____
   Comments:

4. Circle those chapters you liked least:
   1 2 3 4 5 6 7 8 9 10 11 12 13 14 15 16 17 18 19 20
   What was your least favorite chapter? _____
   Comments:

5. List any chapters your instructor did not assign. _____

6. What topics did your instructor discuss that were not covered in the text? _____
   _____

7. Were you required to buy this book? _____ Yes _____ No

   Did you buy this book new or used? _____ New _____ Used

   If used, how much did you pay? _____

   Do you plan to keep or sell this book? _____ Keep _____ Sell

   If you plan to sell the book, how much do you expect to receive? _____

   Should the instructor continue to assign this book? _____ Yes _____ No

8. Please list any other learning materials you purchased to help you in this course (e.g., study guide, lab manual).
   _____

9. What did you like most about this text? _____
   _____

10. What did you like least about this text? _____
    _____

11. General comments:

    May we quote you in our advertising? _____ Yes _____ No

    Please mail to: Boyd Lane
                     College Division, Research Department
                     Box 508
                     1300 Alum Creek Drive
                     Columbus, Ohio 43216

    Thank you!